Methods in Molecular Biology

Series Editor
John M. Walker
School of Life and Medical Sciences,
University of Hertfordshire, Hatfield,
Hertfordshire, UK

For further volumes:
http://www.springer.com/series/7651

For over 35 years, biological scientists have come to rely on the research protocols and methodologies in the critically acclaimed *Methods in Molecular Biology* series. The series was the first to introduce the step-by-step protocols approach that has become the standard in all biomedical protocol publishing. Each protocol is provided in readily-reproducible step-by-step fashion, opening with an introductory overview, a list of the materials and reagents needed to complete the experiment, and followed by a detailed procedure that is supported with a helpful notes section offering tips and tricks of the trade as well as troubleshooting advice. These hallmark features were introduced by series editor Dr. John Walker and constitute the key ingredient in each and every volume of the *Methods in Molecular Biology* series. Tested and trusted, comprehensive and reliable, all protocols from the series are indexed in PubMed.

Angiogenesis

Methods and Protocols

Edited by

Andrew V. Benest

Biodiscovery Institute, University of Nottingham, Nottingham, UK

Humana Press

Editor
Andrew V. Benest
Biodiscovery Institute
University of Nottingham
Nottingham, UK

ISSN 1064-3745 ISSN 1940-6029 (electronic)
ISBN 978-1-0716-2058-8 ISBN 978-1-0716-2059-5 (eBook)
https://doi.org/10.1007/978-1-0716-2059-5

This Humana imprint is published by the registered company Springer Science+Business Media, LLC, part of Springer Nature.
The registered company address is: 1 New York Plaza, New York, NY 10004, U.S.A.

Preface

This is a new edition of this book, and I am honored to have been passed the editing baton by my friends, colleagues, and collaborators, Stewart Martin and Peter Hewett, editors of three earlier editions. I am delighted that the field of angiogenesis continues to develop at pace, and this is reflected in this latest collection.

The nature of angiogenesis research is changing. Whereas previous editions were written at a time where there was a focus on exploring how endothelial cells responded to angiogenic stimuli, the field has now grown to incorporate new ideas. We now know that the dynamic endothelium comprises multiple different phenotypes and behaviors, across developmental, pathological, and physiological settings. This edition builds on the lymphatic protocols included in previous *Methods in Molecular Biology* volumes, and I am delighted to have been able to continue this trajectory.

Not only has our understanding of angiogenesis shifted, but scientists also now have access to huge amounts of data and resources online. While this explosion of freely available data is a huge positive to the field, analysis can seem daunting, so this book includes tools to aid in the interpretation of large datasets. For example, we have included protocols to explore how transcriptional phenotyping can be achieved, in terms of exploring tissue-specific heterogeneity to the use of database mining of existing datasets. Similarly, bioinformatic techniques run alongside protocols for transcript analysis, cell biology, pathology, physiological and developmental angiogenesis models.

As with the previous editions, the focus of this work is to enable anyone in a routinely equipped laboratory to pick up this book and perform the experiments as described. However, it should be noted that experiments relying on animal work (and we have included mammalian and fish models) are reliant on the appropriate measures (legal, ethical, and procedural) being in place.

This book was commissioned, written, edited, and produced entirely against the challenging backdrop of COVID-19, and I will be forever indebted to the contributors who have given their time, expertise, and kindness to the collaborative process. I am delighted to have put together a book of 29 chapters with such esteemed authors, who bring a diverse range of skills and experience to their respective chapters to build a rich learning resource. I would also like to thank my own research team for their support during this time too. Above all, I hope this edition provides a meaningful contribution to angiogenesis and endothelial cell biology research.

Nottingham, UK *Andrew V. Benest*

Contents

Contributors

NADA S. AHMED • *Endothelial Cell Quiescence Group, Centre for Cancer Sciences, Biodiscovery Institute, School of Medicine, University of Nottingham, Nottingham, UK*

CLAIRE L. ALLEN • *Exonate Ltd, Nottingham, UK*

KENTON P. ARKILL • *Tumour and Vascular Biology Laboratories, Centre for Cancer Sciences, School of Medicine, Biodiscovery Institute, University of Nottingham, Nottingham, UK*

PETER BALUK • *Department of Anatomy, Cardiovascular Research Institute, University of California, San Francisco, CA, USA; UCSF Helen Diller Family Comprehensive Cancer Center, University of California, San Francisco, CA, USA*

DAVID O. BATES • *Tumour and Vascular Biology Laboratories, Centre for Cancer Sciences, School of Medicine, Biodiscovery Institute, University of Nottingham, Nottingham, UK*

NICHOLAS BEAZLEY-LONG • *Endothelial Cell Quiescence Group, Centre for Cancer Sciences, Biodiscovery Institute, School of Medicine, University of Nottingham, Nottingham, UK*

ANDREW V. BENEST • *Endothelial Cell Quiescence Group, Centre for Cancer Sciences, Biodiscovery Institute, School of Medicine, University of Nottingham, Nottingham, UK*

SOHNI RIA BHALLA • *Tumour and Vascular Biology Laboratories, Centre for Cancer Sciences, School of Medicine, Biodiscovery Institute, University of Nottingham, Nottingham, UK*

JOSHUA H. BOURNE • *Institute of Cardiovascular Science, Medical and Dental School, University of Birmingham, Birmingham, UK*

MAIRI BRITTAN • *Centre for Cardiovascular Science, The Queen's Medical Research Institute, University of Edinburgh, Edinburgh, UK*

FINNIUS A. BRUTON • *Centre for Cardiovascular Science, The Queen's Medical Research Institute, University of Edinburgh, Edinburgh, UK*

YAN CHEN • *Department of Infection, Immunity and Cardiovascular Disease, University of Sheffield, Sheffield, UK*

EVA M. CLAVANE • *Discovery & Translational Science Department, Leeds Institute of Cardiovascular and Metabolic Medicine, University of Leeds, Leeds, UK*

RICHARD M. CUBBON • *Discovery & Translational Science Department, Leeds Institute of Cardiovascular and Metabolic Medicine, University of Leeds, Leeds, UK*

GABRIELA D'AMICO • *UCL Institute of Ophthalmology, University College London, London, UK*

MARLENE E. DA VITORIA LOBO • *UCL Institute of Ophthalmology, University College London, London, UK*

CHRIS DENNING • *Department of Stem Cell Biology, Biodiscovery Institute, University of Nottingham, Nottingham, UK*

MARTIN A. DENVIR • *Centre for Cardiovascular Science, The Queen's Medical Research Institute, University of Edinburgh, Edinburgh, UK*

SARAH DE VAL • *Department of Physiology, Anatomy and Genetics, University of Oxford, Oxford, UK; Nuffield Department of Medicine, Ludwig Institute for Cancer Research Ltd, University of Oxford, Oxford, UK*

LOUBNA ELSABONI • *Tumour and Vascular Biology Laboratories, Centre for Cancer Sciences, School of Medicine, Biodiscovery Institute, University of Nottingham, Nottingham, UK*

PAUL C. EVANS • *Department of Infection, Immunity and Cardiovascular Disease, University of Sheffield, Sheffield, UK*

ALESSANDRO FANTIN • *Department of Biosciences, University of Milan, Milan, Italy*
KEVIN GASTON • *Biodiscovery Institute and Division of Translational Medical Sciences, School of Medicine, University of Nottingham, Nottingham, UK*
ZOE L. GRANT • *The Walter and Eliza Hall Institute of Medical Research, Parkville, VIC, Australia; Department of Medical Biology, University of Melbourne, Parkville, VIC, Australia*
KATHRYN R. GREEN • *Endothelial Cell Quiescence Group, Centre for Cancer Sciences, Biodiscovery Institute, School of Medicine, University of Nottingham, Nottingham, UK*
TARA L. HAAS • *School of Kinesiology and Health Science and Department of Biology, York University, Toronto, ON, Canada*
LYDIA HARDOWAR • *School of Science and Technology, Nottingham Trent University, Nottingham, UK*
JACQUELINE HEROLD • *Department of Biomedicine, Aarhus University, Aarhus, Denmark*
NICHOLAS A. HODGES • *Crayton Pruitt Family Department of Biomedical Engineering, University of Florida, Gainesville, FL, USA*
JOSEPH L. HORDER • *Endothelial Cell Quiescence Group, Centre for Cancer Sciences, Biodiscovery Institute, School of Medicine, University of Nottingham, Nottingham, UK*
JAMES B. HOYING • *Advanced Solutions Life Sciences, Manchester, NH, USA*
RICHARD P. HULSE • *School of Science and Technology, Nottingham Trent University, Nottingham, UK*
ELENA IOANNOU • *UCL Institute of Ophthalmology, University College London, London, UK*
PADMA-SHEELA JAYARAMAN • *Biodiscovery Institute and Division of Translational Medical Sciences, School of Medicine, University of Nottingham, Nottingham, UK*
LOUISE A. JOHNSON • *MRC Human Immunology Unit, Radcliffe Department of Medicine, MRC Weatherall Institute of Molecular Medicine, John Radcliffe Hospital, University of Oxford, Oxford, UK*
JACINTA ISABELLE KALISCH-SMITH • *Department of Physiology, Anatomy, and Genetics, BHF Centre of Research Excellence, University of Oxford, Oxford, UK*
JOANNA KALUCKA • *Department of Biomedicine and Aarhus Institute of Advanced Studies (AIAS), Aarhus University, Aarhus, Denmark*
JUMANA KARIM • *Tumour and Vascular Biology Laboratories, Centre for Cancer Sciences, School of Medicine, Biodiscovery Institute, University of Nottingham, Nottingham, UK*
ADAM J. KATZ • *Crayton Pruitt Family Department of Biomedical Engineering, University of Florida, Gainesville, FL, USA*
JOHN R. KING • *Faculty of Science, School of Mathematics, University of Nottingham, Nottingham, UK*
PHILIP KITCHEN • *College of Health and Life Sciences, Aston University, Birmingham, UK*
KONSTANTINOS KLAOURAKIS • *Burdon-Sanderson Cardiac Science Centre and Department of Physiology, Anatomy and Genetics, University of Oxford, Oxford, UK; British Heart Foundation - Oxbridge Centre of Regenerative Medicine, CRM, University of Oxford, Oxford, UK*
ARINOLA O. LAMPEJO • *Crayton Pruitt Family Department of Biomedical Engineering, University of Florida, Gainesville, FL, USA*
ZIWEN LI • *Centre for Cardiovascular Science, The Queen's Medical Research Institute, University of Edinburgh, Edinburgh, UK*
BANKS M. LOMEL • *Crayton Pruitt Family Department of Biomedical Engineering, University of Florida, Gainesville, FL, USA*

Gonzalo Gómez López • *Bioinformatics Unit, Spanish National Cancer Research Centre (CNIO), Madrid, Spain*
Amy P. Lynch • *Endothelial Cell Quiescence Group, Centre for Cancer Sciences, Biodiscovery Institute, School of Medicine, University of Nottingham, Nottingham, UK*
Maria J. C. Machado • *Tumour and Vascular Biology Laboratories, Centre for Cancer Sciences, School of Medicine, Biodiscovery Institute, University of Nottingham, Nottingham, UK*
Dima Majbour • *Crayton Pruitt Family Department of Biomedical Engineering, University of Florida, Gainesville, FL, USA*
Naseeb K. Malhi • *Tumour and Vascular Biology Laboratories, Centre for Cancer Sciences, School of Medicine, Biodiscovery Institute, University of Nottingham, Nottingham, UK*
Aarren J. Mannion • *Department of Oncology-Pathology, Karolinska Institute, Stockholm, Sweden*
Stewart G. Martin • *Nottingham Breast Cancer Research Centre, Biodiscovery Institute, School of Medicine, University of Nottingham, Nottingham, UK*
Donald M. McDonald • *Department of Anatomy, Cardiovascular Research Institute, University of California, San Francisco, CA, USA; UCSF Helen Diller Family Comprehensive Cancer Center, University of California, San Francisco, CA, USA*
Alan McIntyre • *Hypoxia and Acidosis Group, Center for Cancer Sciences, Biodiscovery Institute, University of Nottingham, Nottingham, UK*
Paul J. Meakin • *Discovery & Translational Science Department, Leeds Institute of Cardiovascular and Metabolic Medicine, University of Leeds, Leeds, UK*
Nigel P. Mongan • *School of Veterinary Medicine and Science, Biodiscovery Institute, University of Nottingham, Nottingham, UK; Department of Pharmacology, Weill Cornell Medicine, New York, NY, USA*
Walter L. Murfee • *Crayton Pruitt Family Department of Biomedical Engineering, University of Florida, Gainesville, FL, USA*
Alice Neal • *Department of Physiology, Anatomy and Genetics, University of Oxford, Oxford, UK; Nuffield Department of Medicine, Ludwig Institute for Cancer Research Ltd, University of Oxford, Oxford, UK*
Chris R. Neal • *Wolfson BioImaging, Biomedical Sciences, University of Bristol, Bristol, UK*
Zubair Ahmed Nizamundeen • *Tumour and Vascular Biology Laboratories, Centre for Cancer Sciences, School of Medicine, Biodiscovery Institute, University of Nottingham, Nottingham, UK*
Adam F. Odell • *School of Health Sciences, York St. John University, York, UK*
Guillermo Díez Pinel • *Neuronal and Vascular Biology Group, UCL Institute of Ophthalmology, University College London, London, UK*
Alexandra Pislaru • *Department of Biology, York University, Toronto, ON, Canada*
Emad Rakha • *Nottingham Breast Cancer Research Centre, Biodiscovery Institute, School of Medicine, University of Nottingham, Nottingham, UK*
Kavita Raniga • *Department of Stem Cell Biology, Biodiscovery Institute, University of Nottingham, Nottingham, UK*
Elijah W. Rice • *Department of Plastic and Reconstructive Surgery, Wake Forest School of Medicine, Winston-Salem, NC, USA*
Paul R. Riley • *Burdon-Sanderson Cardiac Science Centre and Department of Physiology, Anatomy and Genetics, University of Oxford, Oxford, UK; British Heart Foundation - Oxbridge Centre of Regenerative Medicine, CRM, University of Oxford, Oxford, UK*

FEDERICA RIU • *Tumour and Vascular Biology Laboratories, Centre for Cancer Sciences, School of Medicine, Biodiscovery Institute, University of Nottingham, Nottingham, UK*
HELENA RODRIGUEZ-CARO • *Department of Physiology, Anatomy and Genetics, University of Oxford, Oxford, UK*
KATHERINE M. ROSS STEWART • *Centre for Cardiovascular Science, The Queen's Medical Research Institute, University of Edinburgh, Edinburgh, UK*
MARTINA RUDNICKI • *School of Kinesiology and Health Science and Department of Biology, York University, Toronto, ON, Canada*
CHRISTIANA RUHRBERG • *UCL Institute of Ophthalmology, University College London, London, UK*
HULAN SHANG • *Crayton Pruitt Family Department of Biomedical Engineering, University of Florida, Gainesville, FL, USA*
SARAH J. STORR • *Nottingham Breast Cancer Research Centre, Biodiscovery Institute, School of Medicine, University of Nottingham, Nottingham, UK*
HANNAH A. STROBEL • *Advanced Solutions Life Sciences, Manchester, NH, USA*
ARIANA D. SUAREZ-MARTINEZ • *Crayton Pruitt Family Department of Biomedical Engineering, University of Florida, Gainesville, FL, USA*
ZARAH B. TABRIZI • *Endothelial Cell Quiescence Group, Centre for Cancer Sciences, Biodiscovery Institute, School of Medicine, University of Nottingham, Nottingham, UK*
HANNAH A. TAYLOR • *Discovery & Translational Science Department, Leeds Institute of Cardiovascular and Metabolic Medicine, University of Leeds, Leeds, UK*
NIKITA VED • *Department of Physiology, Anatomy, and Genetics, BHF Centre of Research Excellence, University of Oxford, Oxford, UK*
GRACE VETCH • *Tumour and Vascular Biology Laboratories, Centre for Cancer Sciences, School of Medicine, Biodiscovery Institute, University of Nottingham, Nottingham, UK*
JOAQUIM MIGUEL VIEIRA • *Burdon-Sanderson Cardiac Science Centre and Department of Physiology, Anatomy and Genetics, University of Oxford, Oxford, UK; British Heart Foundation - Oxbridge Centre of Regenerative Medicine, CRM, University of Oxford, Oxford, UK*
NGUYEN T. N. VO • *Department of Stem Cell Biology, Biodiscovery Institute, University of Nottingham, Nottingham, UK*
EMMA C. WATSON • *The Walter and Eliza Hall Institute of Medical Research, Parkville, VIC, Australia*
LACHLAN WHITEHEAD • *The Walter and Eliza Hall Institute of Medical Research, Parkville, VIC, Australia*
ROBERT N. WILKINSON • *School of Life Sciences, Medical School, Queens Medical Centre, University of Nottingham, Nottingham, UK*

Part I

Developmental Biology

Chapter 1

The Embryonic Mouse Hindbrain Model to Study Sprouting Angiogenesis In Vivo

Gabriela D'Amico and Christiana Ruhrberg

Abstract

Blood vessel growth is a fundamental process for organ development and wound healing but is also associated with ischemic diseases and cancer. The growth of new blood vessels from preexisting vasculature, termed sprouting angiogenesis, is the predominant mode of blood vessel growth in central nervous system vascularization and pathological vessel growth. Accordingly, studying the molecular and cellular mechanisms of angiogenesis holds the promise to find novel therapeutic targets to stimulate new vessel formation in ischemic tissues or inhibit pathological vessel growth in disease. The embryonic mouse hindbrain provides an excellent model to study sprouting angiogenesis in vivo by histochemical or fluorescent wholemount immunolabeling, thus allowing high-resolution image capture of nascent vasculature and subsequent quantification of relevant angiogenic parameters. This chapter describes how to use the mouse embryonic hindbrain as a model to study physiological angiogenesis, including detailed protocols for hindbrain dissection, wholemount staining, and angiogenic parameters analysis.

Key words Sprouting angiogenesis, Hindbrain model, Wholemount immunolabeling

1 Introduction

The growth of new blood vessels is a fundamental process required for organogenesis during development [1]. After birth, new blood vessel growth is also required during the female reproductive cycle as well as for wound healing and exercise-induced muscle growth. New blood vessel growth, however, is also associated with different diseases such as cancer and neovascular eye diseases, where it has been linked to disease progression [2]. The main mechanism driving new blood vessel growth in these conditions is termed sprouting angiogenesis; this process involves the formation of endothelial tip-cell led migratory sprouts that extend from preexisting vessels in response to growth factor gradients and extracellular matrix cues [3]. Therefore, elucidating the molecular and cellular mechanisms of sprouting angiogenesis is crucial for identifying new therapeutic targets to stimulate new vessel formation in

Andrew V. Benest (ed.), *Angiogenesis: Methods and Protocols*, Methods in Molecular Biology, vol. 2441,
https://doi.org/10.1007/978-1-0716-2059-5_1,

ischemic tissues or to inhibit pathological vessel growth in disease. This chapter describes briefly how the mouse embryonic hindbrain is vascularized during normal development and how it can be used as a model to study physiological sprouting angiogenesis. Specifically, we provide step-by-step protocols for tissue dissection, sample staining, subsequent imaging, and quantitative analysis.

1.1 Mouse Hindbrain Vascularization

The mammalian brain is vascularized during embryogenesis by sprouting angiogenesis in a stereotypical and precise spatiotemporal manner to form a well-defined vascular network [4, 5]. At embryonic day (E) 9.5 in the mouse, vessels sprout from a pre-formed perineural vascular plexus into the brain parenchyma, then extending perpendicularly as so-called radial vessels toward the ventricular zone (Fig. 1a) [6]. This process occurs in response to vascular endothelial growth factor A (VEGF-A), which is secreted by neural progenitor cells [7]. By around E10.0, the radial vessels reach the subventricular zone and turn at near right angles to grow laterally (Fig. 1a) [6]. From E10.5 onwards, these lateral vessels continue to sprout but also begin to anastomose, thereby giving rise to a honeycomb-like vascular network known as subventricular vascular plexus (SVP), which nurtures the neural progenitor zone and regulates neural progenitor proliferation in the hindbrain (Fig. 1a) [8, 9]. At E11.5, most of the vascular anastomosis takes place, facilitated by the help of tissue-resident macrophages termed microglia, so that the SVP is largely complete by E12.5 [10]. Throughout this period, the hindbrain vasculature undergoes remodeling and maturation, which involves the recruitment of perivascular cells, initially pericytes, and subsequently also astrocytes to ensure the acquisition of especial vascular barrier properties that characterize most of the vessels in the central nervous system, known as the blood–brain barrier [11]. Owing to this unique spatiotemporal process, and because the roof-plate can be opened for hindbrain flatmounting, the mouse embryonic hindbrain has become an instrumental and accessible model system for studying the molecular and cellular mechanisms of sprouting angiogenic in vivo.

1.2 Advantages and Disadvantages of the Hindbrain Model

The embryonic mouse hindbrain offers several advantages as a model system to study sprouting angiogenesis. The first is that hindbrain vascularization occurs early during development and thus allows the study of blood vessel growth and patterning in genetically modified mice that show embryonic lethality, such as NRP1 knockout mice [12, 13]. This is an important consideration, because such mouse mutants cannot be examined with the other widely used angiogenic organ model, the postnatal mouse retina [14, 15].

A second advantage is that the hindbrain is exquisitely suited for wholemount imaging after flatmounting and labeling of

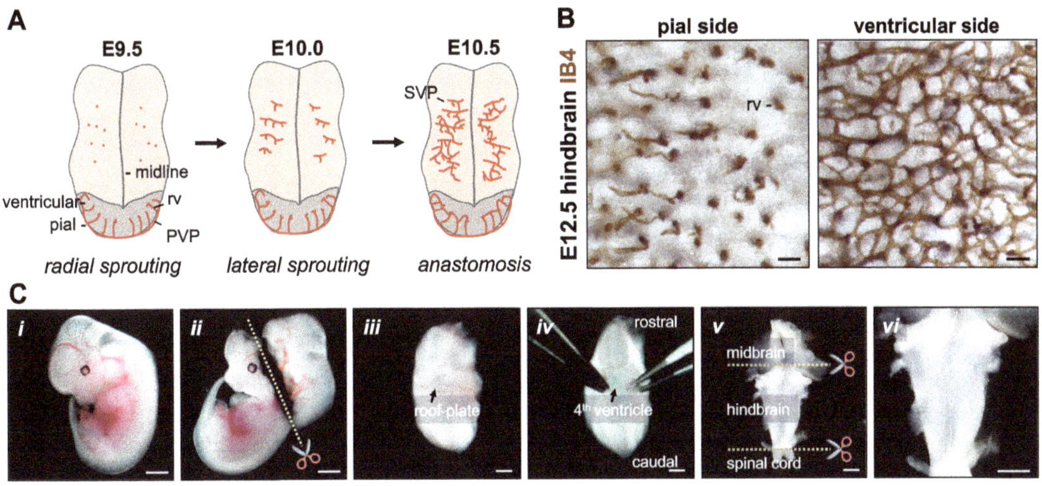

Fig. 1 Mouse hindbrain vascularization and dissection. (**a**) Schematic illustration of vascular development in the embryonic hindbrain. Vessel sprouting takes place in a temporally and spatially well-established sequence of events from E9.5 onwards in the mouse. Abbreviations: *E* embryonic day, *PVP* perineural vascular plexus, *rv* radial vessel, *SVP* subventricular vascular plexus. (**b**) Brightfield stereomicroscope images (zoom 8×) of selected areas from the pial and ventricular sides of a flatmounted E12.5 wild-type mouse hindbrain, histochemically stained for the endothelial cell marker IB4. Scale bar: 50 μm. (**c**) Step-by-step hindbrain dissection from a wild-type mouse embryo at E12.5 (insets *i*–*vi*). Scale bars: 1 mm (*i, ii*) and 0.5 mm (*iii*–*vi*)

endothelial cells and their interacting cells (Fig. 1b). Accordingly, the hindbrain vascular plexus in a flatmount shows vessels with a one-dimensional growth pattern (the radial vessels on the pial side) and vessels with a two-dimensional network structure (the SVP vessels on the ventricular side). This geometric architecture of vasculature in the flatmounted hindbrain enables easy quantification of a wide range of vascular parameters, including vessel numbers, vessel diameter, branchpoints, tip cell numbers, and endothelial cell filopodia characteristics, thus making the hindbrain system one of the most reliable tools to characterize vascular patterning (Figs. 2 and 3) [8, 15].

A third advantage is that the preparation of embryonic hindbrains for wholemount imaging is fairly simple and can be adapted for versatile applications. For example, this method permits staining and subsequent analysis immediately after dissection and fixing, but also much later on, after the hindbrain tissue has been stored. This allows pooling of samples, for example to account for genotype, and thus examining a sufficiently large sample number at the same time to minimize the possibility of intrinsic technical variability and to increase statistical power (*see* Subheading 3).

However, the hindbrain model may not be suitable in some circumstances. For example, genetically modified mice with embryonic lethality before E10.5 cannot be examined with this model, because vessel sprouting in the hindbrain begins only after E9.75. This limitation can be overcome by utilizing Cre-Lox technologies,

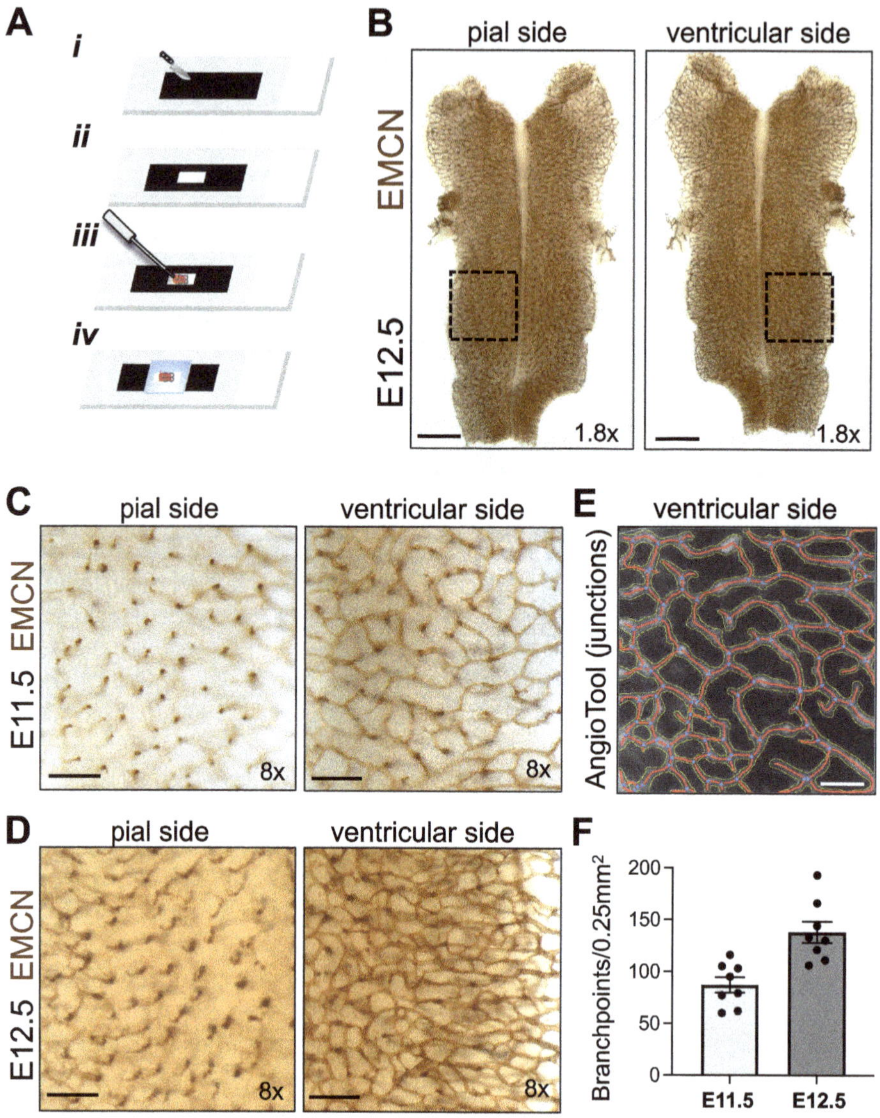

Fig. 2 Analysis of flatmounted, vessel-stained mouse hindbrains. (**a**) Schematic presentation of the mounting procedure for a wholemount stained hindbrain onto a microscope glass slide (insets *i–iv*). (**b**) Brightfield stereomicroscope images (zoom 1.8×) of an entire E12.5 wild-type mouse hindbrain, histochemically immunostained for the capillary/venous endothelial cell marker endomucin (EMCN). Scale bar: 0.5 mm. The stippled boxes indicate the areas shown in (**c**). (**c**,**d**) Brightfield stereomicroscope images (zoom 8×) of the SVP in E11.5 (**c**) and E12.5 (**d**) wild-type mouse hindbrains, stained for EMCN. Scale bars: 100 μm. (**e**) Representative image of the SVP in an EMCN-stained, flatmounted E12.5 hindbrain (zoom 8×), including rendering with the AngioTool macro in Image J. The vessel outlines (yellow), the vessel "skeletons" (red), and vascular branching points (blue dots) are shown, as used for automated quantification of vascular parameters. Scale bar: 100 μm. (**f**) Quantification of vessel branchpoints in the wild-type SVP, measured as junctions with AngioTool at E11.5 and E12.5. Analysis performed on eight individual images of each hindbrain to illustrate the technical variation between different fields. The average is the representative value for the hindbrain. Data are shown as mean ± SE

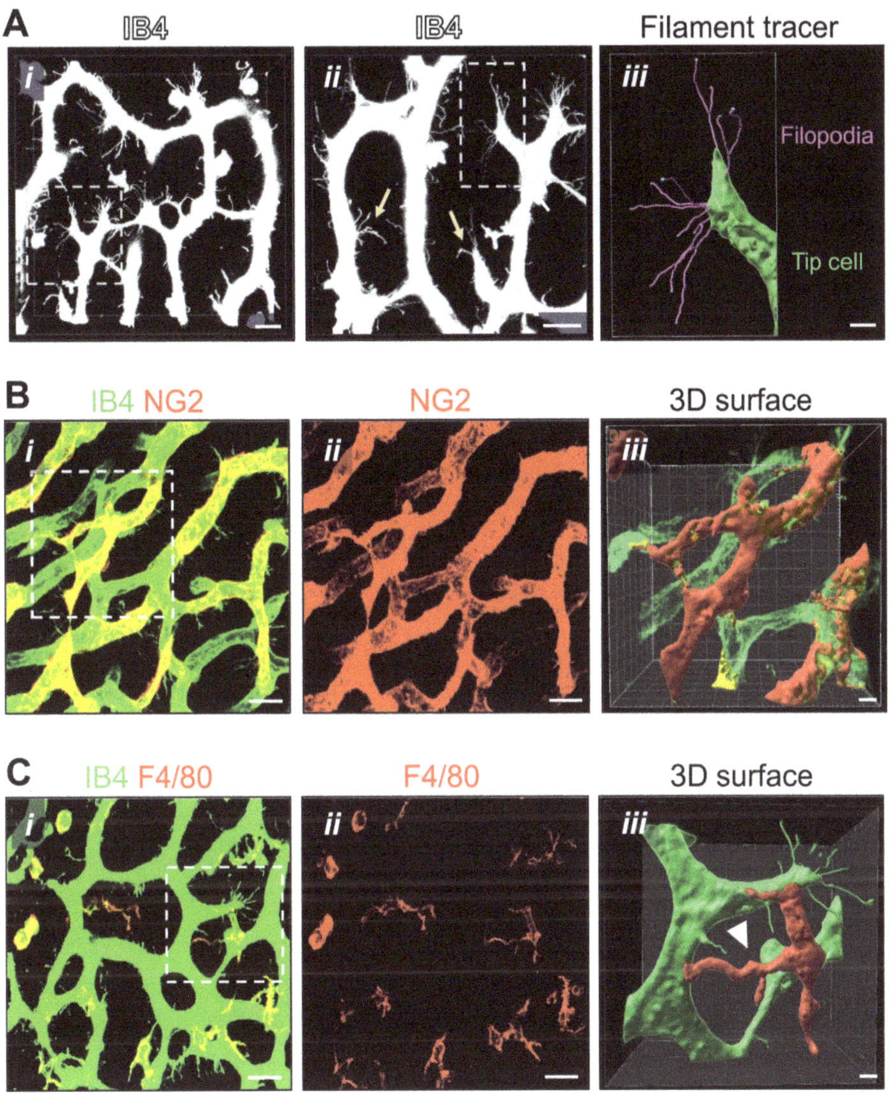

Fig. 3 Endothelial cell features and interaction with surrounding cell types. (**a**) Maximum intensity projection of a confocal z-stack from an E11.5 wild-type hindbrain that is fluorescently stained for IB4. (*i*) SVP, imaged at 40×; the dotted box in (*i*) is shown at higher magnification, imaged at 63×; in a 3D orthogonal view in (*ii*) to visualize endothelial cell filopodial bursts (examples are indicated with arrows). The dotted box in (*ii*) is shown in (*iii*) after processing with the Filament tracer module in IMARIS to analyze, e.g., filopodia numbers (green dots) and lengths (purple segments). Scale bars: 20 μm (*i, ii*) and 5 μm (*iii*). (**b**) Maximum intensity projection of a confocal z-stack from (*i*) an E12.5 wild-type hindbrain, stained for IB4 and NG2 to identify SVP endothelium and pericytes, respectively. The single channel for NG2 is shown in (*ii*). A higher magnification of the area indicated with a dotted square in (*i*) is shown in *(iii)* as a 3D surface view, generated with IMARIS software, to show that pericytes cover the endothelial surface. Scale bars: 50 μm (*i, ii*) and 5 μm (*iii*). (**c**) Maximum intensity projection of a confocal z-stack from (*i*) from an E12.5 wild-type hindbrain, stained for IB4 and ADGRE1 (F4/80 antigen) to identify SVP endothelium and monocyte/macrophages, respectively; (*ii*) shows only the ADGRE1 channel. A higher magnification of the area indicated with a dotted square in (*i*) is shown in (*iii*) as a 3D surface view generated with IMARIS software to analyze endothelial cell–macrophage interactions. As reported [10], tissue macrophages interact with neighboring endothelial tip cell (*arrowhead*). Scale bars: 50 μm (*i, ii*) and 5 μm (*iii*)

which allows refinement of gene targeting to specific cell types or temporally (reviewed by [16]). Indeed, the ablation of VEGF-A results in early embryonic lethality by E9.5 [17, 18], but its role in the developing brain has been elucidated with a constitutive Cre-Lox approach that inactivates VEGF-A in neural progenitors only (Fig. 3) [7]. Another disadvantage of the hindbrain model is its poor suitability for local injections of drugs or viruses for evaluating candidate molecules with the potential of inhibiting or promoting vascularization during development. Although the in utero manipulation of embryos can be achieved in principle, e.g., via injection or electroporation after transient, surgical exposure of the embryos, both are extremely challenging techniques that are more suitable to late gestational stages, such as E15.5 embryos [19–21], and therefore after the time window when angiogenic sprouting in the hindbrain is usually assessed (*see* Subheading 3).

2 Materials

2.1 Sample Collection and Hindbrain Preparation

2.1.1 Equipment

1. Micro-dissecting scissors (e.g., Sigma).
2. Watchmaker forceps no. 5 (e.g., Dumont).
3. Watchmaker forceps no. 55 (e.g., Dumont).
4. Scalpels (no. 10; e.g., Swann-Morton).
5. 2.0-mL round-bottom safe-lock tubes (e.g., Eppendorf).
6. Plastic cell culture dishes (60 cm diameter; e.g., BD Falcon).
7. Plastic Pasteur pipettes (e.g., Alpha Laboratories).
8. 50-mL reagent tubes (e.g., Falcon tubes from Corning).
9. Orbital/plate shaker (e.g., SciQuip).
10. Tabletop balance (e.g., ED Precision balance, Sartorius).
11. Water bath (e.g., Stuart SWB15D, Stuart).
12. Stereomicroscope with bright light illumination (e.g., Nikon SMZ1270).

2.1.2 Reagents

1. Phosphate-buffered saline tablets (PBS, Sigma-Aldrich).
2. Paraformaldehyde (PFA, e.g., Sigma-Aldrich). CAUTION: It is harmful if inhaled and swallowed. It is irritating to eyes, respiratory system, and skin and may cause sensitization of skin upon contact. Wear protective goggles, clothing, and gloves. Use it in a chemical fume hood/fume cabinet.
3. Absolute methanol (e.g., Fisher Scientific). CAUTION: It is toxic and can cause various serious irreversible effects through inhalation, in contact with skin and if swallowed. It is irritating to eyes and skin. Wear protective goggles, clothing, and gloves.

4. Triton X-100 (e.g., Sigma-Aldrich). CAUTION: It is harmful if ingested and can cause serious damage to the eyes. Wear protective goggles, clothing, and gloves.

2.1.3 Equipment and Reagent/Buffer Setup

1. Plastic Pasteur pipettes (used to transfer embryos and hindbrains): Slice the bulb of the pipette to create a spoon-shaped opening suitable to transfer embryos. For hindbrain transfer, cutoff the tip with a scalpel or scissors to create an appropriately wide opening.
2. PBS: Dissolve five tablets of PBS in 1 L of distilled water. Store at room temperature indefinitely.
3. PBT: Add 50 μL Triton X-100 to 50 mL PBS in a 50-mL tube. Store at room temperature indefinitely.
4. Formaldehyde solution (4%, w/v): Under a chemical fume hood, weigh 2 g of paraformaldehyde (PFA) powder into a 50-mL tube and add 50 mL of PBS. Dissolve the PFA by heating up at 60 °C in a water bath, mixing frequently until the PFA powder has dissolved. Store in aliquots at −20 °C for up to 1 year. Thaw at room temperature or at 37 °C in a water bath and cool it down on ice prior to use.
5. Methanol gradient: To prepare 50 mL of 25%, 50%, or 75% methanol in PBT, mix 12.5, 25, or 37.5 mL of absolute methanol with 37.5, 25, or 12.5 mL of PBT, respectively, in 50-mL tubes. Store at room temperature indefinitely.

2.2 Hindbrain Staining, Mounting, and Imaging

2.2.1 Equipment

1. 2.0-mL round-bottom safe-lock tubes (as above).
2. Plastic Pasteur pipettes (as above).
3. Orbital/plate shaker (as above).
4. Microscope glass slides (e.g., VWR International).
5. Microscope glass coverslips (18 × 18 mm; e.g., VWR International).
6. Black electric tape (e.g., SCOTCH33, 3 M).
7. Stereomicroscope with bright light illumination (as above) equipped with a digital camera (e.g., Nikon DS-Fi3) and data acquisition software (e.g., NIS-Elements Microscope Imaging software, Nikon).
8. Confocal laser scanning microscope (e.g. LSM710, Zeiss) with associated data acquisition software (e.g. ZEN Microscope Imaging software, Zeiss).

2.2.2 Reagents

1. PBS (as above).
2. Triton X-100 (as above).
3. Absolute methanol (as above).

4. Hydrogen peroxide (e.g. Sigma-Aldrich). CAUTION: It is harmful if ingested and can cause serious damage to the eyes. Wear protective goggles, clothing, and gloves.
5. Heat-inactivated normal goat serum (NGS, Sigma-Aldrich).
6. Heat-inactivated normal rabbit serum (NRS, Sigma-Aldrich).
7. Ready-to-use serum-free protein block solution (Dako).
8. Biotinylated lectin from *Bandeiraea simplicifolia* BS-I isolectin B4 (IB4; Sigma-Aldrich).
9. Rat anti-mouse PECAM1 monoclonal antibody (CD31, also known as MEC13.3; BD Pharmingen).
10. Rat anti-mouse endomucin monoclonal antibody (V.7C7; Santa Cruz Biotechnology).
11. HRP-tagged streptavidin (Dako).
12. HRP-tagged rabbit anti-rat secondary antibody (Dako).
13. Diaminobenzidine and urea hydrogen peroxide tablets (Sigma-Aldrich).
14. AlexaFluor488-conjugated goat anti-rat secondary antibody (Invitrogen).
15. AlexaFluor488-conjugated streptavidin (Invitrogen).
16. PFA (as above).
17. SlowFade Antifade reagent (e.g., Life Technologies).

2.2.3 Equipment and Reagent/Buffer Setup

1. Plastic Pasteur pipettes (used to transfer hindbrains from round-bottom tubes to microscope slides as previously described).
2. Orbital/plate shaker in a cold room or cold cabinet, because antibody incubations should be performed in tubes that are gently agitated at 4 °C.
3. PBS (as previously described).
4. PBT (as previously described) (*see* **Note 1**).
5. Methanol gradient (to rehydrate hindbrains that were fixed and stored, as previously described).
6. Blocking buffer (10%, v/v): For 1 mL, mix 100 μL of serum with 900 μL of PBT.
7. Diaminobenzidine solution: Dissolve 1 DAB tablet in 5 mL of distilled water at room temperature. Adjust tablet amount to final volume required. Protect from light. Prepare fresh for each immunostaining experiment.
8. Diaminobenzidine and urea/hydrogen peroxide solution: Dissolve 1 DAB and 1 urea hydrogen peroxide tablet in 5 mL of

distilled water at room temperature. Adjust tablet amount to final volume required. Protect from light. Prepare fresh for each immunostaining experiment.

9. Formaldehyde solution (4%, w/v) to post-fix wholemount stained hindbrains, as previously described.

3 Methods

3.1 Day 1: Dissection of Hindbrains from Mouse Embryos

1. Euthanize a timed-mated pregnant female mouse using an ethically approved procedure, e.g., cervical dislocation, at the intended gestational stage (*see* **Note 2**). Dissect both uterine horns containing the embryos using sharp scissors. Transfer the excised uterus into a 6-cm plastic dish containing ice-cold PBS and keep it on ice until moving to **step 2**.
2. Perform all dissections using a stereomicroscope. Using watchmaker forceps (number 5), peel-off the uterine muscle wall, open the yolk sac to expose the embryo, and release each embryo by cutting the umbilical cord (*see* **Note 3**).
3. Transfer each embryo into a clean plastic dish containing ice-cold PBS (Fig. 1c, *i*).
4. Using sharp scissors or a scalpel, slice off the embryo head (Fig. 1c, *ii*) (*see* **Note 4**).
5. Position the head dorsal side up and identify the fourth ventricle and the thin epithelial covering known as the roof-plate (Fig. 1c, *iii*).
6. Pierce the roof-plate using watchmaker forceps (number 5 or 55) and peel off the tissue, moving rostrally along the midline over the midbrain and then caudally over the posterior hindbrain and spinal cord (Fig. 1c, *iv*).
7. Using watchmaker forceps (number 5 or 55), carefully peel away any remaining head mesenchyme and meninges attached to the pial side of the hindbrain. Remove the midbrain and spinal cord tissue (Fig. 1c, *v*) (*see* **Note 5**).
8. The hindbrain should now be exposed as an open-book preparation (Fig. 1c, *vi*).
9. Transfer hindbrains to 2-mL round-bottom safe-lock tubes using a plastic Pasteur pipette with a wide-bore tip, aspirate the PBS and fix for 2 h at 4 °C with gentle agitation in freshly made (or thawed from frozen aliquots) cold 4% formaldehyde solution in PBS.
10. Rinse fixed hindbrains three times with PBS and proceed with wholemount immunolabeling.

PAUSE STEP: Fixed hindbrains can be stored at 4 °C in PBS if immunolabeling will begin within 2–3 days. Alternatively, replace PBS with an increasing methanol gradient (e.g., *25%, 50%, and 75% absolute methanol in PBT; for 5 min each at room temperature) before transferring to absolute methanol for longer storage at– 20 °C.* For some antibody staining, e.g., to detect PECAM1, avoid fixation for longer than 2 h or methanol storage [15].

3.2 Days 2–3: Hindbrain Wholemount Staining

3.2.1 Hindbrain Wholemount Immunolabeling (e.g., Isolectin B4, PECAM1, or Endomucin)

1. If hindbrain was stored in methanol, replace absolute methanol with a decreasing methanol gradient (e.g., 75%, 50%, and 25% absolute methanol in PBT for 5 min each at room temperature) before transferring to PBT alone (*see* **Note 6**).
2. Place the hindbrains into new 2.0-mL round-bottom safe-lock tubes before proceeding with either histochemical or fluorescent wholemount immunostaining.

3.2.2 Hindbrain Wholemount Histochemical Staining

1. Quench endogenous peroxidase activity by incubating in 1% hydrogen peroxide in PBS for 30 min at room temperature with gentle agitation.
2. Wash samples in PBT two times for 5 min each at room temperature.
3. Incubate in blocking solution for 30 min at room temperature with gentle agitation. For biotinylated isolectin B4 (IB4), PECAM1, or endomucin, use 10% heat-inactivated normal goat or rabbit serum in PBT as blocking solution (*see* **Note 7**).
4. Incubate overnight at 4 °C with gentle agitation in blocking solution containing IB4 or primary antibody at the appropriate dilution (e.g., use biotinylated IB4 at 1:100, rat anti-mouse PECAM1 at 1:200, or rat anti-mouse endomucin at 1:400).
5. Wash thoroughly the hindbrains five times for 20–30 min each in PBT at room temperature.
6. Incubate overnight at 4 °C with gentle agitation in blocking solution containing secondary antibodies, diluted as recommended by the manufacturer (typical range 1:200 to 1:500). For biotinylated IB4 detection, use HRP-tagged streptavidin, and for PECAM1 or endomucin detection, use HRP-tagged anti-rat antibodies.
7. Wash the hindbrains as described in **step 5**.
8. Incubate the hindbrains in a solution of diaminobenzidine for 30 min at room temperature with gentle agitation, and then in a solution of diaminobenzidine and urea/hydrogen peroxide for ~3–5 min or until color develops (*see* **Note 8**).
9. To stop the reaction, rinse the hindbrains in distilled water, wash twice in PBS for 5 min each and then post-fix in 4%

formaldehyde for 5 min at room temperature for long-term preservation of antibody binding. Briefly rinse twice in PBS.

PAUSE STEP: Samples can either be processed for imaging or be stored at 4 °C in PBS for up to 1 week before imaging without loss of signal.

3.2.3 Hindbrain Wholemount Fluorescent Staining

1. Incubate in blocking solution for 30 min with gentle agitation at room temperature. For biotinylated IB4, PECAM1, or endomucin labeling, use 10% heat-inactivated normal goat serum in PBT as blocking solution.
2. Incubate overnight at 4 °C with gentle agitation in blocking solution containing IB4 or primary antibody at the appropriate dilution (e.g., use biotinylated IB4 at 1:100, rat anti-mouse PECAM1 at 1:200, or rat anti-mouse endomucin at 1:400).
3. Wash thoroughly the hindbrains five times for 20–30 min each in PBT at room temperature.
4. Incubate overnight at 4 °C with gentle agitation in blocking solution containing fluorophore-tagged streptavidin to detect biotinylated IB4, e.g., AlexaFluor488 streptavidin. For PECAM1 or endomucin detection, use fluorophore-tagged anti-rat antibodies, e.g., AlexaFluor488 goat anti-rat. If combining IB4 and antibody staining, use complementary fluorophores, e.g., AlexaFluor488 and AlexaFluor647 (*see* **Note 9**).
5. Wash the hindbrains as described in **step 3**.
6. Post-fix in 4% formaldehyde for 5 min at room temperature for long-term preservation of antibody binding. Briefly rinse twice in PBS (*see* **Note 10**).

PAUSE STEP: Samples can now be either processed for imaging or stored at 4 °C in PBS for up to a week before imaging.

3.3 Day 4: Hindbrain Mounting and Image Acquisition and Analysis

1. Cover a glass microscope slide with two layers of black electrical tape and use a scalpel to cut a rectangular well big enough to hold one hindbrain (Fig. 2a, *i–ii*).
2. Transfer each hindbrain into the well using a plastic Pasteur pipette with a wide-bore tip (or cut off the very end of a small-bore tip), place the hindbrain ventricular side facing up, and then remove the excess PBS. Add an appropriate aqueous-based mounting medium reagent (e.g., SlowFade Antifade reagent) and cover the well slowly with a glass coverslip to seal it (Fig. 2a, *iii–iv*) (*see* **Notes 11 and 12**).
3. Imaging of histochemical stained hindbrains can be done using a stereomicroscope with bright light illumination and an appropriate zoom range (Fig. 2b–d). For fluorescent stained hindbrains, use an epifluorescent or confocal laser-scanning microscope equipped with lenses suitable for aqueous-based

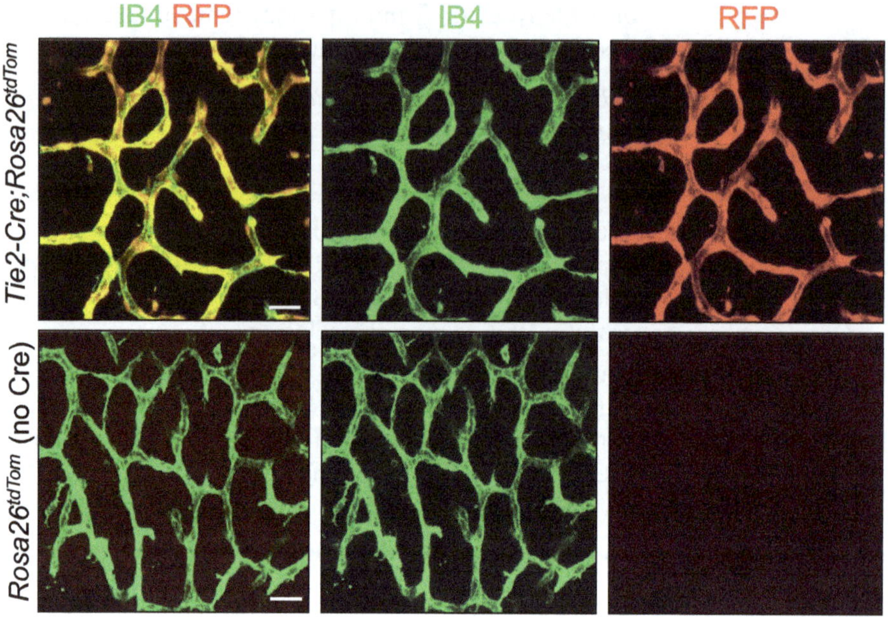

Fig. 4 Cre-mediated labeling of endothelium in embryonic hindbrain vessels. Littermate E12.5 hindbrains from a cross of mice carrying *Tie2-Cre* [25] and *Rosa26*^{tdTom} [26] to activate vascular endothelial tdTomato expression, shown in maximum intensity projections of confocal z-stacks scans after fluorescent staining for IB4 and RFP to identify endothelial cells and Cre-dependent tdTomato, respectively. *Tie2-Cre;Rosa26*^{tdTom} mice express Cre recombinase in endothelial cells under the control of the *Tie2* promoter, and Cre-mediated recombination allows expression of the fluorescent tdTomato reporter due to removal of a floxed stop codon (e.g., [12, 13]). Hindbrains from embryos lacking Cre do not activate the tdTomato reporter. The overlay and individual IB4 and RFP channels are shown. Scale bars: 50 μm

mounted slides and optical filters appropriate for the fluorophores used for immunolabeling (Figs. 3 and 4).

4. Histochemical images for the embryonic hindbrains can be acquired at low magnification (e.g., zoom 0.8×) to visualize the superficial vascular plexus (SVP) on the ventricular side as a whole. Turning the slide upside down allows visualizing the radial vessels entering the brain from the pial side (Fig. 2b). At least two randomly selected areas in the center of each hindbrain half should be acquired at higher magnification (e.g., zoom 8×) to allow manual quantification of the number vessel branchpoints (junctions) or radial vessels in the ventricular or pial side, respectively (Fig. 2c,d). The values obtained for these areas should be averaged for each hindbrain to obtain the value for that hindbrain to decrease technical error. Additionally, these angiogenic parameters can be automated and identified utilizing freely available software, e.g., AngioTool [22] or Angiogenesis Analyser [23], macros in Image J platform (Fig. 2e,f).

5. Fluorescence hindbrain images are best documented with maximum intensity projections of z-stacks scans acquired with a laser scanning confocal microscope (Figs. 3 and 4) (*see* **Note 13**). High-magnification images (e.g., acquired with 40× or 63× lenses) are required to visualize individual cells and their features, such as endothelial tip cell cells and their filopodia extensions (Fig. 3a), or to observe endothelial cell interactions with other hindbrain cell types such as pericytes and macrophages (Fig. 3b,c) or endothelial cell-lineage tracing (Fig. 4). Freely available (see above) or commercially available software packages (e.g., IMARIS from Bitplane or Volocity from Improvision) may be used for semi-automated and automated quantification of angiogenic parameters, including filopodia numbers, length, and branchpoints (Fig. 3a, *iii*).

4 Notes

1. Use PBS for non-permeabilized tissue staining, e.g., for visualization of extracellular rather than cytoplasmic epitopes, and PBT for permeabilization of the tissue to detect both extracellular and intracellular epitopes.
2. The success of tissue wholemount immunolabeling procedures usually relies on good antibody penetration. This could be a limitation at later gestational stages, thus alternative methods and protocols for hindbrain processing have been developed, e.g., staining of vibratome or cryostat preparations, and are elsewhere described [15, 24].
3. A small piece of yolk can be kept for genomic DNA extraction and subsequent PCR-based genotyping of genetically modified mice.
4. A small piece of embryonic tissue (e.g., 2 mm of tail snip) can be kept for genomic DNA extraction and subsequent PCR-based genotyping of genetically modified mice, instead of using yolk sac tissue.
5. This step should be followed from E11.5 onwards, when the meningeal tissue is consolidated, as it increases antibody penetration into the hindbrain neuroepithelium. For E9.5 or E10.5, the hindbrain neuroepithelium is thin and fragile, therefore avoid this step so not to damage the hindbrain tissue. Consider using alternative methods for hindbrain analysis at these early gestational stages, e.g., vibratome or cryostat preparations, which are elsewhere described [15, 24].
6. The graded series of methanol ensures gently hindbrain rehydration to prevent tissue damage.

7. Use serum from the host species that the secondary antibodies were raised in. For primary antibodies raised in goat, incubate in serum-free protein blocking solution (e.g., 5% bovine serum albumin in PBS or a suitable commercial alternative such a Dako serum-free protein block). This will reduce non-specific staining often observed when using primary antibodies raised in goat.
8. Samples need to be light-protected, as diaminobenzidine and urea/hydrogen peroxide solutions are photosensitive. Is critical to observe the samples periodically under a stereomicroscope to avoid overdeveloping the reaction.
9. Samples need to be light-protected to prevent photobleaching. When including primary antibodies raised in goat in the protocol, use anti-goat Fab fragments of secondary antibodies to reduce non-specific staining.
10. The described protocol can be adapted to the multi-labeling of endothelial cells, vessel-associated cells, and surrounding extracellular matrix in the hindbrain as desired. This is achieved by combining primary antibodies raised in different species, or using such antibodies together with IB4 (e.g., IB4 and NG2 or IB4 and F4/80, *see* Fig. 3). For this purpose, apply the primary antibodies, or IB4 and the primary antibodies, together in the blocking solution; after washing, apply the appropriate secondary antibodies (and streptavidin in the case of IB4) together in the blocking solution.

    Additional antibodies specific for.

    Endothelial cells (EC): goat anti-mouse VEGFR2.

    Dilution (1:200).

    R&D Systems.

    Various cell types, including EC: goat anti-mouse/ratNRP1.

    Dilution (1:200).

    R&D Systems.

    Pericytes: rabbit anti-mouse NG2 chondroitin sulfate proteoglycan.

    Dilution (1:200).

    Millipore.

    Macrophage: rat anti-mouse F4/80.

    Dilution (1:400).

    Bio-Rad.

    Basement membrane: rabbit anti-mouse collagen IV.

    Dilution (1:400).

    Bio-Rad.

11. Trapping of air bubbles in mounted samples can occur when placing the coverslip over the sample. Lower the coverslip carefully (e.g., applying a right to left motion) onto the slide to avoid trapping air bubbles; if bubbles persist, remove the coverslip carefully and repeat the mounting procedure.
12. If the hindbrain was fluorescently labeled, store the slide at 4 °C in the dark until image acquisition to avoid photobleaching of the fluorophores.
13. If fluorescent speckles are seen, the reason could be that the secondary antibodies have formed precipitates. To avoid this problem, always centrifuge the fluorescently labeled antibody solutions in a refrigerated benchtop centrifuge at top speed and carefully remove the supernatant without disturbing any precipitates.

Acknowledgments

G. D'A. and C.R. are supported by a Wellcome Trust Award [205099/Z/16/Z] to C. R. The authors thank Laura Denti for mouse husbandry.

References

1. Ruhrberg C (2003) Growing and shaping the vascular tree: multiple roles for VEGF. BioEssays 25(11):1052–1060. https://doi.org/10.1002/bies.10351
2. Potente M, Gerhardt H, Carmeliet P (2011) Basic and therapeutic aspects of angiogenesis. Cell 146(6):873–887. https://doi.org/10.1016/j.cell.2011.08.039
3. Chappell JC, Wiley DM, Bautch VL (2011) Regulation of blood vessel sprouting. Semin Cell Dev Biol 22(9):1005–1011. https://doi.org/10.1016/j.semcdb.2011.10.006
4. Bar T (1983) Patterns of vascularization in the developing cerebral cortex. Ciba Found Symp 100:20–36. https://doi.org/10.1002/9780470720813.ch3
5. Ruhrberg C, Bautch VL (2013) Neurovascular development and links to disease. Cell Mol Life Sci 70(10):1675–1684. https://doi.org/10.1007/s00018-013-1277-5
6. Tata M, Ruhrberg C, Fantin A (2015) Vascularisation of the central nervous system. Mech Dev 138(Pt 1):26–36. https://doi.org/10.1016/j.mod.2015.07.001
7. Raab S, Beck H, Gaumann A, Yuce A, Gerber HP, Plate K et al (2004) Impaired brain angiogenesis and neuronal apoptosis induced by conditional homozygous inactivation of vascular endothelial growth factor. Thromb Haemost 91(3):595–605. https://doi.org/10.1160/TH03-09-0582
8. Ruhrberg C, Gerhardt H, Golding M, Watson R, Ioannidou S, Fujisawa H et al (2002) Spatially restricted patterning cues provided by heparin-binding VEGF-A control blood vessel branching morphogenesis. Genes Dev 16(20):2684–2698. https://doi.org/10.1101/gad.242002
9. Tata M, Wall I, Joyce A, Vieira JM, Kessaris N, Ruhrberg C (2016) Regulation of embryonic neurogenesis by germinal zone vasculature. Proc Natl Acad Sci U S A 113(47):13414–13419. https://doi.org/10.1073/pnas.1613113113
10. Fantin A, Vieira JM, Gestri G, Denti L, Schwarz Q, Prykhozhij S et al (2010) Tissue macrophages act as cellular chaperones for vascular anastomosis downstream of VEGF-mediated endothelial tip cell induction. Blood 116(5):829–840. https://doi.org/10.1182/blood-2009-12-257832
11. Engelhardt B, Liebner S (2014) Novel insights into the development and maintenance of the blood-brain barrier. Cell Tissue Res 355(3):687–699. https://doi.org/10.1007/s00441-014-1811-2
12. Gerhardt H, Ruhrberg C, Abramsson A, Fujisawa H, Shima D, Betsholtz C (2004)

Neuropilin-1 is required for endothelial tip cell guidance in the developing central nervous system. Dev Dyn 231(3):503–509. https://doi.org/10.1002/dvdy.20148
13. Fantin A, Vieira JM, Plein A, Denti L, Fruttiger M, Pollard JW et al (2013) NRP1 acts cell autonomously in endothelium to promote tip cell function during sprouting angiogenesis. Blood 121(12):2352–2362. https://doi.org/10.1182/blood-2012-05-424713
14. Fruttiger M (2007) Development of the retinal vasculature. Angiogenesis 10(2):77–88. https://doi.org/10.1007/s10456-007-9065-1
15. Fantin A, Vieira JM, Plein A, Maden CH, Ruhrberg C (2013) The embryonic mouse hindbrain as a qualitative and quantitative model for studying the molecular and cellular mechanisms of angiogenesis. Nat Protoc 8(2): 418–429. https://doi.org/10.1038/nprot.2013.015
16. Nagy A (2000) Cre recombinase: the universal reagent for genome tailoring. Genesis 26(2): 99–109
17. Ferrara N, Carver-Moore K, Chen H, Dowd M, Lu L, O'Shea KS et al (1996) Heterozygous embryonic lethality induced by targeted inactivation of the VEGF gene. Nature 380(6573):439–442. https://doi.org/10.1038/380439a0
18. Carmeliet P, Ferreira V, Breier G, Pollefeyt S, Kieckens L, Gertsenstein M et al (1996) Abnormal blood vessel development and lethality in embryos lacking a single VEGF allele. Nature 380(6573):435–439. https://doi.org/10.1038/380435a0
19. Saito T (2006) In vivo electroporation in the embryonic mouse central nervous system. Nat Protoc 1(3):1552–1558. https://doi.org/10.1038/nprot.2006.276
20. Walantus W, Castaneda D, Elias L, Kriegstein A (2007) In utero intraventricular injection and electroporation of E15 mouse embryos. J Vis Exp 6:239. https://doi.org/10.3791/239
21. Szczurkowska J, Cwetsch AW, dal Maschio M, Ghezzi D, Ratto GM, Cancedda L (2016) Targeted in vivo genetic manipulation of the mouse or rat brain by in utero electroporation with a triple-electrode probe. Nat Protoc 11(3):399–412. https://doi.org/10.1038/nprot.2016.014
22. Zudaire E, Gambardella L, Kurcz C, Vermeren S (2011) A computational tool for quantitative analysis of vascular networks. PLoS One 6(11): e27385. https://doi.org/10.1371/journal.pone.0027385
23. Carpentier G, Berndt S, Ferratge S, Rasband W, Cuendet M, Uzan G et al (2020) Angiogenesis analyzer for ImageJ - a comparative morphometric analysis of "endothelial tube formation assay" and "fibrin bead assay". Sci Rep 10(1):11568. https://doi.org/10.1038/s41598-020-67289-8
24. Tata M, Ruhrberg C (2018) The mouse hindbrain as a model for studying embryonic neurogenesis. J Vis Exp (131). https://doi.org/10.3791/56793
25. Kisanuki YY, Hammer RE, Miyazaki J, Williams SC, Richardson JA, Yanagisawa M (2001) Tie2-Cre transgenic mice: a new model for endothelial cell-lineage analysis in vivo. Dev Biol 230(2):230–242. https://doi.org/10.1006/dbio.2000.0106
26. Madisen L, Zwingman TA, Sunkin SM, Oh SW, Zariwala HA, Gu H et al (2010) A robust and high-throughput Cre reporting and characterization system for the whole mouse brain. Nat Neurosci 13(1):133–140. https://doi.org/10.1038/nn.2467

Chapter 2

A Workflow to Track and Analyze Endothelial Migration During Vascular Development in Zebrafish Embryos Using Lightsheet Microscopy

Yan Chen, Paul C. Evans, and Robert N. Wilkinson

Abstract

Zebrafish allow unrivalled in vivo imaging of vascular development due to their optical translucency and the availability of transgenic lines which fluorescently label cells and tissues of interest. Advances in light sheet fluorescence microscopy allow longer and faster imaging of live embryos at higher resolutions than previously possible, which facilitates study of dynamic cellular and molecular mechanisms underlying vessel formation and function. Here we describe a workflow using lightsheet microscopy to quantify endothelial cell (EC) migration dynamics during vascular development. Tracking movement of EC nuclei and analyzing the properties of EC migration trajectories permit detailed studies of angiogenesis and vascular remodeling in different contexts.

Key words Zebrafish, Endothelial cell, Lightsheet microscopy, LSFM, Image processing, Cell tracking, Cell migration, Vessel remodeling, TrackMate, MATLAB

1 Introduction

Early studies on the formation of vertebrate vasculature networks using genetic and lineage tracing approaches yielded detailed information on how embryonic vessels are formed [1]. Historically, these studies have relied heavily on static image analysis; however, the development of a functional vasculature is a concert orchestrated by highly dynamic cellular behaviors. Thus, it is unsurprising that current researchers have come to increasingly rely on time-lapse imaging to study the spatiotemporal dynamics of vascular development [2–4]. Endothelial cells (ECs) line the inner lumen of blood vessels and are among the earliest differentiated cell types to develop in the embryo [5]. ECs form the barrier between vessels and tissue, and during embryonic development, these cells represent the surface for metabolic exchange at stages when organs that

Andrew V. Benest (ed.), *Angiogenesis: Methods and Protocols*, Methods in Molecular Biology, vol. 2441,
https://doi.org/10.1007/978-1-0716-2059-5_2,

will ultimately perform these functions are yet to develop. The zebrafish is a particularly useful system to study vascular development and EC behavior in vivo since its cardiovascular system develops rapidly and displays a high degree of genetic conservation with humans [6]. Furthermore, the optical translucency of zebrafish embryos permits in vivo imaging of even the deepest of tissues [3, 7]. Recent advances in fluorescent microscopy and the ever-growing availability of mutant and transgenic lines have enhanced the utility of zebrafish to investigate cardiovascular development at cellular and subcellular resolutions [3]. One such advance in light-sheet fluorescence microscopy (LSFM), also known as Selective Plane Illumination Microscopy (SPIM), illuminates the sample through a thin sheet of laser light and detects emitted light perpendicular to the sample being imaged (Fig. 1). In comparison to confocal microscopy, LSFM reduces phototoxicity and fluorophore bleaching since only the plane in focus is illuminated, thus LSFM is excellent for long-term live imaging. LSFM also facilitates high imaging speed and resolution, allowing observation of dynamic processes and rare events [10–12].

Live imaging experiments often generate a substantial volume of time-lapse images which are frequently terabytes of data in scale and are difficult to fully exploit using simple observation [13]. Consequently, there is an increasing need for computerized image processing to generate quantitative results and interrogate complex biological imaging data. A wide variety of image analysis software has been developed to handle these data [14, 15]; however, most zebrafish cell tracking studies focus on lineage tracing and clonal analysis during gastrulation [15], while studies on EC migration behavior focus on observation of cell migration trajectories in a relatively small region or short time periods [16–19]. This is in part a reflection of the technical limitations of efficient cell detection and trajectory analysis. Here we propose a workflow (Fig. 1) to track migratory ECs in zebrafish using freely available TrackMate [8] plugins in Fiji [9] and provide customized MATLAB scripts to analyze migration trajectory properties including distance, displacement, and angle of migration. These approaches facilitate detailed interrogation of EC behavior during the formation of vascular networks.

2 Materials

1. E3 solution (5 mM NaCl, 0.17 mM KCl, 0.33 mM $CaCl_2$, 0.33 mM $MgSO_4$).
2. Low melting point agarose (Thermo Scientific™).
3. Tricaine (3-amino benzoic acidethylester, Merck).
4. Lightsheet microscope.

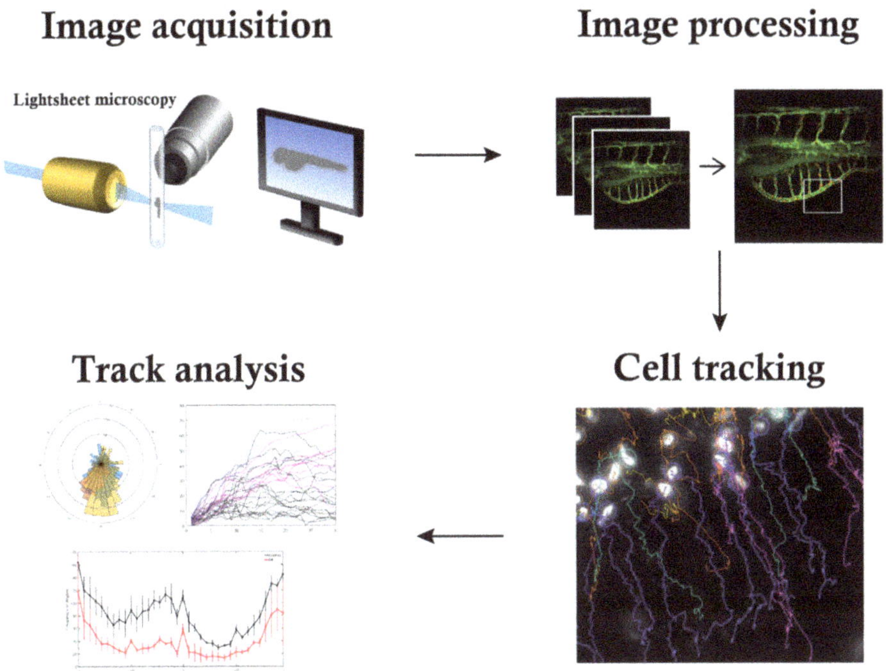

Fig. 1 Endothelial cell tracking workflow. The workflow of this protocol includes four steps: (1) image acquisition introduces how to mount and image zebrafish embryos using lightsheet microscopy; (2) image processing is recommended to rectify vibration- and gravity-induced cell migration noise; (3) cell migration trajectories are traced using TrackMate plugin [8] in Fiji [9]; (4) the track statistics are analyzed using customized MATLAB (Mathworks) scripts and plotted using Prism (GraphPad)

5. Fiji [9] (installed with *Linear Stack Alignment with SIFT* plugin [20] and *TrackMate* plugin [8]).
6. MATLAB 2019a (Mathworks).

3 Methods

3.1 Sample Preparation and Image Acquisition

1. Dissolve the low melting point agarose (LMP) (Thermo Scientific™) to a final concentration of 1% in E3 solution. Make 1 ml of aliquots of this solution in 1.5-ml Eppendorf tubes. Place these tubes in a bench-top heating block to equilibrate the LMP to approximately 42 °C (*see* **Note 1**).
2. Incubate dechorionated embryos in Tricaine working solution (162 mg/L) in a petri dish for at least 5 min to reduce embryo movement.
3. Transfer embryo to a new dry petri dish and remove surplus E3 medium using a Pasteur pipette. Immediately, cover the embryo with a drop of 1% LMP (*see* **Note 2**).

4. Before the gel is set, quickly mount the embedded embryo in LMP agarose as appropriate for the lightsheet microscope in use. If using capillaries or fluorinated ethylene propylene (FEP) tubes, it is easier to draw the embryo into the capillary or tube headfirst (*see* **Note 3**).
5. Fill the lightsheet imaging chamber or cover embedded embryos with a mixture of E3 and 1× tricaine to ensure embryos remained anesthetized throughout imaging. If possible, set the temperature in the lightsheet imaging chamber to 28.5 °C.
6. Set desired parameters (e.g., laser power, interval time of time series) in lightsheet software and commence imaging according to desired experimental parameters (*see* **Note 4**).

3.2 Image Processing

1. Once time-lapse images are acquired, generate a maximum intensity projection (MIP) to confirm time-lapse imaging has proceeded as expected, and export images as a series of TIFF files.
2. Load the TIFF images into Fiji (ImageJ) (by *Shift* select all images and drag to Fiji) for processing. Make the images as stacks by choosing *Image > Stacks > Images to Stack* and select *Use Titles as Labels.* Check the time series and if it is not correctly aligned in time, sort it by choosing *Image > Stacks > Tools > Stack Sorter > Sort by label.*
3. In Fiji, set image scale by choosing *Image > Properties* (channels, slices, frames, and voxel: *x*, *y*, *z*, unit). The pixel size is stored in the metadata of the acquisition file, e.g., Zeiss .czi file.
4. Register the image stacks by aligning them using the Fiji plugin, *Linear Stack Alignment with SIFT* [20]. Go to *Plugins > Registration > Linear Stack Alignment with SIFT* (*see* **Note 5**).
5. If the SIFT registration is used, run the algorithm with the following parameters (*see* **Note 6**): *1.6px* Gaussian blur, *10* steps per scale, octave, *64px* minimum image size, *1920px* single plane field of view, *15* feature descriptors, *8* orientation bins, *0.98* closest/next ratio, *3px* maximum alignment error (global alignment with 10px), *0.05* inlier ratio, *rigid* transform and without interpolation.
6. Save the aligned stacks as a new TIFF file, named appropriately. If multi-channel images are used e.g. EC nuclei labeled with RFP and EC membrane labeled with GFP), go to *Image > Color > Split Channels*, and save each channel as a separate TIFF file (*see* **Note 7**).

3.3 Cell Tracking

1. EC nuclei are labeled and tracked in a semi-automated manner using TrackMate Plugin, implemented in Fiji [8]. Go to *Plugins* > *Tracking* > *TrackMate* (*see* **Note 8**).
2. The image properties including *Pixel width*, *height*, and *depth* and *Time interval* should be automatically captured in calibration settings. Click *Next* button.
3. *Select a detector* > *LoG detector* (*see* **Note 9**) > *Next.*
4. *LoG detector* > *Estimated blob diameter*: 10 μm (auto) (*see* **Note 10**).
5. The detection step may take a few seconds or minutes depending on your computer performance and image sample. Click *Next* once finished.
6. *Initial thresholding* > *(include) Quality Above 0.05* (*see* **Note 11**).
7. *Select a view* > *HyperStack Displayer* > *Next.*
8. *Set filters on spots* > *+ X (Above and Below)* > *+ Y (Above and Below)*, to discard spots outside the ROI. You can also stack *Mean intensity* and *Contrast* filters which are often used in our SIVP samples. Some of the purple circles displayed on the ImageJ stack should disappear after applying filters (*see* **Note 12**). Click *Next* when you are satisfied with the spots.
9. *Select a tracker* > *LAP tracker* > *Next.* In LAP tracker settings, enable Track segment *gap closing* and Track segment *splitting* (*see* **Note 13**), and leave the remaining settings as default.
10. *Set filters on tracks* > *Next.* No filters are used in this step since unwanted tracks will be deleted manually in the following steps (*see* **Note 14**).
11. On *Display options*, click *TrackScheme* which opens a map of all tracks, displayed vertically over time. This map allows users to delete or add tracks manually. An experienced user should be able to correct all tracks within a single day. For more details, follow the *TrackMate manual* [21]. Here we list a few common actions:

    (note that the TrackScheme and the HyperStack viewer are interlinked, e.g., selecting one spot in either panel will highlighted it in green in both panels).

    (a) *Delete a spot*: select one spot on either the HyperStack or the TrackScheme, press *Delete* on keyboard.

    (b) *Delete a whole track*: select one spot on the HyperStack, right click anywhere on TrackScheme, click Select whole track and press *Delete* key.

    (c) *Delete a link*: select a link between two spots on the TrackScheme and press Delete key.

(d) *Create a new spot*: double-click on the HyperStack anywhere free of any existing spots.

(e) *Link two spots*: on HyperStack, select a spot in frame T, move to the next frame T + 1 by scrolling the mouse, hold the *Shift* key, and click on the second spot (both cells/spots now are highlighted in green). Go to TrackScheme and right click anywhere, select Link 2 spots.

(f) *Link several newly added spots*: on the TrackScheme, newly created spots will appear on the far right of the panel, draw a rectangle to select all spots or hold *Shift*, and click individual spots. Right click Link *N* spots to link all spots.

(g) *Refresh the TrackScheme*: once spots and links are manually re-arranged, press the Layout button on the toolbar to refresh the track layout.

12. On the *Display options*, click *Save* on the bottom of the panel to save the tracking file. It is possible to edit the tracks in this file at a later date.

13. Once all tracks are correctly labeled, export these using *Display options*, select *Analysis* (next to *TrackScheme*), and three dialogues will appear. Select *File* > *Save as* on each dialogue will output the file as an Excel file. Three Excel files will be exported, including *Track statistics*, *Spots in tracks statistics*, and *Links in tracks statistics*. The latter two files are required for analysis in the following steps.

14. To present the tracks as a movie, click *Next* and reach the *Select an action* panel, select *Capture overlay* > *Execute*. On the Fiji toolbar, select *File* > *Save as* > *AVI* to save the TrackMate capture as a movie file.

3.4 Data Analysis

1. Locate the exported files, namely the *Spots in tracks statistics* and the *Links in tracks statistics*, in the same folder as the MATLAB scripts, including *script 1 track analysis* and *script 2 direction analysis*. These scripts are available on GitHub with the following link: https://github.com/yanc0913/SIVP_cell_tracking.

2. To run script 1 (*see* **Note 15**), users are asked to type in *input file names* (i.e., the names given to the *Spots in tracks statistics* and the *Links in tracks statistics* files) and an *output file name*. The output file includes values analyzed from the cell trajectory, including *track distance*, *track displacement*, *distance/displacement ratio*, *migration velocity of each step*, as well as *track* and *step delta x* and *delta y*. These results will be output in an Excel file with several spreadsheets.

3. Copy and paste the output data to any statistical analysis tool, for example, GraphPad Prism, to perform statistical analysis and plotting.
4. Script 2 includes two sections.
5. The first section generates scatter plots of cell steps in one control and one experimental repeat. To run this section, values of *step delta x* and *delta y* are required, which are obtained from *script 1*. Users are asked to type in *input* and *output file names*, as well as *column locations* in the output spread sheet for data to be written in. The second section generates scatter plots and polar histogram plots of cell migration steps in all repeats.

4 Notes

1. LMP can be maintained molten in a heating block for several weeks at 42 °C, but do not re-melt LMP once solidified since this reduces clarity.
2. The agarose must have cooled to 42 °C to avoid burning the embryo.
3. Depending on experiment and microscope, it may be possible to mount experimental and control embryos within the same agarose to allow simultaneous imaging.
4. We have employed the zebrafish transgenic lines *Tg(fli1a:* EGFP*)^{y1}* [22] and *Tg(fli1a:nls*-mCherry*)^{SH550}* (to be published elsewhere) to image EC cytoplasm in the green channel and EC nuclei in the red channel. We find that an imaging interval of less than 30 min is required for clear EC nuclei tracking. A shorter interval (e.g.,10 min) allows better image registration and tracking accuracy but has associated increases in computational cost.
5. This step is *optional* depending upon the mode of embryo mounting and the microscope used, but we find that when imaging using a Zeiss Lightsheet Z.1, mounting the embryo in a column of agarose, leads to movement of embryo out of the ROI during long-time imaging due to gravity. Embryos can also twitch due to muscle contraction under insufficient anesthesia. Registration of these deviations with SIFT improves EC tracking in subsequent steps.
6. The SIFT plugin identifies a set of points of interest in two images by difference of Gaussian detector which generates similarity-invariance relative to local features. The parameters have been explained on the Fiji-Wiki https://imagej.net/Feature_Extraction. The parameters used in this method is recommended for zebrafish endothelial cells.

7. The series of TIFF files containing the EC nuclei channel will be used for cell tracking. We recommend keeping the ROI in the orientation you wish to present and crop extraneous regions from the files to reduce the computational cost of cell tracking.
8. The following parameters are used to detect ECs within the zebrafish sub-intestinal vein plexus (SIVP). For more details to optimize your image sample, please follow the *TrackMate Manual* [21].
9. The LoG detector is optimal for intermediate spot sizes (5–20 pixels in diameter), and the DoG detector for small sizes (below 5 pixels in diameter). Downsample LoG detector is for large spot sizes above 20 pixels in diameter, and the Block LoG detector splits the image in small blocks. In our samples, the spots (nuclei) are around 10 μm (5–10 pixels), so we used LoG detector.
10. If your sample has a high noise background, you can check *Use median filter*, which will apply a 3 × 3 median filter and reduces spurious spots. *Do sub-pixel localization* detects fine locations and usually generates more spurious spots.
11. The quality of all detected spots is displayed in a histogram, and the peak at low quality (on the left) usually represents spurious spots. All spots with a quality value below the threshold will be *discarded* and, therefore, will not be tracked in the following steps. This step is *irreversible*, thus leaves the threshold close to 0 to keep all target spots if your sample does not too many spots (>20,000).
12. The spots are not discarded after filtering (these are not displayed) and you can always come back to this step to include/exclude more spots if required. However, the filtered spots are not included in the following particle linking step; therefore, you should include as many desired EC nuclei as possible at this step.
13. *Gap-closing* allows the tracks to recuperate when a spot disappears for one or a few frames and reconnects with reappearing spots later. *Splitting* allows a track to split into two, for instance when a cell undergoes mitosis.
14. We do not use any filters on tracks when tracking the EC nuclei of the SIVP because these ECs display complex migration movements and applying filters may delete partial tracks which are desired for analysis. It is easier to delete a track at a later stage than it is to add one manually.
15. The script is for one experimental repeat at a time; therefore, users would need to change file names accordingly for different repeats. To understand the basic idea of script 1, *see* Fig. 2.

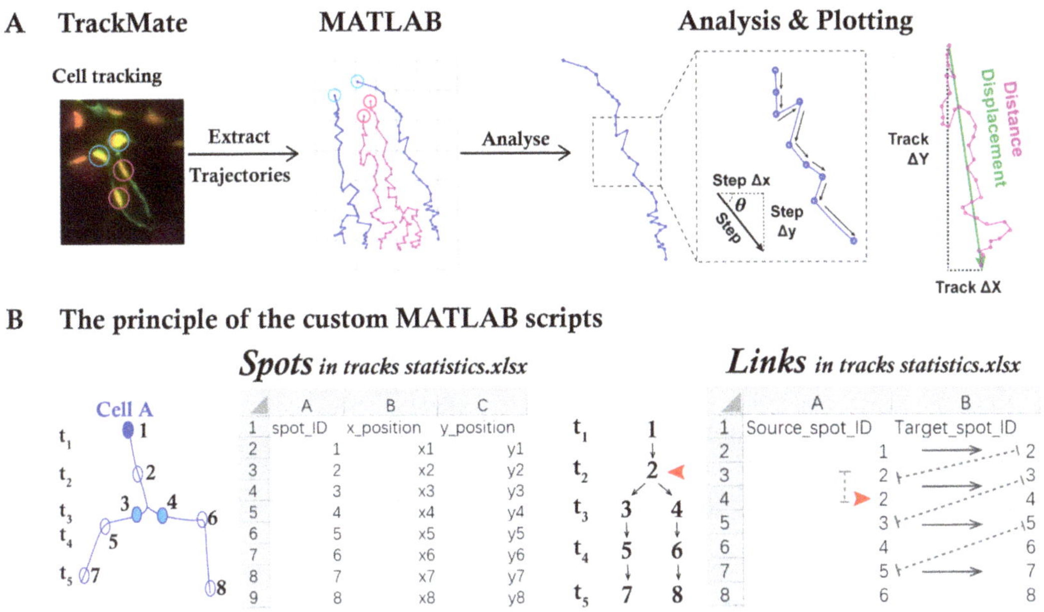

Fig. 2 The results and principle of the custom MATLAB scripts. (**a**) Following cell tracking using TrackMate, cell migration trajectories are extracted from Excel files using customized MATLAB scripts and are analyzed to quantify migration step parameters, including *Step* Δx *and* Δy, *Step displacement*, *migration velocity*, and *polar histogram of migration angles* (θ); and track parameters, including *Track* ΔX *and* ΔY, *Track distance*, *Track displacement*, and *the ratio of Distance/Displacement*. (**b**) Diagrams show the core functions of the MATLAB scripts to detect individual tracks and links of a cell between two consecutive time points. An endothelial cell A (in dark blue) migrates from t_1 to t_5 and undergoes mitosis to generate two daughter cells (in light blue) at t_3. The cell (also known as a spot in TrackMate) at every time frame is given a unique spot ID (1–8), as shown in *Spots in tracks statistics.xlsx* file. The links of a cell between two consecutive time frames is stored in *Links in tracks statistics.xlsx* file. As shown in the examples, a link is established between a *Source* spot (1) in t_1 and a *Target* spot (2) in the following time frame t_2. A condition is set in the scripts where a *Source* spot ID in next ($i + 1$) row equals to a *Target* spot ID; if the condition is true, then a new link is established (e.g., 2→3). If a cell undergoes mitosis (2), it will be flagged (red arrowheads), and the scripts will proceed through the steps of one daughter cell first (3→7) and return to the flagged position to proceed through the steps of the other daughter cell (2→8)

Users can use the core function of detecting links between spots in the MATLAB scripts and customize the outcome to one's own need.

Acknowledgments

We are grateful to Katie Fisher for help with custom MATLAB code and release from debugging hell. This work was supported by a Chinese Scholarship Council studentship (YC) and Nottingham Research Fellowship (RNW).

References

1. Herbert SP, Stainier DYR (2011) Molecular control of endothelial cell behaviour during blood vessel morphogenesis. Nat Rev Mol Cell Biol 12:551–564
2. Meijering E, Dzyubachyk O, Smal I (2012) Methods for cell and particle tracking. Methods Enzymol 504:183–200
3. Okuda KS, Hogan BM (2020) Endothelial cell dynamics in vascular development: insights from live-imaging in zebrafish. Front Physiol 11:842
4. Savage AM et al (2019) tmem33 is essential for VEGF-mediated endothelial calcium oscillations and angiogenesis. Nat Commun 10:732
5. Adams RH, Alitalo K (2007) Molecular regulation of angiogenesis and lymphangiogenesis. Nat Rev Mol Cell Biol 8:464–478
6. Bowley G, Kugler E, Wilkinson R, Lawrie A, Eeden F, Chico TJA, Evans PC, Noël ES, Serbanovic-Canic J (2021) Zebrafish as a tractable model of human cardiovascular disease. Br J Pharmacol
7. Wilkinson RN, Van Eeden FJM (2014) The zebrafish as a model of vascular development and disease. Prog Mol Biol Transl Sci 124:93–122
8. Tinevez JY, Perry N, Schindelin J, Hoopes GM, Reynolds GD, Laplantine E, Bednarek SY, Shorte SL, Eliceiri KW (2017) TrackMate: an open and extensible platform for single-particle tracking. Methods 115:80–90
9. Schindelin J et al (2012) Fiji: an open-source platform for biological-image analysis. Nat Methods 9:676–682
10. Keller PJ, Dodt H-U (2012) Light sheet microscopy of living or cleared specimens. Curr Opin Neurobiol 22:138–143
11. Keller PJ, Schmidt AD, Wittbrodt J, Stelzer EHK (2008) Reconstruction of zebrafish early embryonic development by scanned light sheet microscopy. Science 322:1065–1069
12. Huisken J, Swoger J, Del Bene F, Wittbrodt J, Stelzer EHK (2004) Optical sectioning deep inside live embryos by selective plane illumination microscopy. Science 305:1007–1009
13. Höckendorf B, Thumberger T, Wittbrodt J (2012) Quantitative analysis of embryogenesis: a perspective for light sheet microscopy. Dev Cell 23:1111–1120
14. Xia S, Zhu Y, Xu X, Xia W (2013) Computational techniques in zebrafish image processing and analysis. J Neurosci Methods 213:6–13
15. Mikut R et al (2013) Automated processing of zebrafish imaging data: a survey. Zebrafish 10: 401–421
16. Rochon ER, Menon PG, Roman BL (2016) Alk1 controls arterial endothelial cell migration in lumenized vessels. Development 143:2593–2602
17. Udan RS, Vadakkan TJ, Dickinson ME (2013) Dynamic responses of endothelial cells to changes in blood flow during vascular remodeling of the mouse yolk sac. Development 140: 4041–4050
18. Goi M, Childs SJ (2016) Patterning mechanisms of the sub-intestinal venous plexus in zebrafish. Dev Biol 409:114–128
19. Hen G et al (2015) Venous-derived angioblasts generate organ-specific vessels during embryonic development. Development 142:4266–4278
20. Lowe DG (2004) Distinctive image features from scale-invariant keypoints. Int J Comput Vis 60:91–110
21. Tinevez J-Y (2016) TrackMate documentation. https://imagej.net/_images/8/85/TrackMate-manual.pdf
22. Lawson ND, Weinstein BM (2002) In vivo imaging of embryonic vascular development using transgenic zebrafish. Dev Biol 248:307–318

Chapter 3

Analyzing Vessel Regression and Endothelial Apoptosis as a Component of Angiogenic Vessel Remodeling

Lachlan Whitehead, Emma C. Watson, and Zoe L. Grant

Abstract

Angiogenic vessel remodeling is a critical step in establishing a hierarchical vessel network. Vessel networks rapidly expand through angiogenesis in response to pro-angiogenic factors. This leads to an initially dense vessel network that requires selective regression of vessel branches to establish a hierarchical conduit for blood flow, a process known as pruning. This involves migration of endothelial cells from low-flow vessels to adjacent high-flow vessels and generally occurs independently of cell death. Vessels may also regress in response to other stimuli, including reduced metabolic demand, redundancy, and pathological stimuli. In these contexts, widespread vessel regression typically occurs and involves loss of endothelial cells by apoptotic cell death. Thus, vessel remodeling occurs via both apoptosis independent and dependent vessel regression. In this chapter, we outline a semi-automated method for quantifying vessel regression using the neonatal model of angiogenesis. We further provide instruction on analyzing endothelial apoptosis in this model.

Key words Angiogenesis, Vessel regression, Vessel remodeling, Apoptosis, Cell death

1 Introduction

Blood vessels form a tubular network that facilitates the circulation of blood. This ensures the distribution of oxygen, nutrients, and cells, as well as the removal of waste and metabolic by-products from almost all tissues in the body. This is achieved by hierarchical organization of blood vessels throughout the body. Blood vessel lumens are lined by endothelial cells (ECs) that are surrounded by perivascular support cells and layers of basement membrane [1]. During development, vessel networks expand by angiogenesis to meet the metabolic demands of growing tissues. While angiogenesis is generally regarded as a process of vessel growth, the production of mature, efficient blood vessel networks by angiogenesis is just as dependent on vessel removal as it is on new vessel formation [2].

Andrew V. Benest (ed.), *Angiogenesis: Methods and Protocols*, Methods in Molecular Biology, vol. 2441,
https://doi.org/10.1007/978-1-0716-2059-5_3,

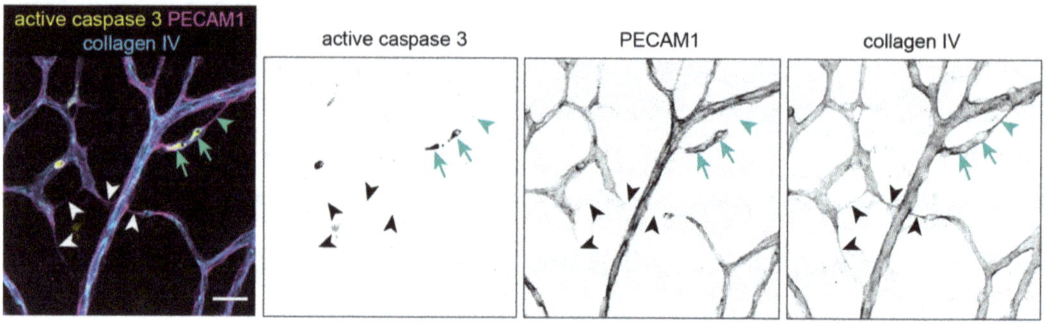

Fig. 1 Apoptosis and vessel regression during angiogenic vessel remodeling. Endothelial cell apoptosis may accompany vessel regression. Green arrowhead indicates regressing vessel associated with apoptotic EC (green arrow); white arrowhead indicates regressing vessel not associated with apoptosis. Scale bar: 30 μm

Ongoing endothelial sprouting and proliferation during angiogenesis creates an initially dense and immature vessel network incapable of efficient blood delivery [2]. The removal of selected vessel segments from this immature plexus through vessel "pruning" shapes it into a mature, hierarchical network that allows efficient blood flow [3]. Vessel regression may also serve to remove vessels as the metabolic demand of the tissue changes, including removal of redundant vessel networks in their entirety, for example, the hyaloid and pupillary vessel networks present in the developing eye [4, 5]. In patent, perfused vessels, the endothelium is surrounded by a collagen IV–rich basement membrane; however, when a vessel regresses, the ECs are lost and a "sleeve" of collagen IV is left behind where the vessel once was [2]. Vessel regression may be influenced by changes in blood flow [3, 6–8], vessel stability [9], and disturbances in tissue physiology [10–14]. As vessels regress, the ECs that formed them must be removed. Depending on the context, this may or may not require apoptosis [15] (Fig. 1).

Apoptosis is a programmed form of cell death that removes superfluous or damaged cells to facilitate normal development and to maintain homeostasis in adults [16, 17]. Apoptosis can be triggered by both intrinsic and extrinsic cell stimuli causing the activation of a cascade of cysteine-aspartic proteases, or caspases, ultimately activating effector caspases 3, 6, and 7 that orchestrates the destruction of the cell [16]. Each regressing vessel network has a different requirement for apoptosis. In the retina, only 5% of pruned vessels in the retina are associated with an apoptotic EC [3], and EC apoptosis is dispensable in this process [18]. Instead, the primary mechanism of vessel pruning is the migration of ECs from low-flow vessels to adjacent high-flow vessels [3, 19, 20]. While EC apoptosis is infrequent and dispensable for the pruning of individual vessels during angiogenesis, it becomes highly active when large numbers of vessels regress together or when whole networks involute. In contexts of complete network regression/involution, EC apoptosis appears to have an indispensable role [18]. This includes in models of pathological vessel loss, such as in the mouse oxygen-induced retinopathy model [10]. The existing model suggests that EC-migration drives vessel pruning, while apoptosis clears away redundant ECs when they are unable to relocate to the protective environment of neighboring vessels.

In this chapter, we provide a detailed protocol for the semi-automated analysis of vessel regression using freely available code written for FIJI [21] as used in our previous publications [10, 18]. This automated program requires user input to choose the threshold for segmenting immunofluorescence images but performs all other analysis steps automatically. Benefits of the semi-automated method of analyzing vessel regression include faster analysis and a reduction in the introduction of human error and bias. Manually selecting the segmentation threshold is an essential step due to variations in staining intensity and background signal between images. The analysis method described here determines the ratio between an endothelial marker, such as PECAM1, and the basement membrane marker collagen IV as a measure of vessel regression. We also present quantification data showing differences in vessel regression ratios across different vascular regions of the postnatal retina. This quantification highlights an advantage of our method; by using ratios between vascular and basement membrane signals, we are able to detect changes in vessel regression between regions with different vessel density. Given that vessel regression may be accompanied by EC apoptosis, we also provide an overview for manually quantifying EC apoptosis.

2 Materials

This protocol details the whole-mount immunofluorescence staining and imaging steps necessary to perform quantification of vessel regression and apoptosis in dissected mouse retinas. We refer to previously published, detailed protocols for enucleating mouse eyes and retina dissection as described by Gammons et al. and Pitulescu et al. [22, 23].

2.1 Retinal Dissection and Flat-Mounting

1. Eyes fixed in 4% PFA for 2 h at 4 °C.
2. Dulbecco's phosphate-buffered saline (DPBS).
3. Tube roller.
4. 1% Triton X-100 in DPBS.
5. Normal donkey serum (NDS).
6. Active (cleaved) caspase 3 antibody (we used rabbit anti-cleaved caspase 3 from Cell Signaling Technology (clone5A1E) at 1/50).

7. Anti- PECAM1 antibody (we used rat anti-PECAM1 from BD Pharmingen (clone MEC13.3) at 5 μg/ml).
8. Anti-collagen IV antibody (we used goat anti-collagen IV from Millipore at 4 μg/ml).
9. Anti-rat, anti-rabbit, and anti-goat fluorescent secondary antibodies (in any combination of minimally overlapping fluorochromes).
10. Blocking buffer: 2% NDS, 1% Triton X-100 in DPBS.
11. Wash buffer: 0.01% Triton X-100 in DPBS.
12. Glass microscope slides.
13. Scotch tape.
14. Scalpel.
15. Coverslips.
16. Prolong Diamond mounting medium (Invitrogen, P36961).

2.2 Imaging/Quantitative Image Analysis

1. Confocal microscope.
2. ImageJ/FIJI.

3 Methods

3.1 Whole-Mount Retina Immunofluorescence Staining (See Notes 1 and 2)

1. Following fixation and washing in DPBS, dissect eyes in DPBS as shown by others [22, 23].
2. Transfer one or both retinas per mouse to a 1.5-ml tube.
3. Block and permeabilize retinas by adding 50 μl blocking buffer. Incubate at room temperature for 1 h with gentle rolling. Multiple 1.5-ml tubes can be put into a 50-ml tube for rolling.
4. Prepare primary antibody mix by diluting anti-PECAM1, anti-collagen IV, and anti-active caspase 3 antibodies to working concentration in blocking buffer. Carefully remove as much blocking buffer from retinas as possible and replace with 50 μl of diluted antibodies mix. Incubate at 4 °C overnight with gentle rolling.
5. Wash retinas with 1 ml wash buffer (0.01% Triton X-100 in DPBS) at room temperature, changing the wash buffer each hour for at least 5 h.
6. Prepare secondary antibody mix by diluting anti-rat, anti-goat, and anti-rabbit secondary antibodies in blocking buffer. Carefully remove wash buffer from retinas and add 50 μl to each retina tube. Incubate at 4 °C overnight with gentle rolling.
7. Wash retinas with 1 ml wash buffer at room temperature, changing the wash buffer each hour for at least 5 h.

3.2 Flat-Mounting Stained Retinas

1. Transfer a washed retina to a dish containing DPBS and use forceps to carefully pull away any debris that may have attached to the retina during the staining procedure. Make four radial cuts into the retina by pressing a syringe needle into the retina lengthways, each cut the length of half the retina. This produces four retina petals that allow the retina to lay flat.
2. Transfer cut retina to a glass microscope slide with pre-cut tape spacer on it. Tape spacers can be made by sticking a piece of tape to a glass slide and using a scalpel to cut out a square slightly larger than the retina. This ensures that the retina is not crushed between the slide and coverslip during mounting.
3. Use forceps to position retina with the vessel surface facing upwards.
4. Remove excess wash buffer from around retina using 200-µl pipette and use forceps to carefully lay each petal flat.
5. Add a suitable amount of Prolong Diamond mounting medium (about 10–20 µl) on a coverslip and around the retina and place the coverslip on top. Store flat, in the dark and allow to dry overnight before imaging.

3.3 Whole-Mount Retina Imaging

Imaging should be performed on a confocal microscope with sufficient resolution and magnification to visualize vessels. We use the Leica TCS SP8 confocal microscope with 20×/0.75 NA water objective. This is suitable for quantifying both vessel regression and apoptosis. All images should be obtained as *Z*-stacks through the entire region of interest. 1-µm step *Z*-slices are sufficient for detecting apoptotic ECs. We generally image the whole retina at the relevant timepoint and crop specific regions of interest for further analysis.

3.4 Generating Images for Analysis

The vessel regression analysis uses maximum intensity projection images. For best results, analysis should be performed on as many regions (technical replicates) as possible per retina (at least three regions). These can be cropped from a whole retina overview (*see* **Notes 3** and **4**) (Fig. 2a). Cropping smaller images also allow for testing changes in vessel regression across different regions of the retina, e.g., around arteries, veins, or the sprouting plexus (Fig. 2b,c).

3.5 Automated Analysis of Vessel Regression

The code for the automated vessel regression analysis can be found at the following website: https://github.com/DrLachie/VesselRegression

We recommend downloading the vesselRegression.ijm macro and README.md files. The vesselRegression.ijm file can be opened in FIJI to run the analysis. The details for each analysis step are embedded within the code.

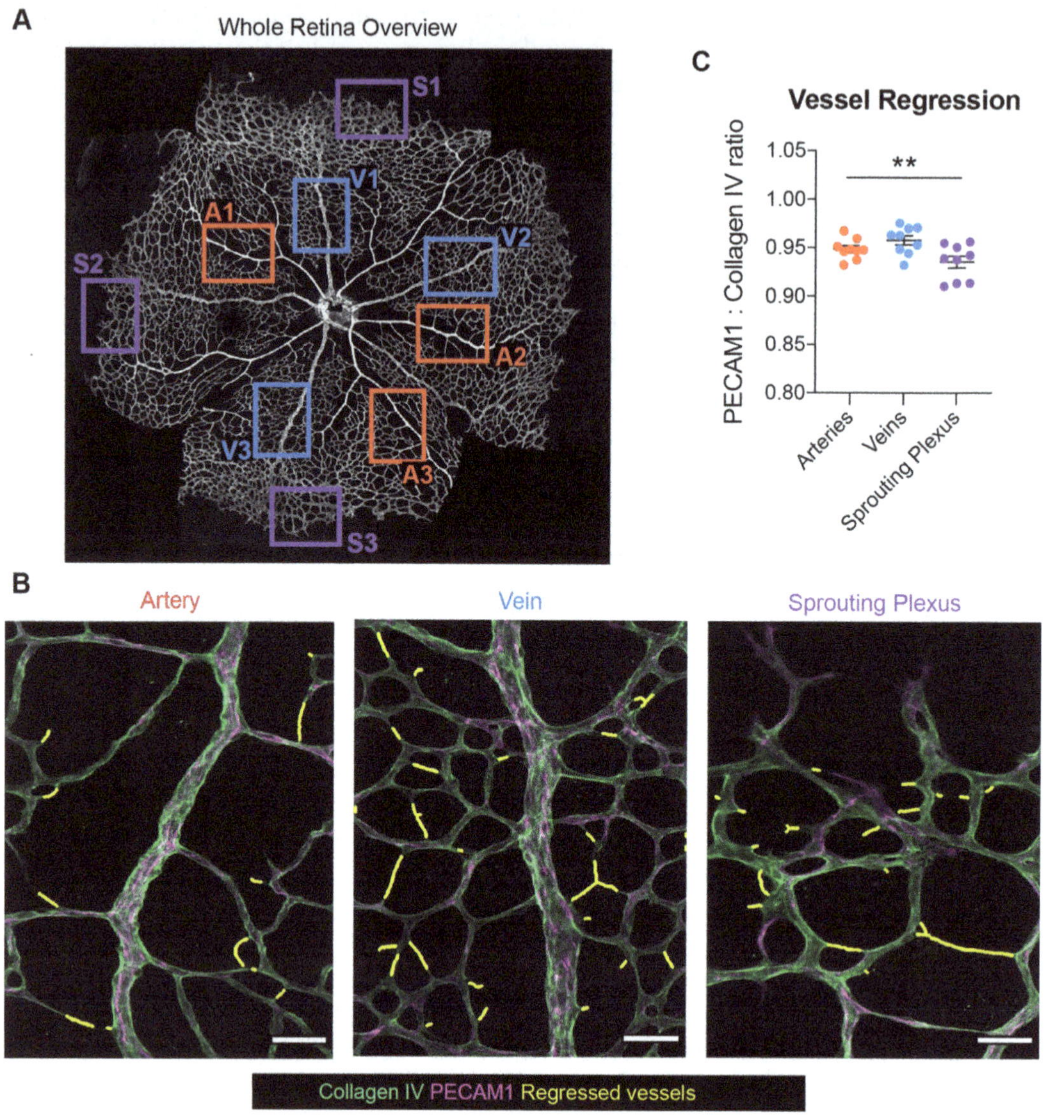

Fig. 2 Vessel regression analysis across different vascular regions of the postnatal retina. (**a**) Whole retina overview from wild-type P6 mouse with examples of three replicates for vessel regression analysis highlighted by boxes in red (artery, A), blue (vein, V), and purple (sprouting plexus, S). (**b**) Examples of vessel regression analysis output for regions surrounding an artery, vein, and sprouting plexus. Yellow lines on output image highlight regressing vessels detected by our software, see details in "Automated analysis of vessel regression". Scale bar: 50 μm. (**c**) Quantification of vessel regression analysis across arterial, venous, and sprouting plexus regions using our software. Data are mean ± SEM, statistical testing by one-way ANOVA, $^{**}p < 0.01$

1. Open the vesselRegression.ijm macro file in FIJI. This analysis code requires the MorphoLibJ plugin [24]. Ensure this is installed by following the Installation instructions as per https://imagej.net/MorphoLibJ.

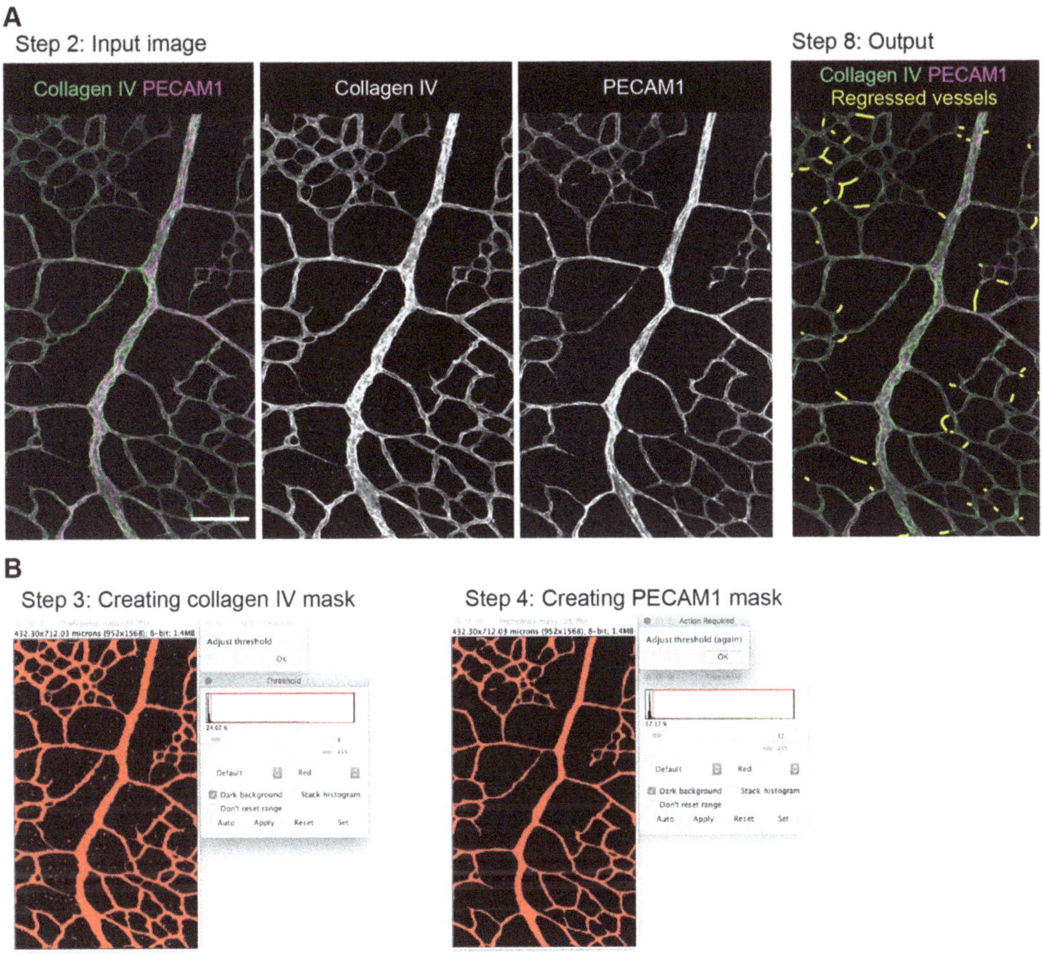

Fig. 3 Semi-automated vessel regression analysis. (**a**) Example of input and output image showing vasculature around an artery, stained for PECAM1 (magenta) and collagen IV (green). Yellow lines on output image highlight regressing vessels detected by our software. The width of the yellow lines has been increased for visualization purposes, see step 8 of "Automated analysis of vessel regression." Scale bar: 100 μm. (**b**) Example of thresholding steps for collagen IV and PECAM1

2. Click "Run". This opens up a panel with multiple options. First, select the source directory to the location of the files to be analyzed (*see* **Note 5**). Next, select the appropriate PECAM1 and collagen IV channel numbers. Click "OK." The code will automatically run (Fig. 3a).
3. It will provide a prompt to manually adjust the threshold for the collagen IV channel. Slide the upper threshold track until it masks the collagen IV signal with minimal nonvascular regions detected. Click "OK." See example in Fig. 3b.
4. Next, it will provide a prompt to manually adjust the threshold for the PECAM1 channel. Slide the upper threshold track until

it masks the PECAM1 signal with minimal nonvascular regions detected. Click “OK.” See example in Fig. 3b.

5. Continue until all images in the directory have been analyzed.
6. The program generates a results table that gives the file name, the total length of collagen IV skeleton, the total length of PECAM1 skeleton, and the ratio between the two as PECAM1:collagen IV. This table can be saved and opened in Microsoft Excel.
7. The vessel regression score per retina can be calculated as the average PECAM1:collagen IV ratio across each technical replicates.
8. The macro will also generate an image output of the difference between the collagen IV and PECAM1 skeletons, highlighting those sections that are collagen IV positive and PECAM1 negative, i.e., regressed vessels (*see* **Note 6**) (Fig. 3a). The thickness of the mask line can be altered by setting the value as greater than zero before running the macro, see line 26 of the macro code (var increaseSizeOfMaskForVisibility = 0;).
9. Output images should be verified individually to ensure the analysis has not included non-regressing vessels. Any images with too little or too many regressing vessels highlighted may be reanalyzed with slight corrections in thresholding (*see* **Note 7**).

3.6 Analyzing Apoptotic Endothelial Cells

Apoptotic ECs can be defined as active-caspase 3+, PECAM1+, and enclosed within collagen IV. Many apoptotic ECs have an elongated appearance. Using whole retina overviews or smaller fields of view, search for apoptotic ECs and label with a “region of interest” marker (*see* **Note 8**). As some cells that are attached to the endothelium may be apoptotic, it is important to use *Z*-stack images in order to determine whether active-caspase 3 signal is within the PECAM1 signal (as opposed to on top or below, examples shown in Fig. 4a). *See* Fig. 4b for an example of the number of ECs counted as apoptotic (*see* **Note 9**). The number of apoptotic ECs should be normalized to vessel area to account for any differences in vessel density between conditions. Total vessel area is best calculated using a *Z* maximum projection image of PECAM1 signal for the area where apoptosis was analyzed. To determine the association between apoptosis and vessel regression, each apoptotic EC can be manually assigned as within a regressing vessel (PECAM1- collagen IV+) or not (PECAM1+ collagen IV+).

The following steps can be used to remove noise for clean vessel area analysis:

1. Despeckle image.
2. Median filter (1–2 pixels).

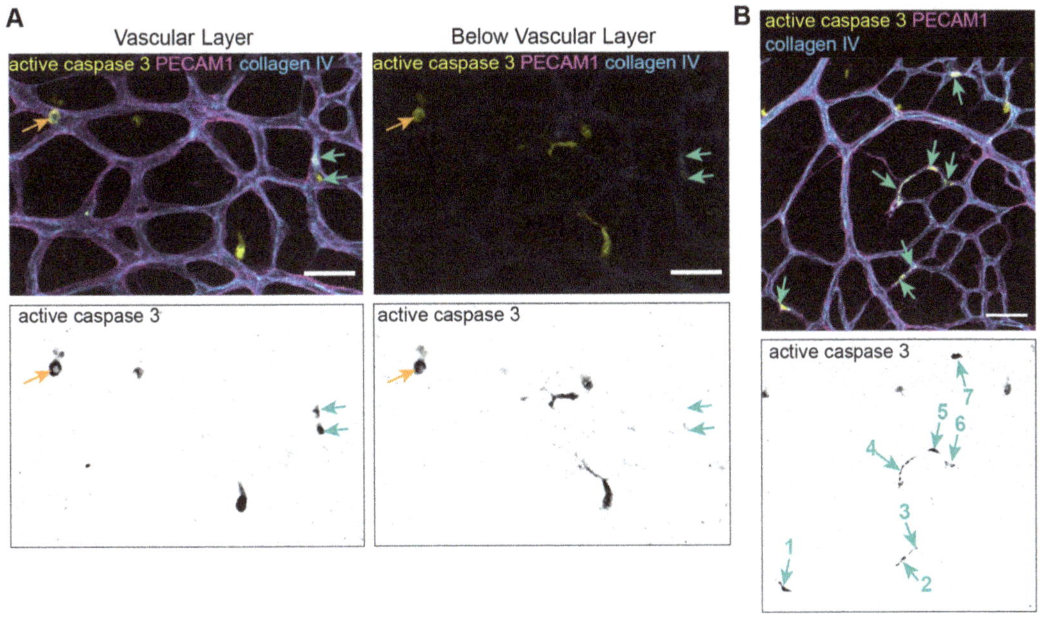

Fig. 4 Endothelial cell apoptosis. (**a**) Using Z-slices for quantification is important for determining whether an apoptotic cell is endothelial or not. Left hand image shows maximum projection image within vascular layer, where the orange arrow indicates an apoptotic cell that appears endothelial. Right hand image shows the cell (orange arrow) is actually deeper into the retina (below vasculature). Green arrows indicate bona fide apoptotic ECs. Scale bar: 30 μm. (**b**) Examples of varying morphology of apoptotic ECs (green arrows). All images show PECAM1 (cyan), collagen IV (magenta), and active caspase 3 (yellow) staining around artery in P6 retina. Numbers refer to each individual EC as an example of quantification. Scale bar: 50 μm. Some active caspase 3 images have been despeckled for visual clarity

3. Remove outliers.
4. Adjust threshold such that it masks PECAM1 signal with minimal nonvascular regions detected (ensure dark background is checked).
5. Measure area (in 'Set Measurements' ensure 'Limit to threshold' is ticked).

Convert area measurement to mm^2 and divide the total number of apoptotic ECs per region by mm^2 vessel area.

4 Notes

1. There are many variations in whole-mount retina staining protocols. The one described here is reliable with these antibodies in our hands but can be modified to suit your own preferred protocol. Antibody concentrations may require titrating.
2. The combination of the three primary antibodies allows direct comparison between vessel regression and apoptosis if

required. The combination of rabbit, rat, and goat antibodies ensures the same retinas can be stained with all three antibodies. If only staining to study vessel regression, omit the active-caspase 3 antibody.

3. Within a large image such as a whole retina overview, different portions may require differences in threshold when segmenting. Therefore, cropping to smaller regions minimizes artefacts due to poor segmentation.
4. If analyzing retinas with multiple vessel layers (i.e., superficial, middle, and deep retinal vasculature) be careful in generating maximum intensity projection images of each layer. Signals from vessels in other layers that show in projection will affect segmentation of PECAM1 or collagen IV signals and may give spurious results.
5. Saving all files to be analyzed in the same directory will enable all images to be systematically analyzed in one sitting.
6. Output images are generated in color-blind-friendly green for PECAM1 and magenta for collagen IV, regardless of color settings in original image. Regressing vessels are visualized with a yellow line.
7. Slightly over or undersetting the threshold for each channel will change the segmentation and skeletonization of the vessel channels and may benefit the accuracy of the vessel regression output. Before analyzing a large number of images, test the segmentation accuracy on a few to identify the optimal threshold settings that result in the identification of true regressing vessels by comparing the output images to the originals.
8. A region of interest (ROI) can be assigned with the point tool and pressing "t" (or click "Add [t]" with ROI manager open) in FIJI. This enables the analyzer to have a definitive list of cells counted. ROIs per image can be saved using FIJI's ROI manager into a RoiSet, which can be reopened in the future.
9. As the appearance of active caspase 3 staining may look different for each apoptotic EC, person to person variation in quantification for the same image will occur. It is therefore essential to outline strict criteria for defining apoptotic ECs at the outset and remain consistent. Figure 4b gives an example of the number of cells that we counted in that region.

References

1. Potente M, Makinen T (2017) Vascular heterogeneity and specialization in development and disease. Nat Rev Mol Cell Biol 18(8):477–494
2. Korn C, Augustin HG (2015) Mechanisms of vessel pruning and regression. Dev Cell 34(1):5–17
3. Franco CA, Jones ML, Bernabeu MO, Geudens I, Mathivet T, Rosa A et al (2015) Dynamic endothelial cell rearrangements drive developmental vessel regression. PLoS Biol 13(4):e1002125. https://doi.org/10.1371/journal.pbio.1002125. PMID: 25884288;

PMCID: PMC4401640 Refer: https://doi.org/10.1371/journal.pbio.1002125
4. Ito M, Yoshioka M (1999) Regression of the hyaloid vessels and pupillary membrane of the mouse. Anat Embryol (Berl) 200(4):403–411
5. Saint-Geniez M, D'Amore PA (2004) Development and pathology of the hyaloid, choroidal and retinal vasculature. Int J Dev Biol 48(8–9):1045–1058
6. Lenard A, Daetwyler S, Betz C, Ellertsdottir E, Belting HG, Huisken J et al (2015) Endothelial cell self-fusion during vascular pruning. PLoS Biol 13(4):e1002126
7. Kochhan E, Lenard A, Ellertsdottir E, Herwig L, Affolter M, Belting HG et al (2013) Blood flow changes coincide with cellular rearrangements during blood vessel pruning in zebrafish embryos. PLoS One 8(10): e75060
8. Lobov IB, Cheung E, Wudali R, Cao J, Halasz G, Wei Y et al (2011) The Dll4/notch pathway controls postangiogenic blood vessel remodeling and regression by modulating vasoconstriction and blood flow. Blood 117(24):6728–6737
9. Schimmel L, Fukuhara D, Richards M, Jin Y, Essebier P, Frampton E et al (2020) c-Src controls stability of sprouting blood vessels in the developing retina independently of cell-cell adhesion through focal adhesion assembly. Development 147(7):dev185405
10. Grant ZL, Whitehead L, Wong VH, He Z, Yan RY, Miles AR et al (2020) Blocking endothelial apoptosis revascularizes the retina in a model of ischemic retinopathy. J Clin Invest 130(8):4235–4251
11. Korn C, Scholz B, Hu J, Srivastava K, Wojtarowicz J, Arnsperger T et al (2014) Endothelial cell-derived non-canonical Wnt ligands control vascular pruning in angiogenesis. Development 141(8):1757–1766
12. Rao S, Chun C, Fan J, Kofron JM, Yang MB, Hegde RS et al (2013) A direct and melanopsin-dependent fetal light response regulates mouse eye development. Nature 494(7436):243–246
13. Gale NW, Thurston G, Hackett SF, Renard R, Wang Q, McClain J et al (2002) Angiopoietin-2 is required for postnatal angiogenesis and lymphatic patterning, and only the latter role is rescued by Angiopoietin-1. Dev Cell 3(3):411–423
14. Hackett SF, Wiegand S, Yancopoulos G, Campochiaro PA (2002) Angiopoietin-2 plays an important role in retinal angiogenesis. J Cell Physiol 192(2):182–187
15. Watson EC, Grant ZL, Coultas L (2017) Endothelial cell apoptosis in angiogenesis and vessel regression. Cell Mol Life Sci 74(24):4387–4403
16. Czabotar PE, Lessene G, Strasser A, Adams JM (2014) Control of apoptosis by the BCL-2 protein family: implications for physiology and therapy. Nat Rev Mol Cell Biol 15(1):49–63
17. Voss AK, Strasser A (2020) The essentials of developmental apoptosis. F1000Res 9:148
18. Watson EC, Koenig MN, Grant ZL, Whitehead L, Trounson E, Dewson G et al (2016) Apoptosis regulates endothelial cell number and capillary vessel diameter but not vessel regression during retinal angiogenesis. Development 143(16):2973–2982
19. Chen Q, Jiang L, Li C, Hu D, Bu JW, Cai D et al (2012) Haemodynamics-driven developmental pruning of brain vasculature in zebrafish. PLoS Biol 10(8):e1001374
20. Zhang Y, Xu B, Chen Q, Yan Y, Du J, Du X (2018) Apoptosis of endothelial cells contributes to brain vessel pruning of zebrafish during development. Front Mol Neurosci 11:222
21. Schindelin J, Arganda-Carreras I, Frise E, Kaynig V, Longair M, Pietzsch T et al (2012) Fiji: an open-source platform for biological-image analysis. Nat Methods 9(7):676–682
22. Gammons MV, Bates DO (2016) Models of oxygen induced retinopathy in rodents. Methods Mol Biol 1430:317–332
23. Pitulescu ME, Schmidt I, Benedito R, Adams RH (2010) Inducible gene targeting in the neonatal vasculature and analysis of retinal angiogenesis in mice. Nat Protoc 5(9):1518–1534
24. Legland D, Arganda-Carreras I, Andrey P (2016) MorphoLibJ: integrated library and plugins for mathematical morphology with ImageJ. Bioinformatics 32(22):3532–3534
25. Erratum in: PLoS Biol. 2015 May;13(5): e1002163. PMID: 25884288; PMCID: PMC4401640

Chapter 4

Whole-Mount Immunofluorescence Protocol for 3D Imaging, Reconstruction, and Quantification of Fourth Pharyngeal Arch Artery Formation

Elena Ioannou and Christiana Ruhrberg

Abstract

Pharyngeal arch arteries (PAA) are formed early during mouse embryogenesis and remodel soon thereafter into the aortic arch arteries. Failure of these vessels to form or remodel results in congenital heart defects. This protocol is designed to study the formation of the PAA using whole-mount immunofluorescence staining, followed by tissue clearing with benzyl alcohol/benzyl benzoate (BAAB) and imaging by confocal microscopy. The fine cellular resolution obtained with this technique allows the embryonic vasculature of the pharyngeal arch artery endothelium to be visualized by surface rendering and quantitatively analyzed by counting the number of endothelial cells in both the PAA and the vascular plexus surrounding them.

Key words Pharyngeal arch arteries, Whole-mount immunofluorescence, Endothelial cells, Tissue clearing, 3D reconstruction, Surface rendering

1 Introduction

During mouse embryogenesis, the pharyngeal arch arteries (PAA) form as five pairs of symmetrical structures in a rostro-caudal fashion to connect the aortic sac and the paired dorsal aorta [1]. The arteries take their name from the pharyngeal arches (PA) they form in and are termed the first, second, third, fourth, and sixth pair. In the mouse embryo, the formation of these arteries starts around embryonic day (E) 8.5 and is completed by E10.5. The first and second PAA regress early, whereas the third, fourth, and sixth undergo asymmetric remodeling to form the aortic arch arteries. The third PAA gives rise to the common carotid arteries. The fourth left PAA contributes to the aortic arch, and the fourth right PAA becomes a proximal segment of the subclavian artery. The sixth left PAA becomes the ductus arteriosus and the sixth right PAA regresses [2]. The PAA pairs 3, 4, and 6 form via

Andrew V. Benest (ed.), *Angiogenesis: Methods and Protocols*, Methods in Molecular Biology, vol. 2441,
https://doi.org/10.1007/978-1-0716-2059-5_4,

vasculogenesis, a process of blood vessel formation in the embryo that involves the de novo production of endothelial cells and their assembly into a plexus [3]. Failure of PAA formation or remodeling causes congenital heart defects, for example, similar to those seen in patients with DiGeorge Syndrome [4, 5]. Comprehensive knowledge of the mechanisms that regulate PAA development therefore identifies causes congenital heart disease and, in the long run, will improve molecular diagnosis and identify suitable strategies for intervention.

A plethora of approaches have been used to visualize and analyze PAA formation, such as vascular casts, India ink injection, high-resolution episcopic microscopy, immunofluorescence of tissue sections, and whole-mount immunohistochemistry [1, 4–11]. Although all these techniques have provided invaluable information on the timing and mechanisms of PAA formation, these approaches do not provide information about how PAA pairs are formed. The protocol below is an adaptation of the technique developed in Sophie Astrof's lab to observe PAA formation [3]. This approach involves whole-mount immunofluorescence staining for the vascular endothelial growth factor receptor 2 (VEGFR2, also known as KDR or FLK1) and the nuclear endothelial marker ETS-related gene (ERG) as cell surface and nuclear endothelial markers, respectively, together with staining for islet 1 and islet 2 (ISL1/2) to mark the PA endoderm and ectoderm, followed by tissue clearing with benzyl benzoate/benzyl alcohol (BABB). Subsequent high-resolution imaging of embryos using laser confocal microscopy and 3D image rendering with appropriate software such as Imaris. This workflow allows the embryonic vasculature of the PAA endothelium to be visualized by surface rendering and quantitatively analyzed by counting the number of endothelial cells in both the PAA and the PA vascular plexus surrounding them. This protocol can be adapted for the analysis of other vascular networks.

2 Materials

Prepare all solutions using ultrapure water and analytical grade reagents.

1. Phosphate-buffered saline (PBS): to make 1 L of 1× PBS isotonic solution, add 100 mL of 10× PBS to 900 mL of water in a glass bottle and mix well. 1× PBS solution contains 137 mM NaCl, 2.7 mM KCl, 10 mM Na_2HPO_4, and 1.8 mM KH_2PO_4. Store at 4 °C.
2. Phosphate-buffered saline containing 0.1% Triton-X-100 (PBST): to make 1 L of 0.1% PBST, prepare 900 mL of 1× PBS and add 1 mL of Triton-X-100 using a 1-mL syringe,

because Triton-X-100 is viscous. Adjust final volume to 1 L with 1× PBS. Filter-sterilize the final solution prior to using it. Store at 4 °C.

3. Blocking buffer: to make 300 μL of 10% (v/v) serum-free protein block in PBST, add 30 μL of serum-free protein blocking solution (Dako) into a 1-mL reagent tube and add PBST to a final volume of 300 μL. Make this solution fresh each time.
4. Methanol (MeOH) dilutions: to make 25%, 50%, 75% (v/v) MeOH in PBS and 100% MeOH, in a fume hood, use 50-mL disposable tubes, and add 12.5 mL of MeOH and PBS to a final volume of 50 mL to make 25% MeOH/PBS, 25 mL of MeOH and PBS to a final volume of 50 mL to make 50% MeOH/PBS, 37.5 mL MeOH and PBS to a final volume of 50 mL to make 75% MeOH/PBS, or 50 ml of MeOH for 100% MeOH. Vortex to mix well. Store at room temperature (RT). CAUTION: MeOH is toxic and flammable, wear gloves.
5. Benzyl alcohol-benzyl benzoate (BABB): to make 100% BABB, use two parts benzyl benzoate and one part benzyl alcohol. In a fume hood, add 32 mL of benzyl benzoate to 16 mL of benzyl alcohol in a 100-mL glass bottle. Cover the glass bottle with aluminum foil to protect it from light. Store at RT. CAUTION: BABB is a toxic, corrosive solution. Handle in a fume hood while wearing appropriate protective clothing.
6. Sodium azide (NaN_3): to make 0.02% NaN_3, add 2 μL of 10% NaN_3 stock solution to a final volume of 1 mL water. To make the 10% stock solution of NaN_3, weigh 10 g of NaN_3 and dissolve in 100 mL of distilled water. Store at RT. CAUTION: NaN_3 is toxic, wear gloves.
7. Formaldehyde (4%): to make 100 mL of 4% formaldehyde, pre-heat 50 mL 1× PBS at 65 °C. Add 4 g paraformaldehyde in a fume hood and dissolve for 15 min at 65 °C using the flow hood. Add 12.5 μL 10 M NaOH to the solution, then adjust the final volume to 100 mL. Swirl the bottle gently to dissolve the PFA. Filter-sterilize the final solution through a 0.45-μm membrane filter to remove any particulate matter and aliquot in 15-mL tubes. Store at −20 °C. CAUTION: Formaldehyde is known to be toxic, allergenic, and carcinogenic—avoid contact with skin and inhalation (*see* **Note 1**).
8. Ethanol (70%): to make 100 mL of 70% ethanol (v/v), to 70 mL of absolute ethanol add water to a final volume of 100 mL. Store at RT.

   CAUTION: ethanol is toxic and flammable, wear gloves.
9. Low-melting agarose (1%): to make 200 mL of 1% low-melting agarose solution, weigh 2 g of low-melting agarose and transfer to a 500-mL glass bottle. Add 200 mL of distilled water. Swirl the bottle to suspend the powder and microwave for 1 min or

until agarose is dissolved—become a clear solution (*see* **Note 2**). Be careful not to overheat the solution, to prevent the agarose from spilling in the microwave. Cool to 55 °C before use. Store the remaining 1% low-melting agarose at 4 °C.

10. Fab fragment affinity-purified antibody: unconjugated Fab fragment donkey anti-mouse IgG (H + L). Store at 4 °C.
11. Primary antibodies: mouse anti-mouse ISL1/2, goat anti-mouse VEGFR2, and rabbit anti-mouse ERG. Store at −20 °C.
12. Secondary antibodies: Alexa Fluor® 488 FAB donkey anti-mouse, Alexa Fluor® 647 FAB donkey anti-goat, Alexa Fluor® 555 FAB donkey anti-rabbit. Store at 4 °C.

3 Methods

All animal work described below should be carried out according to institutional ethical and legal guidelines, such as those from the Animal Welfare Ethical Review Body (AWERB) and Home Office in the United Kingdom.

3.1 Mouse Embryo Generation

To isolate E10.5 embryos, perform timed matings. The morning after the mating, check for the presence of a vaginal (copulation) plug which signifies a potential pregnancy. Consider the day of finding the vaginal plug as E0.5 and harvest embryos 10 days later. To account for the rapid changes during early development, stage match the embryos by carefully counting the number of somite pairs that have formed [12]. Embryos on day 10.5 of gestation usually have a range of 34–38 somite pairs.

3.2 Embryo Dissection and Fixation

1. To dissect E10.5 mouse embryos, you will need a stereoscope, Watchmaker's forceps (size 50 or 55), 60-mm disposable dishes, Pasteur pipettes (cut tip), a 24-well plate, 2-mL safe-lock tubes, a bucket full of ice, cold 1× PBS, and 4% formaldehyde.
2. Euthanize the pregnant female by cervical dislocation and confirm death with a secondary method.
3. Spray the abdominal area with 70% ethanol to prevent fur from entering the abdominal cavity. Use forceps to lift the skin of the lower abdomen and make a "v" shape incision using scissors to expose the abdominal cavity (*see* **Note 3**). Excise the gravid uterine horns and transfer them to a fresh dish filled with cold 1× PBS. Place the dish with the uterine horns on ice (Fig. 1a).
4. Use fine forceps to make perpendicular cuts between each individual embryo still surrounded by the uterus and transfer

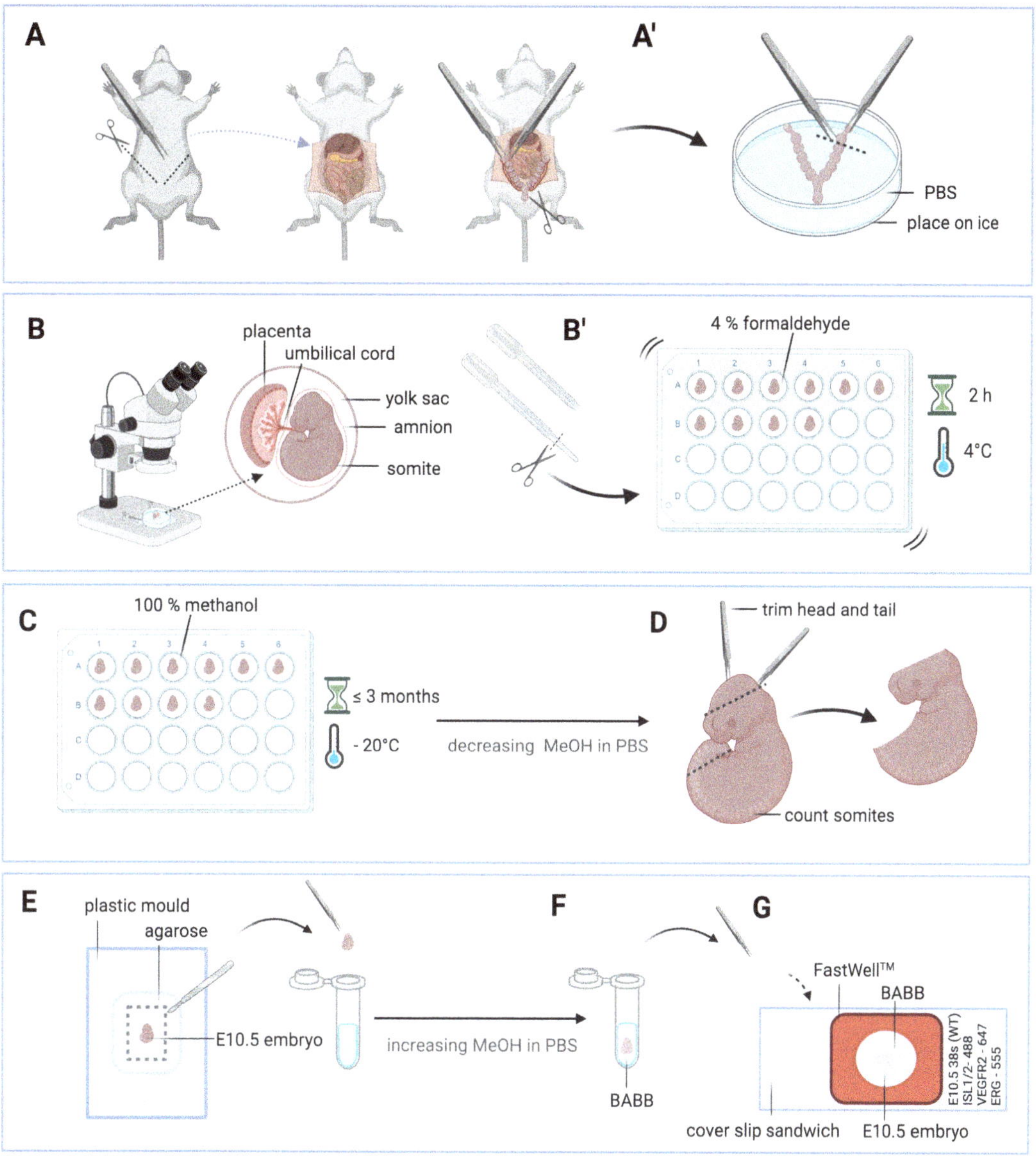

Fig. 1 Embryo immunofluorescent staining. (**a**) Place the culled mouse on its back and use forceps to lift the skin of the lower abdomen, then make a "v" shape incision (dotted lines). Excise the gravid uterine horns and transfer them to a dish with cold PBS (**a**′), placed on ice. Use forceps to separate the embryos still in their uterus (**a**′). (**b**) Dissect the embryos in a fresh dish with cold PBS under the stereoscope to remove the indicated tissue layers surrounding the embryo. Use a Pasteur pipette with a cut tip to transfer the embryos to a 24-well plate with cold PBS (**b**′); when all embryos are dissected, replace the PBS with cold 4% formaldehyde and incubate for 2 h at 4 °C with gentle agitation. (**c**) Embryos can be dehydrated through an increasing methanol gradient and stored in methanol in −20 °C but should be rehydrated before use in a decreasing methanol gradient and then transferred into PBS. (**d**) Count the number of somites, then use forceps to trim the embryo along the dotted lines. Perform immunostaining (not shown). (**e**) Use a Pasteur pipette (cut tip) to transfer the immunostained embryo into a plastic mold (sagittal position). Remove residual liquid and add warm agarose to fill the mold. Once set, trim excess agarose away and use blunt forceps to transfer the agarose block into a 2-mL tube containing 1 mL of 25% methanol in PBS and dehydrate through an increasing methanol gradient. (**f**) In a fume hood, replace methanol with 1 mL of 100% BABB. (**g**) Mount the

them one by one into a fresh dish with cold 1× PBS for further dissection under the stereoscope (Fig. 1a').

5. Use a pair of forceps to remove the uterine tissue that surrounds the decidua. Locate the placenta attached to the embryo and remove it. Remove Reichert's membrane from the yolk sac. Finally, remove the yolk sac and amniotic sac and excise the allantois and umbilical vein, so that only the embryo proper remains (Fig. 1b).

   Tip: the yolk sac can be used for genotyping.

6. Use a Pasteur pipette with a cut tip end (to fit the embryo without damaging it) to collect each embryo and transfer it to a 24-well plate filled with 1× PBS where it is essential to identify individual embryos, for example, where a litter has different genotypes (Fig. 1b'). Repeat the process for the rest of the embryos in the litter.

7. Replace 1× PBS with freshly thawed, cold 4% formaldehyde and incubate for 2 h at 4 °C with gentle agitation, e.g., in an ice box on a rocking platform (*see* **Note 4**).

8. Following fixation, wash embryos twice with 1× PBS for 5 min at room temperature to remove any residual formaldehyde. Embryos can be stored in 1× PBS at 4 °C (≤2 weeks). For long-term storage, embryos can be dehydrated down a methanol gradient comprised of 25%, 50%, 75% (v/v) MeOH in 1× PBS and then stored in 100% MeOH in −20 °C (*see* **Note 5**) (Fig. 1c).

3.3 Embryo Staining

This protocol is optimized for the following antibodies: mouse anti-mouse ISL1/2, goat anti-mouse VEGFR2, and rabbit anti-mouse ERG. You may need to alter conditions when using other antibodies. Use 0.02% NaN_3 when incubating to avoid microbial growth that will damage the antibody and/or tissue.

Prepare embryos for the staining protocol:

- Embryos stored in MeOH should be rehydrated using a MeOH-gradient comprised of 100%, 75%, 50%, and 25% (v/v) MeOH and finally 1× PBS.
- Use a Pasteur pipette (cut tip) to transfer the embryo from the 24-well plate to a 30-mm dish filled with 1× PBS.
- Count the number of somites and record it in your lab book.
- Use forceps to trim off the top of the head, without damaging the PA area; remove the limbs; remove the trunk after the

Fig. 1 (continued) embryo between two coverslips using the FastWells™ reagent barrier with a small drop of 100% BABB on top to cover the agarose area only

hindlimbs (*see* **Note 6**). This step aids antibody penetration (Fig. 1d).

- Use a Pasteur pipette (cut tip) to transfer the embryo to a labeled 2-mL tube.

Attention: Perform all steps carefully and avoid touching the embryo while pipetting solutions in and out of the tube. Always leave a small amount of liquid surrounding the embryo—do not let the embryo dry at any point.

1. To permeabilize the embryo, use a pipette to remove 1× PBS and add 1 mL of PBST containing 0.02% NaN_3. Leave the tube shaking gently at 4 °C overnight.
2. To prevent nonspecific binding of antibodies, prepare blocking solution with 0.02% NaN_3 and include unconjugated Fab Fragment Donkey Anti-Mouse IgG (H + L) 1:200 to prevent background staining caused by the secondary antibody binding to endogenous mouse IgG.
3. Centrifuge blocking solution at >10,000 × *g* before use to remove particulates. Remove PBST from the tube and add 300 μL of the prepared blocking solution. Block with gentle agitation at 4 °C overnight.
4. Prepare primary antibody solution by adding the antibodies in the blocking solution with 0.02% NaN_3 and include unconjugated Fab Fragment Donkey Anti-Mouse IgG (H + L) 1:200. To stain and quantify endothelial cells, use antibodies against VEGFR2 (early endothelial marker) and ERG (endothelial cell nuclear marker). To stain for and identify ectoderm/endoderm use ISL1/2. Dilute anti-VEGFR2 and ERG antibodies 1:200 and anti-ISL1/2 1:50. Centrifuge primary antibodies at >10,000 × *g* in a microfuge before adding them to the blocking solution to remove particulates. Vortex to mix the solution well.
5. Remove blocking solution and replace with 300 μL of blocking solution with the primary antibodies. Incubate with gentle agitation at 4 °C for 72 h.
6. Remove primary antibody solution and wash the embryo five times for 1 h with 1 mL of fresh PBST at room temperature with gentle agitation.
7. Prepare secondary antibody solution comprised of 1:200 anti-mouse Alexa Fluor 488, anti-goat Alexa Fluor 647, and anti-rabbit Alexa Fluor 555 in blocking buffer. Centrifuge secondary antibodies at >10,000 × *g* before use to remove particulates. Remove PBST and replace with 300 μL of blocking solution with the secondary antibodies. Incubate embryo with gentle agitation at 4 °C for 48 h. From this point onwards, protect the samples from light (either keep the 2-mL tube with

the embryo in a container that does not allow light to reach the embryo or wrap the tubes in aluminum foil).

8. Remove secondary antibody solution and wash the embryo every hour with 1 mL of fresh PBST with gentle agitation at room temperature five times.

3.4 Embedding Embryos in Agarose

This step is particularly important. It allows handling of the embryo without damaging it after it becomes transparent following the BABB clearing.

1. Place 1% low-melting agarose solution in the microwave for 45 s or until it is melted. It will become a clear bubbling solution when it is hot. Cool it down until is still warm enough to be liquid but not hot, approximately 55 °C (*see* **Note 7**).
2. Use a Pasteur pipette (cut tip) to carefully transfer the embryo to a plastic paraffin mold. Place the embryo in a sagittal position (Fig. 1e). Carefully remove PBST from around the embryo using a pipette. Quickly, add enough amount of warm agarose to the mold to cover the embryo and fill the mold. Ensure that there are no air bubbles around the embryo (*see* **Note 8**).
3. Carefully place the mold on ice until the agarose solidifies. Cover the ice box to protect the fluorescently stained embryo from light exposure.

3.5 Dehydration and Tissue Clearing

3.5.1 Methanol Dehydration

1. Use a clean scalpel to cut the agarose from around the embryo. Cut close to the embryo but leave enough agarose to be able to pick the sample up without touching the embryo. Use blunt forceps to gently grab the agarose containing the embedded embryo and place it into a 2-mL tube with 1 mL of 25% MeOH.
2. Incubate embryo with gentle agitation for 10 min in the dark at RT.
3. Remove 25% MeOH and add 1 mL of 50% MeOH. Incubate embryo with gentle agitation for 10 min in the dark at RT.
4. Remove 50% MeOH and add 1 mL of 75% MeOH. Incubate embryo with gentle agitation for 10 min in the dark at RT.
5. Remove 75% MeOH and add 1 mL of 100% MeOH. Incubate embryo with gentle agitation for 10 min in the dark at RT.
6. Repeat 100% MeOH wash twice (*see* **Note 9**).

3.5.2 Clearing with BABB

Reminder: BABB is toxic and corrosive. Take extra care when using it (*see* **Note 10**).

1. In a fume hood, remove 100% MeOH from the tube and add 1 mL of 100% BABB. Incubate embryo with gentle agitation for 10 min.

2. In a fume hood safety, replace with fresh 100% BABB and incubate for 2 h in the dark at room temperature or at 4 °C overnight (Fig. 1f).

Note: Optional stopping point. Embryos can remain in 100% BABB in tubes for up to a week.

3.6 Mounting Embryos for Imaging

Mount the embryo between two coverslips in order to allow imaging the embryo on both sides (*see* **Note 11**). It is important to use a divider between the coverslips to prevent the embryo from being squashed. Make sure the divider does not melt when in contact with the corrosive reagent BABB, e.g., use FastWells™ reagent barriers (Fig. 1g).

1. Peel off the adhesive tape from a FastWell™ reagent barrier and attach it to a 24 mm × 60 mm number 1.5 glass cover slip. Tap with your fingers around the barrier and apply gentle pressure to make sure it is attached properly (*see* **Note 12**). Label the coverslip with important information about the staining such as embryo age (number of somites) and antibodies used.
2. Remove the 100% BABB from the 2-mL tube containing the embryo and discard it according to your institutional guidelines for safe handling. Locate the embryo and use forceps to pick it up and place it in the middle of the barrier on the coverslip. Add a small drop of 100% BABB on top of the embryo to cover the agarose area only.
3. Remove the adhesive tape from the other side of the barrier and attach carefully the second coverslip on top. Press gently with your finger around the barrier area only and be careful not to break the glass. Coverslips with the sample can be stored flat in the dark at RT for up to a year if the seal is tight.

3.7 Data Acquisition

To image the stained and cleared embryo, use a confocal microscope equipped with a 10× water immersion objective, numerical aperture 0.95, working distance 0.95 mm, and appropriate software (e.g., NIS-Elements AR 5.11.01 64-bit).

Image the left and right side of the embryo separately by flipping the coverslip sandwich.

1. Place a clear glass slide on the platform and add the coverslip sandwich with the embryo on top of the glass slide. Use a Pasteur pipette to add a drop of distilled water onto the cover slip above the area where the embryo is located. Select a 10× water immersion objective and set the stage position.
2. Select wide-field fluorescence and bring the embryo into the field of view to locate the PA of interest, e.g., the fourth PA. Centre the field of view to the PAA of interest, e.g., the fourth PAA.

3. Set up the acquisition parameters (ND Acquisition tab):
   (a) Pinhole size: 1.0 Airy Unit.
   (b) Frame: X 1024 * Y 1024.
   (c) Speed: 7.
   (d) Adjust zoom and specimen orientation.
   (e) Select the channels: 488, 647, 555.
4. Set up Z-stack (Setup Manager):
   (a) Select 488 channel only and click "live."
   (b) Turn the focus knob on the microscope to adjust the focus. Start from the top of the specimen and move towards the middle. When you see the pharyngeal arch forming, click "set first" to determine the top position. Carry on turning the focus knob until you reach the middle of the embryo (end of the pharyngeal arch) and press "set last" to determine the bottom position.
   (c) Adjust the gain master throughout the *Z*-stack. Set the laser intensity and gain at the middle of the *Z*-stack for each channel (488, 647, and 555).
   (d) Run the *Z* Correction option.
   (e) Click "start experiment."
   (f) Save file as <filename>.lsm, including information of parameters and channels used, date, embryo identification, embryo orientation, etc.
   (g) Repeat this procedure for the other side of the embryo.

3.8 Analysis Using Imaris Software

Imaris is a comprehensive imaging software that allows the visualization, analysis, and interpretation of 3D microscopy images (alternative imaging software may be available that is suitable for this step, but such suitability would need to be established by the reader).

For the analysis of the PAA, e.g., the fourth PAA, three main functions of Imaris are needed:

Surface: to create surfaces of the regions of interest.

Mask: to visually separate the regions of interest.

Spot: to quantify the number of endothelial cells (EC) in each region of interest.

Here, we describe the procedure for fourth PAA analysis.

1. Use Imaris Converted (download for free) to convert <filename>.lsm file to <filename>.ims file.
2. Open Imaris 9.7.0 software and import the <filename>.ims file.

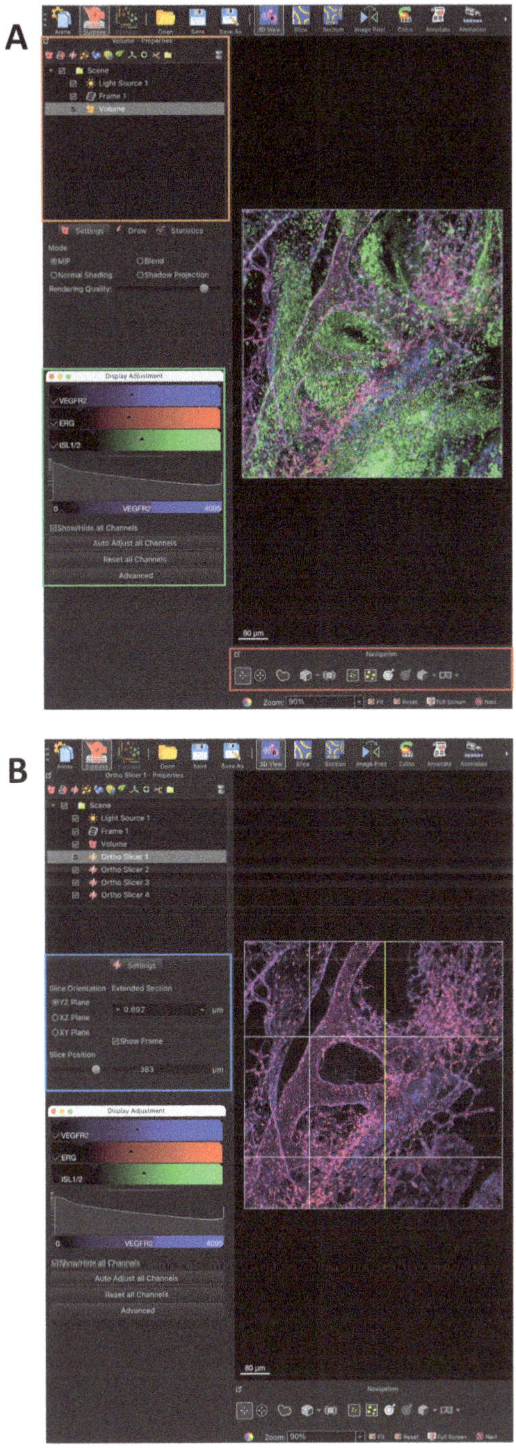

Fig. 2 Analysis using Imaris software. (**a**) Imaris desktop 3D view after the confocal image is imported. The properties box (outlined in orange) allows choosing functions such as volume, slicers, spots, creating surfaces, etc. The navigation box (outlined in red) allows choosing between different camera views. The display adjustment box (outlined in green) allows adjusting the visible

3. The image will open by default with the 3D view and the camera to perspective mode (45 °). Change the camera to the orthogonal view (Fig. 2a) (*see* **Note 13**).
4. Click edit and select "Show Display Adjustment." A new window will pop up with all the three channels: Green (488)—ISL1/2, Red (555)—ERG, and Blue (647)—VEGFR2. You can use the display adjustment window to select which channels you want to be visible and adjust them (*see* **Note 14**).
5. Tick off the ISL1/2 channel and use the pointer selection mode to rotate the image until you locate the fourth PAA parallel to the x-axis in the view of plane (dorsal to ventral).
6. Use the orthoslicer function to define the area around the fourth PAA and exclude the outside volume (Fig. 2b):
   (a) Go to "Properties" and click "Add new Ortho Slicer"  (*see* **Note 15**).
   (b) A new line will appear inside the "Properties" box named "Ortho Slicer 1." Below, a new box named "Settings" for Ortho Slicer 1 will appear with three plane options (YZ, XZ, XY), as well as a bar which allows you to choose the slice position.
   (c) Select "YZ Plane" and move the slice position to the dorsal side until the fourth PAA meets the dorsal aorta. Information beyond this point is not needed to create the fourth PA and PAA area. Click "Add new Ortho Slicer" to create Ortho Slicer 2 and choose "YZ plane." This time move the slice position toward the ventral side and stop at the point where the fourth PAA meets aortic sac.
   (d) Go to "Display Adjustment" and select the ISL1/2 channel. Click "Add new Ortho Slicer" to create Ortho Slicer 3 and choose "XZ plane." Move the slice position toward the cranial position and stop in the middle of the third PAA. Click "Add new Ortho Slicer" to create Ortho Slicer 4 and choose "XZ plane." Move the slice position toward the caudal position and stop in the middle of the sixth PAA (if not yet formed, below the fourth PA).

3.8.1 Surfacing the Fourth Pharyngeal Arch

1. Go to "Properties" and select "Add New Surface"  (Fig. 3a).
2. A new line will appear inside the "Properties" box named "Surfaces 1." Double click on it and rename it "PA." Below, a

Fig. 2 (continued) channels. (**b**) Demonstration of how to use the ortho slicer function to define the area of interest. The settings box (outlined in blue) allows choosing the plane of view, slice position, etc.

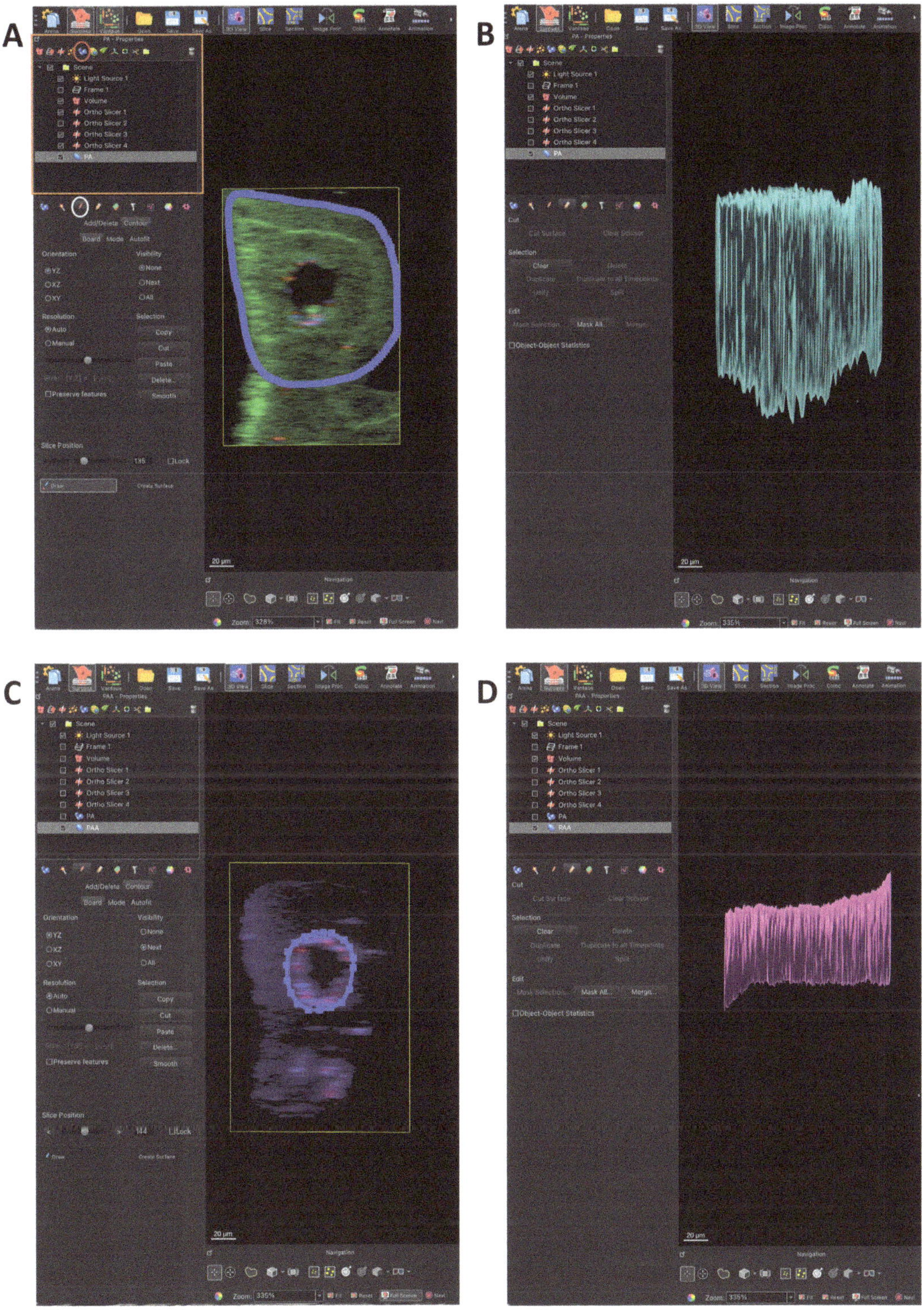

Fig. 3 Surfacing the fourth pharyngeal arch and artery. (**a**,**b**) Manual drawing of the fourth pharyngeal arch. (**a**) From the properties box (outlined in orange), select the function "Add New Surface" (circled in red) and subsequently the function "draw" (circled in white). The blue line (inside the yellow outlined box) illustrates how to draw around the arch for manual creation every 3 μm. (**b**) Final product of the fourth pharyngeal arch

new box will appear named "Create." Click "Skip automatic creation, edit manually" which takes you automatically to the "Draw"  mode → "Contour" → "Board." Click "YZ" under "Orientation" and use the "Slice Position" bar to place the board close to Ortho Slicer 1 to mark the beginning of the manual drawing toward the ventral side. At this point, tick off Ortho Slicer 1 and "Volume" .

3. Use "Pointer Selection Mode" to rotate the image and bring the YZ plane into view.
4. Go to "Draw" → "Contour" → "Mode" and choose "Distance Drawing Mode." Set the vertex spacing to 10 μm.
5. Go back to "Draw" mode → "Contour" → "Board" to begin drawing the surface. Select "Draw" under the "Slice Position" and check that a red dot appears under the brush icon which will indicate that the draw mode is on. Use the "Slice Position" bar to move through the slices. Trace the perimeter of the fourth PA with the mouse cursor and repeat it every three slices (9 μm) until the arch is fully traced.
6. Click "Create Surface" to generate the surface of the traced region (Fig. 3b) (*see* **Note 16**).

   Tip: to change the color of the created surface, go to "Properties" → "PA" → "Color" .
7. Remove all four Ortho Slicers—not needed any more.
8. Go to "Properties" → "PA" → "Statistics"  to get information about the size of the arch.

   Attention: remember to "Export" and "Save" the image (*see* **Note 17**).

3.8.2 Create the Fourth PA Channels

Use the "Mask" function to create the PA channels (*see* **Note 18**):

1. Go to "Properties" → "PA" → "Edit"  and select "Mask All. . . ." A new window will appear.
2. Tick "Duplicate channel before applying mask" and select Green channel. Go to "Mask Settings" → "Constant inside/outside" → "Set voxels outside surface to zero" → "OK."
3. A new channel will appear in the "Display Adjustment" panel named "Masked ISL1/2 channel." Double click to rename "PA ISL1/2 channel." Repeat the same process for the ERG

Fig. 3 (continued) surface after tracing the relevant region in (**a**) through the sections of the confocal *z*-stack; the arch surface is shown after 90 °C rotation relative to (**a**). (**c**, **d**) Manual drawing of the fourth pharyngeal arch artery. (**c**) Use the method shown in (**a**,**b**); the blue line illustrates how to draw around the artery every 1 μm. (**d**) Final product of the fourth pharyngeal arch artery surface after tracing the relevant region of the confocal z-stack in (**c**); the artery surface is shown after 90 °C rotation relative to (**c**)

and VEGR2 channel. The new channels include only information of the region of interest.

4. "Export" and "Save."
5. Tick off ISL1/2, ERG, and VEGR2 channels and keep only the PA ISL1/2, PA ERG, and PA VEGFR2 channels. Tick off "PA" surface and turn "Volume" back on.
6. Go to "Edit" → "Crop 3D" and crop it around the PA volume only (*see* **Note 19**).
7. Export and save the new file as "CROPPED."

3.8.3 Surfacing the Fourth Pharyngeal Arch Artery

1. Go to "Properties" and click "Add New Surface" (Fig. 3c).
2. A new line will appear inside the "Properties" box named "Surfaces 1." Double click on it and rename it "PAA." Below, a new box will appear named "Create." Click "Skip automatic creation, edit manually" which takes you automatically to the "Draw" mode → "Contour" → "Board." Click "YZ" under "Orientation" and use the "Slice Position" bar to place the board at the beginning of the artery and turn off the "Volume." Go to "Display Adjustment" and tick off the ISL1/2 channel (information not useful anymore).
3. Use "Pointer Selection Mode" to rotate the image and bring the YZ plane into view.
4. Go to "Draw" → "Contour" → "Mode" and choose "Distance Drawing Mode." Set the vertex spacing to 10 μm.
5. Go back to "Draw" mode → "Contour" → "Board" to begin drawing the surface. Click "Draw" under the "Slice Position" and check that a red dot appears under the brush icon, which will indicate that the draw mode is on. Use the "Slice Position" bar to move through the slices. Trace the perimeter of the fourth PAA with the mouse cursor and repeat it every slice until the artery is fully traced.
6. Click "Create Surface" to generate the surface of the traced region (Fig. 3d).
7. Go to "Properties" → "PAA" → "Statistics" to get information about the size of the artery.

   Attention: "Export" and "Save" image.

3.8.4 Create the Fourth PAA Channels

Use the "Mask" function to create the PAA channels (*see* **Note 20**):

1. Go to "Properties" → "PAA" → "Edit" and select "Mask All...." A new window will appear.
2. Tick "Duplicate channel before applying mask" and select "PA ISL1/2 channel." Go to "Mask Settings" → "Constant inside/outside" → "Set voxels outside surface to zero" → "OK."

3. A new channel will appear in the "Display Adjustment" panel named "Masked PA ISL1/2 channel." Double click to rename "PAA ISL1/2 channel." Repeat the same process for the PA ERG and PA VEGR2 channel.
4. "Export" and "Save."

3.8.5 Create a Channel for the Plexus

Use the "Mask" function to create the Plexus channel (*see* **Note 21**):

1. Go to "Properties" → "PAA" → "Edit" and select "Mask All. . . ." A new window will appear.
2. Tick "Duplicate channel before applying mask" and select "PA VEGR2 channel." Go to "Mask Settings" → "Constant inside/outside" → "Set voxels inside surface to zero" → "OK."
3. A new channel will appear in the "Display Adjustment" panel named "Masked PA VEGFR2 channel." Double click to rename "Plexus."
4. "Export" and "Save."

3.8.6 Surfacing the Plexus (Automatic Creation)

1. Go to "Properties" and click "Add New Surface."
2. A new line will appear inside the "Properties" box named "Surfaces 1." Double click on it and rename it "Plexus." Below, a new box will appear named "Create." A series of four boxes will appear: first—blue arrowhead (back), second—blue arrowhead (next), third—two green chevrons (finish), and fourth—orange "x" (cancel) which signifies the automatic creation.
3. Click next and go to "Source Channel" and choose "Plexus."
4. Click next and go to "Threshold" → "Threshold (Absolute Intensity)" and set it manually.
5. Click next and set the number of Voxels.
6. Click finish to complete the automatic creation.
7. "Export" and "Save."

    Attention: Automatic creation should not be used for quantification, as it not sufficiently accurate.

3.8.7 Quantification of Endothelial Cell Numbers

For the Number of Fourth Pharyngeal Arch EC

1. Go to "Properties" and click "Add New Spots"  (Fig. 4a).
2. A new line will appear inside the "Properties" box named "Spots 1." Double click on it and rename it "PA spots." Below, a new box will appear named "Create." A series of four boxes will appear: first—blue arrowhead (back), second—blue arrowhead (next), third—two green chevrons

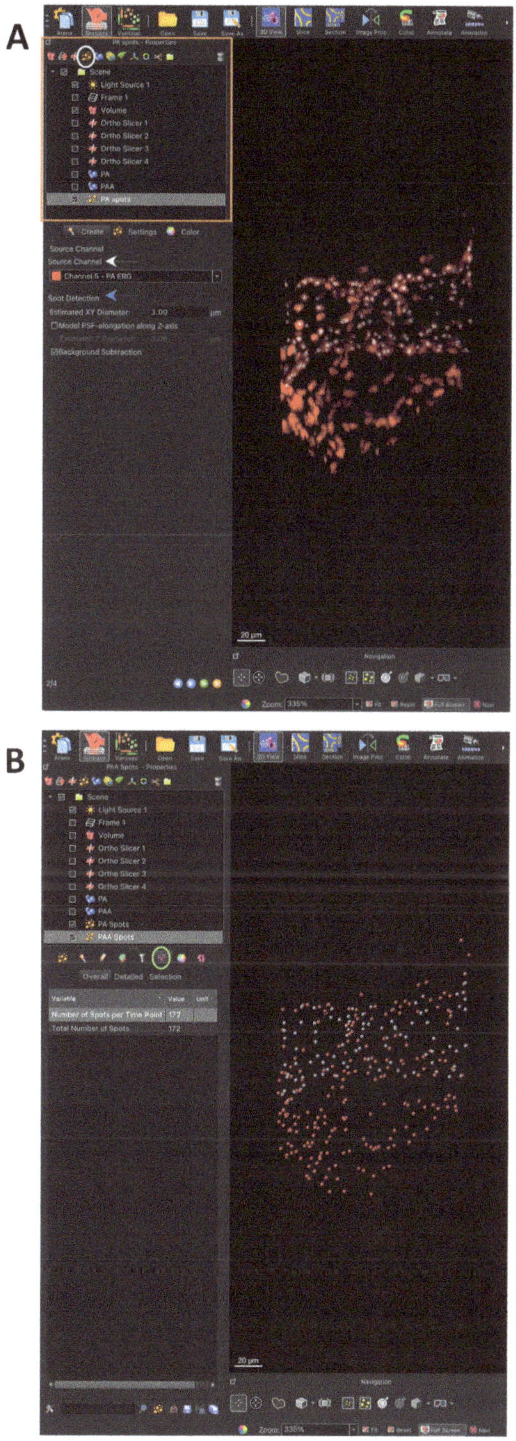

Fig. 4 Quantifying endothelial cell numbers using the spotting function of Imaris. (**a**) From the properties box (outlined in orange), select the function "Spots" (circled in white) and then the source channel (white arrow). The diameter is set to 3 μm (blue arrow). The spots are then created automatically. (**b**) Spots in the pharyngeal arch artery area are shown in white, spots in the pharyngeal arch

(finish), and fourth—orange "x" (cancel) which signifies the automatic creation.

3. Click next and go to "Source Channel" and choose "PA ERG channel." Go to "Spot Detection" → "Estimated XY Diameter" and set it to 3 μm (diameter of an endothelial cell).
4. Click next and go to "Filter Type" → "Quality" and select manually the endothelial cells.
5. Click finish to complete the automatic creation.
6. Go to "Properties" → "PA Spots" → "Statistics" to obtain the number of endothelial cells.
7. "Export" and "Save."

Attention: Make sure that each spot is positive for both EC markers (ERG and VEFGR2). In case the automatic creation has selected a spot that it is not positive for both EC markers, manually delete the spot. To do this, go to "Properties" → "PA Spots" → "Edit." Hold down the "command" key and select the spot that needs to be deleted. The dot will become yellow. Go to "Edit" → "Selection" and click delete. **N.B.** yellow color on Imaris indicates selection.

For the Number of Fourth Pharyngeal Arch Artery Endothelial Cells

1. Create a channel for the fourth PAA endothelial cells:
   (a) Go to "Properties" → "PA Spots" → "Edit" and select "Mask All. . . ." A new window will appear.
   (b) Tick "Duplicate channel before applying mask" and select "PAA ERG channel." Go to "Mask Settings" → "Constant inside/outside" → "Set voxels outside surface to zero" → "OK."
   (c) A new channel will appear in the "Display Adjustment" panel named "Masked PAA ERG channel." Double click to rename "PAA spots."
2. Go to "Properties" and click "Add New Spots" (Fig. 4b).
3. A new line will appear in the "Properties" box named "Spots 1." Double click on it and rename it "PAA spots". Below, a new box will appear named "Create." A series of four boxes will appear: first—blue arrowhead (back), second—blue arrowhead (next), third—two green chevrons (finish), and fourth—orange "x" (cancel) which signifies the automatic creation.

Fig. 4 (continued) area are shown in red. A record of the total number of the spots of interest can be obtained by selecting the "statistics" icon (circled in green)

4. Click "next" and go to "Source Channel" and choose "PAA Spots channel." Go to "Spot Detection" → "Estimated XY Diameter" and set it to 3 μm (approximate diameter of an embryonic endothelial cell in vivo).
5. Click next and go to "Filter Type" → "Quality" and select manually the endothelial cells.
6. Click finish to complete the automatic creation.
7. Go to "Properties" → "PAA Spots" → "Statistics" to get the number of endothelial cells.
8. "Export" and "Save."

4 Notes

1. Transfer the scales and hotplate with magnetic stirrer to the fume hood to avoid exposure to paraformaldehyde. Wear gloves at all times and avoid inhaling it. Use freshly made or freshly thawed formaldehyde. Old formaldehyde solution causes stronger endogenous fluorescence. N.B. formaldehyde is hazardous and must be disposed of in accordance with institutional regulations.
2. Place the lid loosely on top of the glass bottle to keep the bottle from exploding. Pause the microwave halfway through to check that solution has not overspilled and to stir the bottle. Solution is ready when big bubbles are formed, and solution is clear.
3. It is important to make upward movements with the scissors to avoid damaging the embryos.
4. E10.5 mouse embryo fixation for this protocol and antibody combination was optimized to 2 h fixation with 4% formaldehyde at 4 °C. Fixation time should be checked and adjusted to the antibodies being used.
5. Store fixed embryos in methanol for maximum of 3 months. Longer storage may result in degradation of the proteins used for immunostaining.
6. Trimming the head and trunk of the embryo facilitates better antibody penetration. Cutting the limbs off the embryo allows the sample to be embedded flat, which is important for imaging.
7. Prior to adding the 1% low melting agarose, you should be able to touch the bottle with your hand (wearing gloves) without burning; alternatively, place bottle into a water bath or oven set to 55 °C for 30 min. This is particularly important because very hot agarose can damage the embryo.

8. It is important to add just enough agarose to cover the embryo, but not too much, otherwise imaging will be difficult and results not clear for analysis.
9. Make sure the embryos embedded in agarose are fully dehydrated before adding BABB solution to the 2 mL tube. If not properly dehydrated the embryos will become milky instead of transparent. If this does happen, repeat the final step of the dehydration process and try clearing with BABB again.
10. Use glass or polypropylene plastic that will not dissolve when in contact with BABB. BABB is toxic and corrosive, take extra care when using it.
11. Due to the thickness of the embryo, imagining the two sides individually produces clearer results and more accurate analysis. To achieve that, you should mount the embryo between two coverslips instead of using a slide. This way you can simply flip the coverslip sandwich and image either side.
12. Take extra care when trying to seal the coverslip on top of the barrier. Coverslips are very thin and extremely easy to break. Apply gentle pressure on the area of the coverslip that is occupied by the barrier only. If you touch elsewhere, it is very likely that the glass will break.
13. The camera view is a matter of personal preference. The analysis can be carried out either way.
14. To change the name or channel display color, double-click on the current channel's name. A new window will appear with the option to rename the channel and change the color.
15. Orthoslicers do not add information to the image, they just define the area of interest.
16. If you are not happy with the surface created or want to change an area, go to "Properties" → "Surface" → "Creation"  → "Rebuild" → "Skip automatic creation, edit manually" → "Contour" → "Board"→ "Visibility" → "All." When you locate the area that needs modification, select "Delete…" and then repeat drawing procedure. Once modifications are completed, click "Create." Then modified surface will appear.
17. Creating surfaces manually is time consuming; therefore, it is advisable that you save your work immediately after you create them in case the software or computer crashes.
18. To remove unwanted information, you should create a new channel for the pharyngeal arch that will only display the information inside the arch. Use the "Mask" function to select and set everything outside the arch to zero, which deletes it.
19. This step helps the program to run with a smaller volume of information and therefore minimize the risk of the software

crashing. **N.B.** recommended computer specifications to operate Imaris are 16-32 GB RAM, 3.3 GHz CPU (Intel or AMD) 4 to 6 Cores, graphics: NVIDA Quadro P2000 5 GB, Multiple Fast Hard Disks or (SATA) SSDs, monitor: single 1280 × 1024 pixels up to dual 1920 × 1200.

20. To create a new channel that shows information about the pharyngeal arch artery only, use the "Mask" function and set everything outside the artery to zero, which deletes it. **N.B.** you do not need to create a new pharyngeal arch artery channel for ISL1/2 (green) because there is no green staining inside the artery.
21. To see the volume of the plexus for the automatic creation, use the "Mask" function to create a new channel and set everything inside the artery to zero, which deletes it. You should now only have information about the plexus without the artery.

Acknowledgments

We thank Sophie Astrof and her lab members for teaching us this method, Laura Denti and the staff of the Biological Resources Unit for animal husbandry and the Confocal Microscopy Unit for maintaining the imaging equipment. EI is funded by a PhD studentship from the British Heart Foundation, ref. FS/19/29/34367. CR is supported by a Wellcome Investigator Award, ref. 205099/Z/16/Z.

References

1. Hiruma T, Nakajima Y, Nakamura H (2002) Development of pharyngeal arch arteries in early mouse embryo. J Anat 201(1):15–29. https://doi.org/10.1046/j.1469-7580.2002.00071.x
2. Hutson MR, Kirby ML (2007) Model systems for the study of heart development and disease. Cardiac neural crest and conotruncal malformations. Semin Cell Dev Biol 18(1):101–110. https://doi.org/10.1016/j.semcdb.2006.12.004
3. Wang X, Chen D, Chen K, Jubran A, Ramirez A, Astrof S (2017) Endothelium in the pharyngeal arches 3, 4 and 6 is derived from the second heart field. Dev Biol 421(2):108–117. https://doi.org/10.1016/j.ydbio.2016.12.010
4. Jerome LA, Papaioannou VE (2001) DiGeorge syndrome phenotype in mice mutant for the T-box gene, Tbx1. Nat Genet 27(3):286–291. https://doi.org/10.1038/85845
5. Lindsay EA, Vitelli F, Su H, Morishima M, Huynh T, Pramparo T et al (2001) Tbx1 haploinsufficieny in the DiGeorge syndrome region causes aortic arch defects in mice. Nature 410(6824):97–101. https://doi.org/10.1038/35065105
6. Weninger WJ, Maurer-Gesek B, Reissig LF, Prin F, Wilson R, Galli A et al (2018) Visualising the cardiovascular system of embryos of biomedical model organisms with high resolution episcopic microscopy (HREM). J Cardiovasc Dev Dis 5(4):58. https://doi.org/10.3390/jcdd5040058
7. Phillips HM, Stothard CA, Shaikh Qureshi WM, Kousa AI, Briones-Leon JA, Khasawneh RR et al (2019) Pax9 is required for cardiovascular development and interacts with Tbx1 in the pharyngeal endoderm to control 4th pharyngeal arch artery morphogenesis. Development 146(18):dev177618. https://doi.org/10.1242/dev.177618

8. Azaripour A, Lagerweij T, Scharfbillig C, Jadczak AE, Willershausen B, Van Noorden CJ (2016) A survey of clearing techniques for 3D imaging of tissues with special reference to connective tissue. Prog Histochem Cytochem 51(2):9–23. https://doi.org/10.1016/j.proghi.2016.04.001
9. Richardson DS, Lichtman JW (2015) Clarifying tissue clearing. Cell 162(2):246–257. https://doi.org/10.1016/j.cell.2015.06.067
10. Erturk A, Becker K, Jahrling N, Mauch CP, Hojer CD, Egen JG et al (2012) Three-dimensional imaging of solvent-cleared organs using 3DISCO. Nat Protoc 7(11):1983–1995. https://doi.org/10.1038/nprot.2012.119
11. Kuwajima T, Sitko AA, Bhansali P, Jurgens C, Guido W, Mason C (2013) ClearT: a detergent- and solvent-free clearing method for neuronal and non-neuronal tissue. Development 140(6):1364–1368. https://doi.org/10.1242/dev.091844
12. Chan AOK, Dong M, Wang L, Chan WY (2004) Somite as a morphological reference for staging and axial levels of developing structures in mouse embryos. Neuroembryol Aging 3(2):102–110. https://doi.org/10.1159/000089005

Chapter 5

Quantifying and Characterizing Angiogenesis Using the Postnatal Mouse Retina

Alessandro Fantin

Abstract

Angiogenesis refers to the expansion of blood vessels from a preexisting vascular plexus, and it is a fundamental process for organ development, the female reproductive system, and wound healing, but it is also a common denominator in several diseases such as cancer and neovascular eye disease. For these reasons, shedding light on the molecular and cellular mechanisms of angiogenesis has the potential to devise new therapeutic strategies to refrain pathological vessel growth or even promote new vessel formation in ischemic conditions and organ grafts. The mouse postnatal retina provides an excellent and widely adopted model to study physiological angiogenesis in vivo, and this chapter outlines a detailed protocol for its dissection, staining, and analysis of the vasculature.

Key words Angiogenesis, Retina, Image analysis, Quantification, Immunofluorescence

1 Introduction

Angiogenesis is a complex and tightly regulated process that undergoes multiple discrete steps that can be individually evaluated and quantified by several different bioassays. Many different experimental settings have been exploited, ranging from the simplest in vitro models that focus on specific parameters implicated in angiogenesis, to complex in vivo models that recapitulate key features of pathological conditions with a prominent neovascular component, such as murine models of solid cancers or neovascular eye pathology [1]. Therefore, the choice of the right experimental model is critical to uncover molecular and cellular mechanisms that are of physiological relevance. Indeed, much knowledge on the signaling pathways and the cell involved in controlling vascular growth and patterning has been gained by studies of vascular growth in developing organs, such as the mouse perinatal retina (e.g., [2–10]),

Andrew V. Benest (ed.), *Angiogenesis: Methods and Protocols*, Methods in Molecular Biology, vol. 2441,
https://doi.org/10.1007/978-1-0716-2059-5_5,

which, together with the mouse embryo hindbrain (e.g., [2, 7, 8, 11–18]), is among the most widely used in vivo model for physiological angiogenesis.

1.1 The Postnatal Mouse Retina

In the mouse, vascularization of the retina occurs only after birth, concomitant with the regression of hyaloid vessels that supported the inner eye during embryogenesis (reviewed in [19, 20]). Retinal vessels emerge from the optic nerve head at the center of the retina and extend toward the periphery to form a two-dimensional planar network, variably called the primary, superficial, or inner vascular plexus (Fig. 1a,b); these vessels follow a gradient of vascular endothelial growth factor A (VEGF-A) deposited by retinal astrocytes and neurons that migrate in front of the nascent vasculature [4, 21–23]. Beginning on postnatal day (P) 3, the blood vessels closer to the optic nerve head begin to remodel and form an alternating radial pattern of arteries and veins that are connected by capillary beds (Fig. 1b). Thus, some vessel segments increase in diameter as they remodel into arteries and veins, while other vessel segments are pruned to facilitate optimal blood flow in vessels that are retained. Pruning is particularly evident in the proximity of arteries, where capillary free zones emerge. This process can occur via migration and reposition of endothelial cells into feeder vessels or via selective endothelial cell apoptosis [24]. After the primary vascular plexus has reached the periphery of the retina at around P7, vessel sprouts start diving into the tissue underneath to form by the third week of age the deeper and the intermediate vascular plexi that will provide oxygen and nutrients to the internal retinal layers [19, 20].

The main advantage of the developing mouse retina as a model to study angiogenesis is the planar nature of its primary vascular plexus, because it facilitates the visualization of the vasculature under the microscope in flat-mount preparations. Moreover, as described above, retinal angiogenesis proceeds in a stereotypical and well-defined sequence of events. In addition, remodeling in the center of the primary plexus by the end of the first week after birth allows the study of vascular maturation and at the same time the study of vascular extension and sprouting at the vascular front. Since arteriovenous remodeling occurs concomitantly to vessel sprouting, the quantitative analysis of capillary branching complexity has to be carefully performed in areas containing only capillaries, i.e., just behind the vascular front and between arteries and veins (Fig. 1f). Another advantage of using the retina as a model for angiogenesis is its accessibility to experimental manipulation, by either local or systemic injection of drugs, inhibitors, or growth factors [4, 8, 10], which is normally precluded in prenatal models such as the hindbrain unless via an invasive surgical procedure. Additionally, the postnatal retina can be explanted and cultured for live imaging or ex vivo pharmacological treatments

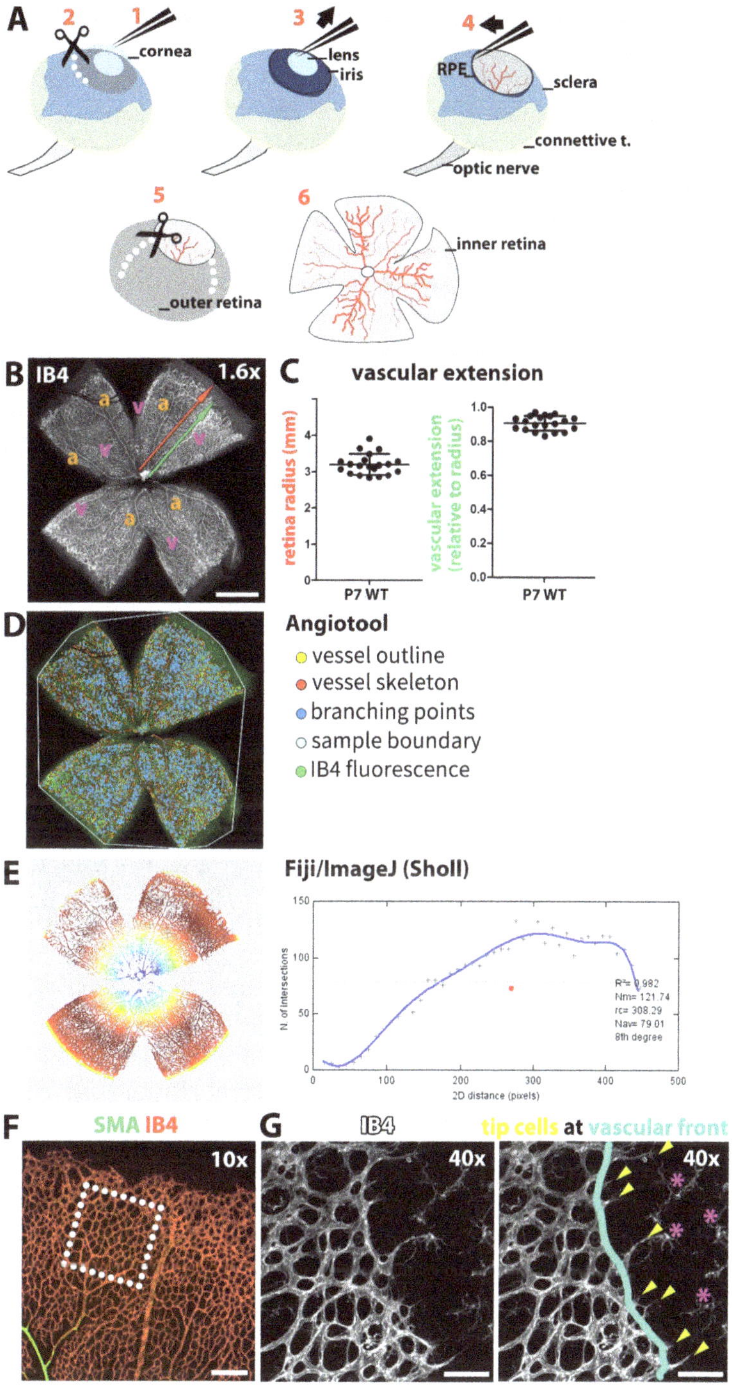

Fig. 1 Retina dissection and quantification of retina angiogenesis. (**a**) Schematic representation of the dissection of a P7 mouse retina. Grab the cornea with watchmaker's forceps (step 1) and cut along the border between the sclera and the iris (white dots) with spring scissors (step 2). After dissecting the cornea, remove the iris together with the lens and hyaloid vasculature with watchmaker's forceps (step 3). Carefully peel sclera and RPE away from the retina

[25]. Furthermore, even though the fact that the retina is vascularized postnatally could be disadvantageous when gene targeting causes embryonic lethality (e.g., the *Vegfa*, *Vegfr1*, *Vegfr2*, and *Nrp1* genes), the development of constitutive or tamoxifen-inducible conditional knockout technology can overcome this limitation [8, 9, 26]. Finally, the dissection and imaging procedures described in this protocol could also be applied to the analysis of vascular changes in the retinas of mouse pups following induction of oxygen-induced retinopathy, a condition that mimics some of the pathological features of retinopathy of prematurity and diabetic retinopathy in humans and is achieved by exposing pups to high oxygen concentration for 3–5 days during the second week after birth [7, 9, 10, 27].

2 Materials

2.1 Equipment

- 96-well flat-bottom tissue culture plate.
- Benchtop centrifuge.
- Benchtop orbital shaker.
- Confocal laser scanning microscope.

Fig. 1 (continued) (step 4). Make four radial incisions with spring scissors (step 5) and position the retina with the inner side facing upward in a small plastic dish for fixation (step 6). (**b**) Fluorescent IB4 labeling (white) of vessels in the P7 mouse retina was imaged with a stereo fluorescent microscope at 1.6× magnification. Orange "a" and purple "v" indicate the alternating pattern of arteries and veins, respectively. Red and green arrows show how to measure the retina radius and the retina vascular extension, respectively. (**c**) Quantification of radius (left hand side) and of vascular extension normalized to retina radius (right hand side) in $n = 20$ P7 wild-type (WT) retinas. (**d**) Skeletization of the vasculature of the retina shown in (**b**) with AngioTool. Blue dots indicate branch points while red lines indicate vessel segments. (**e**) Sholl analysis of the vasculature of the retina shown in (**b**) with Fiji/ImageJ. Rainbow color rendering according to vessel density, i.e., the numbers of skeletonized vessel structures that intersect concentric circles at 10 μm space from each other as shown in the plot on the right hand side. The plot also indicates the average vascular extension. (**f**) Confocal laser scan of a P7 retina stained with IB4 (red) and αSMA (green) with a 10× objective. Note higher αSMA staining in the artery compared to the vein. The stippled square indicates the region of interest to analyze capillary density. (**g**) High-magnification image of the retinal vascular front stained for IB4 (white) and acquired with a 40× objective. The right panel shows the counting of tip cells (yellow arrowhead) to be normalized to the length of the vascular front (cyan line); IB4 positive microglial cells are labeled by purple asterisks. Scale bars: 1 mm (**b**, **d**, **e**), 200 μm (**f**), 50 μm (**g**)

- Dissection instruments: Watchmaker forceps (no. 5), standard surgical scissors, and spring scissors.
- Epifluorescence stereomicroscope equipped with a digital camera.
- Falcon tubes, 50 mL.
- Glass coverslips, 22 × 55 mm.
- Glass slides.
- Parafilm.
- Plastic cell culture dishes, 30 mm diameter.
- Plastic Pasteur pipettes.
- Stereomicroscope equipped with fiber optic gooseneck lamps.
- Tabletop balance.
- Water bath.

2.2 Reagents

- Absolute methanol.
- Alexa Fluor 647-conjugated streptavidin (Life Technologies). It is possible to choose an alternative fluorophore-conjugated streptavidin.
- Biotinylated IB4 from Bandeiraea simplicifolia BS-I (Sigma).
- Blocking solution (0.1% Triton X-100 (v/v) and 10% normal goat or donkey serum (v/v) in PBS).
- Fluorescein isothiocyanate (FITC)-conjugated mouse antibody anti-α smooth muscle actin (SMA, Sigma).
- Mowiol/DABCO antifade (0.6% Mowiol 4-88 (w/v) in 25% glycerol with 2.5% DABCO (w/v)).
- Paraformaldehyde (PFA); dissolve at 4% (w/v) in PBS to yield formaldehyde fixative.
- PBS (phosphate-buffered saline).
- PBT (0.1% Triton X-100 (v/v) in PBS).
- Rising methanol gradient (25%, 50%, and 75% (vol/vol) absolute methanol in PBS, then absolute methanol).

3 Methods

3.1 Retina Dissection and Fixation

1. Mice of the desired age (usually P6-7 pups for retina primary plexus analysis) are culled by cervical dislocation.
2. The eyes are immediately removed with forceps and fixed for 5 min in 100 μL of formaldehyde on ice in the wells of a 96-well plate (one well for both eyes from the same animal).

3. The fixative is then replaced with PBS, and the eyes from one well are moved to a 30-mm plastic dish filled with PBS for further dissection. Catch the eyes by the optic nerve with forceps for the transfer. Avoid removing the optic nerve from the eyeball as its removal will cause the retina to squeeze out when applying pressure to the eye in the next step.
4. Grab the cornea with the forceps (Fig. 1a, step 1) and cut along the cornea border between the sclera and the iris using spring scissors (Fig. 1a, step 2). Alternatively, if failing to grab the cornea, puncture the eye with a needle to insert one tip of the scissors into the eye at the level of the cornea.
5. After dissecting the cornea, remove the iris and pull the lens out with the forceps (Fig. 1a, step 3); this action normally also removes any remaining fragments of the iris, ciliary body, hyaloid vasculature, and vitreous humor. Ensure to completely remove the hyaloid vessels that might still be connected to the optic nerve head region before fixing the retina.
6. Carefully peel the retinal pigmented epithelium (RPE)/choroid/sclera/connective tissue away from the retina avoiding touching the retina with the tips of the forceps (Fig. 1a, step 4). At this stage, double check that hyaloid vessels that might still be connected to the optic nerve head or to any remaining ciliary body at the peripheral edges have been completely removed by removing any floating structure in the eye cup using forceps.
7. To flat-mount the dissected retina, make four radial incisions using spring scissors (Fig. 1a, step 5), position the retina with the inner side facing upward and remove excess PBS from the dish using a plastic Pasteur pipette. If the retina is not completely opened up, flatten the unfolded lobes using curved forceps or by adding PBS by dropping it onto the folded retina with a Pasteur pipette followed again by removal of the excess PBS (Fig. 1a, step 6).
8. Fix the tissue by slowly dropping cold 100% methanol directly at the center of the retina for a few seconds until the tissue appears white and stiff. Alternatively, if the antigen of interest for subsequent procedures is sensitive to methanol fixation, use formaldehyde.
9. Use forceps to carefully lift the retinas and transfer them to a well of a 96-well plate already containing 200 μL of methanol, seal the plate with Parafilm, and incubate at −20 °C for 2 h. Retinas can be stored at −20 °C for many months, provided that methanol is replenished regularly in the wells, as it evaporates over time even in the freezer.

3.2 Fluorescent Labeling

1. For whole-mount staining, wash the samples with 200 μL of PBS twice at room temperature (RT) until retinas sink to the bottom of the wells.
2. Plates containing retinas are placed on an orbital shaker and agitated at moderate speed. *See* **Note 1**, if performing histochemical detection methods, otherwise proceed to next step.
3. The samples are permeabilized and blocked against unspecific interactions by incubating for 30 min to 1 h at RT with 200 μL of the blocking solution (incubation volume can be scaled down to 50 μL if necessary). The choice of the serum species for the blocking solution depends on the secondary antibody species, i.e., if the secondary antibody is raised in goat, then goat serum should be used.
4. Then samples are incubated on an orbital shaker with 100 μL of the blocking solution containing a 1:200 dilution of biotinylated IB4 (or one or more primary antibodies for the protein (s) of interest, also in combination) overnight at 4 °C or for 2 h at RT.
5. Samples are washed three times with 200 μL of PBT at RT for 15 min each on an orbital shaker.
6. Samples are then incubated on an orbital shaker overnight at 4 °C or for 2 h at RT with 100 μL of blocking solution containing a 1:200 dilution of Alexa Fluor 647-conjugated streptavidin (or the secondary antibodies specific for the primaries used at **step 4**) and FITC-conjugated mouse antibody anti-αSMA. Protect samples from light by wrapping plates in aluminum foil.
7. Washed samples as described above at **step 5**. Optional: include DAPI or other nuclear counterstaining in the last wash.
8. Post-fix in formaldehyde at 4 °C for at least 30 min and store in the fridge until mounting.
9. Retinas are gently lifted out of their wells with forceps without touching the inner side of the retina and placed directly on a slide with the inner side facing up. Pay extra care to avoid damaging the tissue at this stage. The samples are then mounted by covering the retinas with a few drops of Mowiol/DABCO antifade and a coverslip.
10. Retinas can be imaged using a fluorescent stereomicroscope with appropriate filters at low magnification to capture the whole retina in a single picture or at higher magnification to better visualize vascular details (confocal imaging followed by maximal intensity projections of *z*-stacks is recommended to obtain more detailed images of the vasculature including subcellular structures such as the filopodia of tip cells that lead the vascular front).

3.3 Quantitation of Angiogenesis

1. Low-magnification images of IB4 staining (alternatively PECAM1 immunostaining can be used) allow to quantify the number of arteries (Fig. 1b, orange "a") and veins (Fig. 1b, purple "v"). Arteries and veins can be discriminated based on their different morphology, with arteries showing a smaller caliber compared to veins and flanked by capillary-free areas. αSMA immunostaining to label vascular smooth muscle cells can also be used to help discriminating arteries from vein, with arteries showing a compact coverage by vascular smooth muscle cells that are oriented orthogonally to the vessel axis, while veins are covered by a reduced number of cells that appear less tightly packed around the vessel (*see also* Fig. 1f for αSMA immunostaining).
2. Low-magnification images of IB4 (or PECAM1) staining can also be used to quantify the radial vascular expansion from the center toward the periphery of the eye. This measurement can be obtained by quantifying the distance from the optic nerve head in the center of the retina to both the retina external border (Fig. 1b, red arrow) and the vascular front (Fig. 1b, green arrow), followed by the calculation of the ratio between the two values (Fig. 1c). The operation is performed using freely available (e.g., Fiji/ImageJ) or commercial software (e.g., Adobe Photoshop) for each of the four leaflets of the mounted retina, and the mean value for each retina is then used for comparisons to other retina samples.
3. The vascular complexity of the whole retina can be measured using freely available software, such as AngioTool [28] or Fiji/ImageJ [29]. AngioTool allows image segmentation to build a vascular skeleton for the quantification of branchpoints and vessel segment length and area, among other parameters (Fig. 1d). The Sholl plugin in Fiji/ImageJ, developed to quantify neuron arborization, calculates both the extension and the complexity of the vascular plexus by creating a series of concentric shells (circles) around the focus (optic nerve head) and counting how many times the skeletonized vascular plexus intersects the sampling shells (Fig. 1e). Alternatively, commercial software packages may be used for automated quantitation of these parameters, such as Imaris (Bitplane).
4. Capillary density can be measured in the plexus behind the vascular front and between the adjacent arteries and veins by selecting a proper region of interest (Fig. 1f) followed by the image analysis procedures explained at **step 3**.
5. Higher magnification imaging of the vascular front allows the manual counting of tip cells extending toward the avascular periphery (Fig. 1g, yellow arrowheads), followed by normalization to the length of the examined vascular front (Fig. 1g, cyan line).

3.4 Troubleshooting

1. Fluorophore-conjugated reagents (streptavidin or antibody) could form precipitates, identifiable as fluorescent speckles in the sample. To avoid artifacts, always briefly centrifuge the reagent in a benchtop centrifuge at top speed before use to avoid precipitates.
2. Weak staining might be due to insufficient tissue penetration by IB4 or photobleaching. Increase incubation time, optimize IB4 concentration, keep samples in the dark (e.g., wrap containers in tin foil), and mount samples in antifade solution. The choice of fluorochromes that are excited and emit at high wave lengths (red or far red) will allow improved detection of IB4 in deeper layers of whole-mounted tissues if retinas older than P7 are analyzed.
3. Trapping of air bubbles in mounted sections can occur when placing the coverslip. Lower the coverslip carefully onto the samples to avoid trapping air bubbles; if bubbles persist, slowly remove the coverslip and repeat the mounting procedure again.

4 Note

1. This protocol can be used for immunostaining the retina with many different antibodies, either alone or in combination, and is compatible with histochemical detection methods (e.g., HRP-conjugated streptavidin), provided that endogenous HRP activity is blocked at the beginning of the staining procedure (*See* Subheading 3.2, **step 3**), and the appropriate blocking reagents and secondary antibodies are used.

References

1. Nowak-Sliwinska P, Alitalo K, Allen E, Anisimov A, Aplin AC, Auerbach R et al (2018) Consensus guidelines for the use and interpretation of angiogenesis assays. Angiogenesis 21(3):425–532. https://doi.org/10.1007/s10456-018-9613-x
2. Fantin A, Schwarz Q, Davidson K, Normando EM, Denti L, Ruhrberg C (2011) The cytoplasmic domain of neuropilin 1 is dispensable for angiogenesis, but promotes the spatial separation of retinal arteries and veins. Development 138(19):4185–4191. https://doi.org/10.1242/dev.070037
3. Benedito R, Roca C, Sorensen I, Adams S, Gossler A, Fruttiger M et al (2009) The notch ligands Dll4 and Jagged1 have opposing effects on angiogenesis. Cell 137(6):1124–1135. https://doi.org/10.1016/j.cell.2009.03.025
4. Gerhardt H, Golding M, Fruttiger M, Ruhrberg C, Lundkvist A, Abramsson A et al (2003) VEGF guides angiogenic sprouting utilizing endothelial tip cell filopodia. J Cell Biol 161(6):1163–1177. https://doi.org/10.1083/jcb.200302047
5. Wang Y, Nakayama M, Pitulescu ME, Schmidt TS, Bochenek ML, Sakakibara A et al (2010) Ephrin-B2 controls VEGF-induced angiogenesis and lymphangiogenesis. Nature 465(7297):483–486. https://doi.org/10.1038/nature09002
6. Hellstrom M, Phng LK, Hofmann JJ, Wallgard E, Coultas L, Lindblom P et al (2007) Dll4 signalling through Notch1

regulates formation of tip cells during angiogenesis. Nature 445(7129):776–780. https://doi.org/10.1038/nature05571

7. Fantin A, Herzog B, Mahmoud M, Yamaji M, Plein A, Denti L et al (2014) Neuropilin 1 (NRP1) hypomorphism combined with defective VEGF-A binding reveals novel roles for NRP1 in developmental and pathological angiogenesis. Development 141(3):556–562. https://doi.org/10.1242/dev.103028
8. Fantin A, Lampropoulou A, Gestri G, Raimondi C, Senatore V, Zachary I et al (2015) NRP1 regulates CDC42 activation to promote filopodia formation in endothelial tip cells. Cell Rep 11(10):1577–1590. https://doi.org/10.1016/j.celrep.2015.05.018
9. Liyanage SE, Fantin A, Villacampa P, Lange CA, Denti L, Cristante E et al (2016) Myeloid-derived vascular endothelial growth factor and hypoxia-inducible factor are dispensable for ocular neovascularization--brief report. Arterioscler Thromb Vasc Biol 36(1):19–24. https://doi.org/10.1161/ATVBAHA.115.306681
10. Raimondi C, Fantin A, Lampropoulou A, Denti L, Chikh A, Ruhrberg C (2014) Imatinib inhibits VEGF-independent angiogenesis by targeting neuropilin 1-dependent ABL1 activation in endothelial cells. J Exp Med 211(6):1167–1183. https://doi.org/10.1084/jem.20132330
11. Ruhrberg C, Gerhardt H, Golding M, Watson R, Ioannidou S, Fujisawa H et al (2002) Spatially restricted patterning cues provided by heparin-binding VEGF-A control blood vessel branching morphogenesis. Genes Dev 16(20):2684–2698
12. Gerhardt H, Ruhrberg C, Abramsson A, Fujisawa H, Shima D, Betsholtz C (2004) Neuropilin-1 is required for endothelial tip cell guidance in the developing central nervous system. Dev Dyn 231(3):503–509
13. Fantin A, Vieira JM, Gestri G, Denti L, Schwarz Q, Prykhozhij S et al (2010) Tissue macrophages act as cellular chaperones for vascular anastomosis downstream of VEGF-mediated endothelial tip cell induction. Blood 116(5):829–840. https://doi.org/10.1182/blood-2009-12-257832
14. Fantin A, Vieira JM, Plein A, Denti L, Fruttiger M, Pollard JW et al (2013) NRP1 acts cell autonomously in endothelium to promote tip cell function during sprouting angiogenesis. Blood 121(12):2352–2362. https://doi.org/10.1182/blood-2012-05-424713
15. Graupera M, Guillermet-Guibert J, Foukas LC, Phng LK, Cain RJ, Salpekar A et al (2008) Angiogenesis selectively requires the p110alpha isoform of PI3K to control endothelial cell migration. Nature 453(7195):662–666. https://doi.org/10.1038/nature06892
16. Lu X, Le Noble F, Yuan L, Jiang Q, De Lafarge B, Sugiyama D et al (2004) The netrin receptor UNC5B mediates guidance events controlling morphogenesis of the vascular system. Nature 432(7014):179–186
17. Tammela T, Zarkada G, Nurmi H, Jakobsson L, Heinolainen K, Tvorogov D et al (2011) VEGFR-3 controls tip to stalk conversion at vessel fusion sites by reinforcing notch signalling. Nat Cell Biol 13(10):1202–1213. https://doi.org/10.1038/ncb2331
18. Plein A, Fantin A, Denti L, Pollard JW, Ruhrberg C (2018) Erythro-myeloid progenitors contribute endothelial cells to blood vessels. Nature 562(7726):223–228. https://doi.org/10.1038/s41586-018-0552-x
19. Fruttiger M (2007) Development of the retinal vasculature. Angiogenesis 10(2):77–88. https://doi.org/10.1007/s10456-007-9065-1
20. Hofmann JJ, Luisa I-AM (2007) Notch expression patterns in the retina: an eye on receptor-ligand distribution during angiogenesis. Gene Expr Patterns 7(4):461–470. https://doi.org/10.1016/j.modgep.2006.11.002
21. Carmeliet P, Tessier-Lavigne M (2005) Common mechanisms of nerve and blood vessel wiring. Nature 436(7048):193–200
22. Okabe K, Kobayashi S, Yamada T, Kurihara T, Tai-Nagara I, Miyamoto T et al (2014) Neurons limit angiogenesis by titrating VEGF in retina. Cell 159(3):584–596. https://doi.org/10.1016/j.cell.2014.09.025
23. Scott A, Powner MB, Gandhi P, Clarkin C, Gutmann DH, Johnson RS et al (2010) Astrocyte-derived vascular endothelial growth factor stabilizes vessels in the developing retinal vasculature. PLoS One 5(7):e11863. https://doi.org/10.1371/journal.pone.0011863
24. Hughes S, Chang-Ling T (2000) Roles of endothelial cell migration and apoptosis in vascular remodeling during development of the central nervous system. Microcirculation 7(5):317–333
25. Sawamiphak S, Ritter M, Acker-Palmer A (2010) Preparation of retinal explant cultures to study ex vivo tip endothelial cell responses. Nat Protoc 5(10):1659–1665. https://doi.org/10.1038/nprot.2010.130

26. Nagy A (2000) Cre recombinase: the universal reagent for genome tailoring. Genesis 26(2):99–109
27. Smith LE, Wesolowski E, McLellan A, Kostyk SK, D'Amato R, Sullivan R et al (1994) Oxygen-induced retinopathy in the mouse. Invest Ophthalmol Vis Sci 35(1):101–111
28. Zudaire E, Gambardella L, Kurcz C, Vermeren S (2011) A computational tool for quantitative analysis of vascular networks. PLoS One 6(11): e27385. https://doi.org/10.1371/journal.pone.0027385
29. Schindelin J, Arganda-Carreras I, Frise E, Kaynig V, Longair M, Pietzsch T et al (2012) Fiji: an open-source platform for biological-image analysis. Nat Methods 9(7):676–682. https://doi.org/10.1038/nmeth.2019

Part II

Adult Physiology and Pathology

Chapter 6

Characterization of Lymphatic Vasculature Using Whole-Mount Immunostaining of Mouse Embryonic Dorsal Skin

Nikita Ved and Jacinta Isabelle Kalisch-Smith

Abstract

Understanding the development of the lymphatic vasculature is essential to the understanding of how these vessels function in health and disease. High-resolution imaging of histological techniques such as immunostaining of sectioned tissue provides a snapshot into lymphatic vessel morphogenesis, patterning, and organization. Whole-mount staining of embryonic dermal vasculature allows for a deeper analysis and characterization of the developing lymphatic vascular network.

Key words Lymphatic vessels, Immunostaining, Characterization, Vessel width, Branching

1 Introduction

The lymphatic vascular network is critical to tissue homeostasis, regulation of interstitial fluid, lipid absorption, fluid drainage, and immune cell trafficking [1]. The blood and lymphatic vascular networks form during embryogenesis through the activation of an intricate series of transcription factors. During development, blood vessels are formed de novo through vasculogenesis from mesoderm- and somite-derived progenitors to form a primitive vascular network, which then matures through angiogenesis [2].

Upon formation of this vascular network, between embryonic days 9.5 and 10.5 (E9.5 and E10.5), endothelial cells in the cardinal vein express lymphatic markers such as lymphatic vascular endothelial hyaluronan receptor-1 (LYVE-1) and prospero-related homeobox domain 1 (PROX-1) [3, 4]. These endothelial cells differentiate to lymphatic endothelial cells. PROX-1-positive cells are necessary for lymphangiogenesis, and these cells will form the early lymph sacs which will ultimately form the lymphatic vascular network [5]. Upon vessel pruning and reshaping, these primary

Andrew V. Benest (ed.), *Angiogenesis: Methods and Protocols*, Methods in Molecular Biology, vol. 2441,
https://doi.org/10.1007/978-1-0716-2059-5_6,

plexuses become complex vascular networks [6]. As blood and lymphatic vasculature are a vital aspect of organ and tissue homeostasis, their developmental and morphogenetic programmes are highly sensitive to local, tissue-specific cues. This can also be detrimental as under pathological conditions, such as hypoxia, altered metabolism, or trauma, these blood vessels can exhibit perturbed angiogenesis/lymphangiogenesis, both during development [7] and in adult tissue [4, 8]. Perturbed lymphatic development can result in subcutaneous edema [9], a phenotype commonly observed in embryonic lethal mouse strains. Many of these mouse strains also express malformations in heart, forebrain, and eye development [10, 11]. To observe the blood and lymphatic vasculature during normal and perturbed development, traditional histological techniques of sectioned tissue can be used, as well as whole-mount imaging of the tissue of interest. Flat-mount imaging of the skin is particularly advantageous as the dermal lymphatics are highly organized in an arborized vascular network and allows for vessel characterization. In mouse embryos, the lymphatic vasculature can be observed in the dorsal skin from E12.5 onwards [12]. Immunofluorescent staining of the developing lymphatic vasculature provides an insight into many aspects of lymphangiogenesis including gene function, signaling pathways, vessel sprouting, migration, and proliferation.

Here we describe whole-mount immunofluorescence staining of these vessels in dorsal embryonic skin, using techniques adapted from [12, 13], to obtain a better understanding of lymphangiogenesis during development.

2 Materials

2.1 Reagents

- 10× Hanks Balanced Salt Solution (HBSS, Sigma).
- Ethylenediaminetetraacetic acid (EDTA, Sigma, EDS).
- 1× Phosphate-Buffered Saline (PBS, Sigma).
- 4% Paraformaldehyde (PFA, Sigma) solution in 1× PBS.
- 0.01% Thimerosal (Sigma) in 1x PBS.
- Triton-X100 (Sigma, X100).
- Bovine Serum Albumin (BSA, Sigma).
- Normal Donkey Serum (NDS, Sigma).
- Vectashield mounting medium (Sigma).

2.2 Buffer Preparation

- 0.5 M EDTA: To prepare EDTA at 0.5 M (pH 8.0): Add 186.1 g of EDTA to 800 mL of H_2O and adjust the pH to 8.0 with sodium hydroxide solution (NaOH).

- Dissection buffer: To prepare 500 mL of dissection buffer, add 50 mL 10× HBSS to 445 mL PBS, followed by 5 mL 0.5 M EDTA.
- 0.01% thimerosal: Add 1 g of thimerosal and add to 10 mL of PBS, then dilute 1 in 1000 in PBS and store at room temperature.
- Blocking solution: To prepare 50 mL blocking solution, add 0.5 g of BSA and 15 μL Triton-X100 to 50 mL PBS and 2.5 mL animal serum from the same species that 2° antibodies are raised in, we used normal donkey serum.
- 0.1% Triton-X100: Add 50 μL Triton-X100 to 50 mL PBS and place on a rotator until fully dissolved.
- Many combinations of fluorophores can be used; however, we used the following:
  - Primary antibodies: rabbit anti-NRP2 (Cell Signaling Technology, 3366), rat anti-CD31 (BD Pharmingen, 553370), goat anti-PROX1 (R&D Systems, AF2727) all diluted 1:500 in blocking solution.
  - Fluorescent-conjugated secondary antibodies: donkey anti-rabbit Cy™3 (Jackson ImmunoResearch, 711-165-152), donkey anti-goat AlexaFluor® 488 (Life Technologies, A-11055), donkey anti-rat DyLight™ 405 (Jackson ImmunoResearch, 712-475-153) all diluted 1:500 in blocking solution.
  - Nuclear staining: TO-PRO®-3 Iodide (Life Technologies, T3605) diluted 1:10,000 in blocking solution.

2.3 Tools

- Vannas scissors/fine scissors.
- Fine forceps.
- 10-cm petri dishes.
- 12-well flat-bottom tissue culture plates.
- 24-well flat-bottom tissue culture plates.
- Aluminum foil.
- Superfrost Plus microscope slides (Thermo Scientific).
- 22 × 32 mm Coverslips (Fisherbrand).
- Nutator/Rocker.
- Dissection Microscope, we used a Leica M80 dissecting microscope.
- Confocal microscope, we used an Olympus FV3000.

3 Methods

3.1 Embryo Collection

- Euthanize pregnant mice according to the desired timepoint and as per institutional ethical regulations. We collected embryos at E14.5. All animal experiments were compliant with the UK Animals (Scientific Procedures) Act 1986 and approved by the University of Oxford animal welfare review board and the Home Office (project license PB01E1FB3).
- Dissect embryos out of intact uteri in a petri dish with room temperature dissection buffer. Place each dissected embryo into a well of a 12-well plate with pre-warmed, fresh dissection buffer (*see* **Note 1**).
- Wash embryos in fresh dissection buffer 2× 10 min.
- Fix embryos overnight in 4% PFA at 4 °C on a nutator (*see* **Note 2**).

3.2 Dissection and Staining of Dorsal Skin

- Wash embryos in ice-cold PBS 3× 10 min each.
- Dissect the dorsal skin in PBS and keep in a 24-well plate: (*see* **Note 3**)
  - Use light pressure to score a rectangle into the dorsal skin, with the vannas or other fine scissors, as shown in Fig. 1.
  - Cut superficially along the scored lines, using the forceps to keep the embryo in place.
  - Use forceps to gently tease the dorsal skin from the embryo (*see* **Note 4**).
- Block tissue overnight in blocking solution at 4 °C on a nutator.
- Discard the blocking solution and incubate samples in 1° antibodies diluted in blocking solution. To investigate blood and lymphatic vasculature, we used CD31, NRP2, and PROX1 (Fig. 2).
- Discard the 1° antibody solution and wash samples in 0.1% Triton-X100 for 6× 1-h washes on ice on a nutator.
- Incubate samples in 2° antibodies in blocking solution a nuclear stain such as TO-PRO®-3 Iodide, on a nutator, overnight at 4 °C. Wrap the plate in aluminum foil to prevent photobleaching (*see* **Note 5**).
- Discard the 2° antibody solution and wash samples in 0.1% Triton-X100 at room temperature for 6× 1-h washes on ice on a rotator (*see* **Note 6**).
- Mount samples by placing tissue on a slide and then by making small incisions at 90° to each other. Using a mounting medium, we used Vectashield, let the coverslips set overnight, again keeping samples shielded from light.

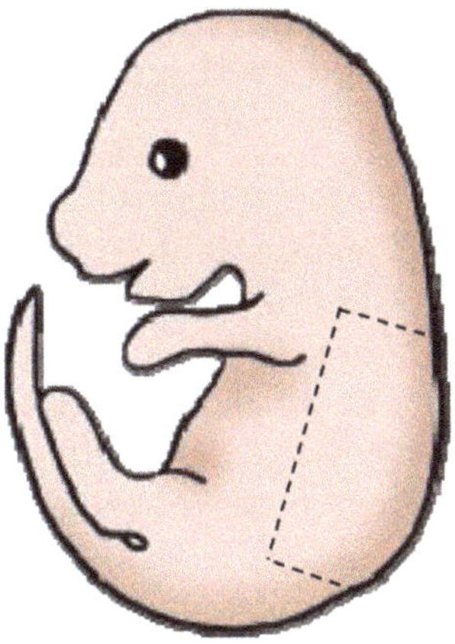

Fig. 1 Dissection schematic. Illustration of a sagittal view of an E14.5 mouse embryo. The dotted line indicates the region of dorsal skin used to immunostain the dermal vasculature, on both sides of the embryo

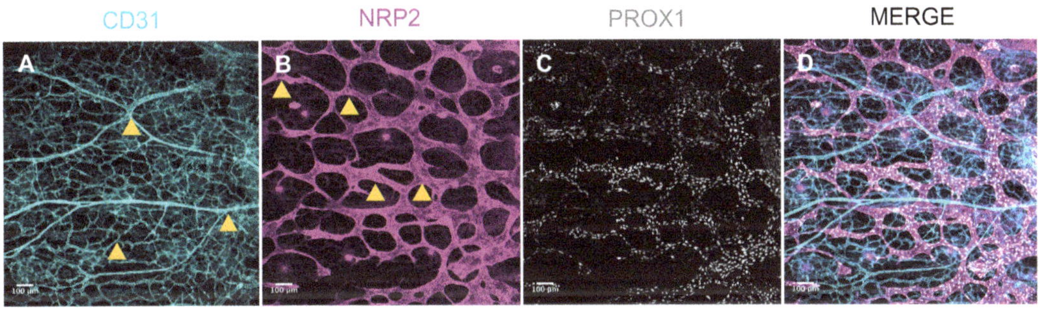

Fig. 2 Whole-mount staining of dorsal skin vessels. Dermal vasculature from E14.5 embryonic dorsal skill stained for (**a**) blood vessel marker CD31 (cyan), (**b**) lymphatic vessel marker NRP2 (magenta), (**c**) lymphatic endothelial cell marker PROX1 (white), and a composite image (**d**). Arrowheads indicate vessel branch points and scale bar represents 100 μM

- Once mounting medium has hardened, tissue can be stored at in the dark at 4 °C until ready for confocal imaging.

3.3 Imaging

- Due the thickness of the immunostained tissue, samples are best analyzed using Z-stacks of tiled images collected by fluorescent confocal microscopy.
- Let samples re-equilibrate at room temperature for 20 min before imaging.
- We performed a resonance scan at 4× magnification to map the dorsal-skin region on the slide, using the nuclear staining. We then used this map at 10× magnification to capture our final images.

- Data presented here have been imaged using an Olympus FV3000 confocal microscope with Fluoview FV31S-SW software.

3.4 Analysis

Quantification of lymphatic vessel parameters such as vessel width and branching index can be measured using FIJI [14], the AngioTool [15] plugin, and other similar software. It is important to map out which region of the dorsal skin is being analyzed as lymphatic vessels closest to the midline are the sprouting front and represent a region of sprouting angiogenesis [12].

4 Notes

1. Embryos/dissected dorsal skin can be stored in 0.01% thimerosal at 4 °C before staining, to prevent fungal growth. This allows for staining in larger batches.
2. If storing tissue long-term (several months), be sure to top up 0.01% thimerosal regularly.
3. We recommend filling 12-well plates with a minimum of 700 μL of solution and 24-well plates with 500μLof solution.
4. When dissecting dorsal skin, take care to avoid cutting too deep into the skin and too far into the abdomen, as the sample may include muscle tissue, which will obscure any staining.
5. The weakest fluorophore should be paired with the strongest primary antibody signal, in this case we paired DyLight™ 405 with CD31.
6. Primary antibodies can be saved following overnight incubation and re-used for multiple batches of dorsal skin immunostaining.

Acknowledgments

NV is supported by a Novo Nordisk postdoctoral fellowship run in partnership with the University of Oxford, and JIK is supported by funding from the British Heart Foundation (FS/17/55/33100 and RE/13/1/30181).

Author Contributions: NV contributed to compilation/writing of the chapter. NV and JIK contributed to the development and modification of whole-mount dorsal skin staining protocol. JIK contributed the final images.

References

1. Tammela T, Alitalo K (2010) Lymphangiogenesis: molecular mechanisms and future promise. Cell 140:460–476
2. Park C, Kim T, Malik AB (2013) Transcriptional regulation of endothelial cell and vascular development. Circ Res 112:1380–1400
3. Gale NW et al (2007) Normal lymphatic development and function in mice deficient for the lymphatic hyaluronan receptor LYVE-1. Mol Cell Biol 27:595–604
4. Wigle JT, Oliver G (1999) Prox1 function is required for the development of the murine lymphatic system. Cell 98:769–778
5. Marcelo KL, Goldie LC, Hirschi KK (2013) Regulation of endothelial cell differentiation and specification. Circ Res 112:1272–1287
6. Risau W (1997) Mechanisms of angiogenesis. Nature 386:671–674
7. Carmeliet P et al (1996) Abnormal blood vessel development and lethality in embryos lacking a single VEGF allele. Nature 380:435–439
8. Wang Y, Oliver G (2010) Current views on the function of the lymphatic vasculature in health and disease. Genes Dev 24:2115–2126
9. Kalisch-Smith JI, Ved N. Szumska D et al (2021) Maternal iron deficiency perturbs embryonic cardiovascular development in mice. Nat Commun 12, 3447. https://doi.org/10.1038/s41467-021-23660-5
10. Wilson R et al (2017) Highly variable penetrance of abnormal phenotypes in embryonic lethal knockout mice. Wellcome Open Res 1:1
11. Perez-Garcia V et al (2018) Placentation defects are highly prevalent in embryonic lethal mouse mutants. Nature 555:463–468
12. Betterman KL, Harvey NL (2018) Histological and morphological characterization of developing dermal lymphatic vessels. In: Oliver G, Kahn ML (eds) Lymphangiogenesis: methods and protocols. Springer, New York, pp 19–35. https://doi.org/10.1007/978-1-4939-8712-2_2
13. Fatima A et al (2015) Whole-mount immunostaining methods to study the blood and lymphatic vasculature in the embryonic mouse skin and adult mouse cornea. In: Slevin M, McDowell G (eds) Handbook of vascular biology techniques. Springer, Dordrecht, pp 245–253. https://doi.org/10.1007/978-94-017-9716-0_21
14. Schindelin J et al (2012) Fiji: an open-source platform for biological-image analysis. Nat Methods 9:676–682
15. Zudaire E, Gambardella L, Kurcz C, Vermeren S (2011) A computational tool for quantitative analysis of vascular networks. PLoS One 6: 27385

Chapter 7

Analyzing Lymphatic Vessel Patterning in Adult Tissue

Louise A. Johnson

Abstract

Whole-mount immunostaining allows intact tissue to be surveyed in three dimensions, avoiding the more restricted fields of view provided by visualizing thin sections. This technique is particularly useful for imaging lymphatic and blood networks by high-resolution confocal microscopy, revealing how such vessels are spatially positioned, the subcellular arrangements of individual antigens, and interactions with individual cells within the interstitium or vessel lumen. The purpose of this chapter is to provide a practical guide for obtaining images of lymphatic vessels following immunofluorescence staining, primarily in mouse skin.

Key words Lymphatic vessels, Blood vessels, Endothelial cells, Whole-mount, Immunofluorescence microscopy, Confocal microscopy, Skin, Antibodies, LYVE-1

1 Introduction

Afferent lymphatic vessels transport lymph fluid, macromolecules, and leukocytes from peripheral tissue to lymph nodes for immune surveillance, returning excess fluid to the blood vasculature via efferent lymphatics [1–3]. Afferent vessels in the tissue are wide, thin, blind-ended capillaries with discontinuous button-like junctions between overlapping oakleaf-shaped endothelial cells [4]. Such junctions are specialized to aid passive uptake of fluid and entry of lymph-migratory leukocytes, which squeeze in through these overlapping flaps. These initial capillaries then merge into larger pre-collector and collector vessels, which are largely impermeable and permit more rapid, leakage-free transport to draining lymph nodes.

Specific markers to distinguish lymphatic vessels from blood capillaries include lymphatic vessel endothelial hyaluronan receptor-1, LYVE-1 [5], the type I transmembrane protein podoplanin [6–8], the transcription factor and lymphatic lineage determinant Prox-1 [9], the tyrosine kinase VEGFR3 [10, 11] and the chemokine CCL21 [12]. Junctional proteins VE-cadherin, CD31,

Andrew V. Benest (ed.), *Angiogenesis: Methods and Protocols*, Methods in Molecular Biology, vol. 2441,
https://doi.org/10.1007/978-1-0716-2059-5_7,

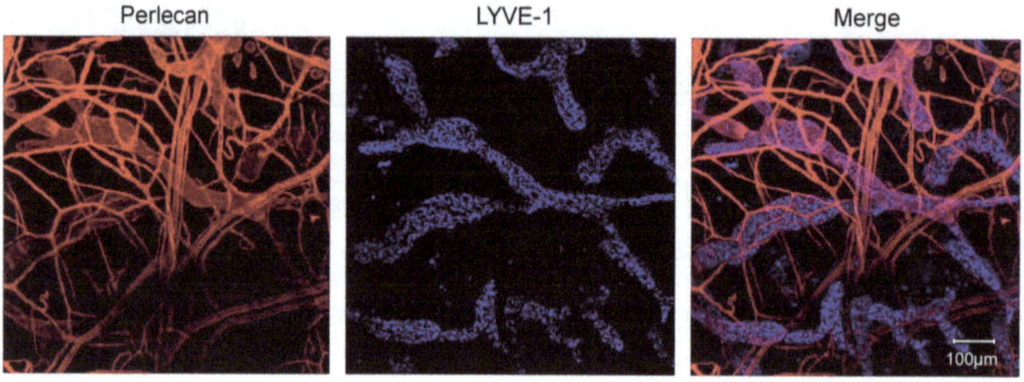

Fig. 1 Whole-mount immunostaining of lymphatic and blood vessels in mouse dermis. Tissue was fixed in PFA at room temperature for 1 h, washed and permeabilized with Triton X-100, prior to immunostaining with rat anti-perlecan and rabbit anti-LYVE-1, detecting with goat anti-rat AlexaFluor®568 and goat anti-rabbit AlexaFluor®647, respectively. Images captured by Zeiss LSM 780 confocal microscope, using a 10×/0.3 DIC M27 Plan-Apochromat (total magnification: 100×)

and ESAM are expressed by both lymphatic and blood vessels, although immunostaining for these molecules is more pronounced on blood capillaries [4, 13]. Similarly, basement membrane proteins such as perlecan, laminin α5, and α-smooth muscle actin (SMA-α) partially surround lymphatic vessels, particularly larger collecting vessels, but immunostaining is more pronounced in association with blood capillaries (Fig. 1) [14, 15]. To irrefutably identify lymphatic vessels, markers should be used in combination as they are also expressed by other cell types, for example, LYVE-1 by macrophages [16], and podoplanin by glomerular podocytes [17] and alveolar epithelial cells [18, 19]. Furthermore, expression of these proteins varies from initial lymphatic capillaries to pre-collector and collector vessels, with LYVE-1 abundantly expressed on initial lymphatic vessels but absent from collector vessels. Also, lymphatic vessels adopt novel expression profiles in response to inflammation and when in proximity to tumor cells, with respect to adhesion and junctional molecules, chemokines, matrix proteins, and immune regulatory molecules [20–23].

Mouse ear skin has been proved very useful for assessing the microanatomical structure of initial lymphatic capillaries, and this book chapter focuses on preparing such tissue for imaging by immunostaining and confocal microscopy. For lymphatic imaging of mouse trachea and lung tissue, the reader is directed to excellent protocols by Baluk and McDonald [24] and also included in this Protocols edition.

2 Materials

Prepare all solutions using ultrapure deionized water and make up fresh stocks of all reagents. Biological and chemical waste materials must be disposed of carefully, in line with local safety regulations.

1. Phosphate-buffered saline (PBS). For 1× concentration, dissolve one tablet into 100 mL deionized water, or according to the manufacturers' instructions; for 2× concentration, dissolve one tablet into 50 mL deionized water. 1× PBS contains: 0.2 g/L KH_2PO_4, 2.16 g/L $Na_2HPO_4 \cdot 7H_2O$, 0.2 g/L KCl, 8.0 g/L NaCl.
2. Paraformaldehyde (PFA) Fixative. To make 100 mL of 1% paraformaldehyde solution, add 1 g powdered paraformaldehyde to a 0.5-L flask with a magnetic stirring bar and add approximately 30 mL deionized water (*see* **Note 1**). Heat the suspension to 55–60 °C on a hot plate with magnetic stirrer in a fume hood. Add 1 M NaOH one drop at a time, until a clear solution forms from the cloudy white suspension. Any remaining white precipitate can be removed by passing the solution through a 0.2-μm filter and syringe, once cooled (if the suspension is heated to over 65 °C, discard and begin again). Add 2× PBS, 50 mL, to the PFA and when the solution is at room temperature, adjust the pH to 7.4 by adding either HCl or NaOH, drop by drop. Add deionized water to attain the final volume of 100 mL. Store 1% PFA at room temperature in a sealed bottle for up to 2 weeks, or discard if precipitate forms (*see* **Note 2**).
3. Permeabilization and wash buffer. Add 1.5 mL Triton X-100 to 500 mL 1× PBS and slowly stir using a magnetic stirring bar (*see* **Note 3**).
4. Blocking Buffer. Dissolve 1 g BSA in 90 mL permeabilization buffer, slowly stirring with a magnetic stirring bar. Then add 10 mL of goat serum or (if using antibodies raised in goat) 10 mL fetal calf serum (FCS) (*see* **Note 4**).
5. VECTASHIELD mounting medium (Vector Laboratories) (*see* **Note 5**).
6. CoverGrip Sealant (Biotium) (*see* **Note 6**).

3 Methods

All incubations should be carried out with gentle shaking/swirling on a plate shaker.

3.1 Immunostaining

1. Initial tissue preparation. Euthanize mice and cut off ears using a pair of sharp dissection scissors. Slide a pair of forceps into the exposed side and gently peel and separate the dorsal and ventral halves of each ear. Lay each ear half within individual wells of a 24-well dish, dermis side down.
2. Fixation. Apply 1% PFA in PBS, 1 mL/well (*see* **Note 7**). Length of fixation should be determined according to individual antibodies as certain epitopes may be more sensitive to the fixation process whereas binding of other antibodies may be improved by prolonged fixation (*see* **Note 8**). For the majority of antibodies, incubate for 16 h at 4 °C. A lighter fixation (1 h, room temperature) may be more appropriate for other antibodies. If the antibody epitope proves too sensitive to PFA, this fixation step may be omitted, although routinely it is recommended that tissue is fixed.
3. Washing and permeabilization. Remove the fixative from each well by aspiration with pipette. To rinse out fixative and permeabilize the tissue, apply Permeabilization and Wash Buffer, 1 mL/well, swirl to mix and then remove and discard. Add more Permeabilization and wash buffer, 1 mL/well and incubate and at room temperature, 1 h, ×3. Best results can be attained with more buffer changes and shorter (30 min) wash steps may prove sufficient.
4. Blocking nonspecific binding. Incubate tissue in Blocking Buffer at room temperature for 2 h, 0.5 mL/well.
5. Primary antibodies. Remove the Blocking Buffer from each well and apply fresh blocking buffer with primary antibodies included, 0.25 mL/well. Combinations of primary antibodies must be raised in different species, e.g., rat, rabbit, goat. Incubate at 4 °C for 16 h.
6. Washing. Remove and discard primary antibody solutions from tissue. Apply Permeabilization and Wash Buffer, 1 mL/well, swirl to mix and then remove and discard. Add more Permeabilization and Wash Buffer, 1 mL/well and incubate at room temperature, 1 h, ×3. Again, best results can be attained with more buffer changes.
7. Secondary antibodies. Remove and discard the Permeabilization and Wash Buffer from each well and apply Blocking Buffer with secondary antibodies included, 0.25 mL/well. AlexaFluor® antibodies (2 mg/mL), 1 in 200, are highly effective, most commonly AlexaFluor®488, AlexaFluor®568, AlexaFluor®647, and Pacific Blue™ (*see* **Note 9**). Incubate at room temperature for 2 h.
8. Washing. Repeat the previous wash step in (6).

9. Fixation. Remove and discard Permeabilization and Wash Buffer and apply 1% PFA in PBS, 1 mL/well, 5 min, to fix the labeled antibodies, thus preserving fluorescence signal.
10. Washing. Remove and discard fixative, then briefly wash tissues by applying PBS, three washes × 5 min.
11. Mounting. Carefully remove mouse ear tissue using curved forceps and place on a glass microscope slide, with the exposed dermis facing upwards, avoiding folds forming in the tissue. Place a coverslip on tissue paper and apply a drop of VECTA SHIELDmounting medium, using the dropper supplied by the manufacturer. Gently lower the slide onto the coverslip so that the tissue comes into contact with the VECTASHIELD. Secure the coverslip into position by painting the edges with Cover-Grip Sealant (Biotium). Allow to dry for approximately 30 min before viewing slides by microscope.
12. Storage. Imaging should be performed soon after staining but slides may be stored in a dark slide box at 4 °C for up to 1 year, with minimal loss of fluorescence signal.

3.2 Pretreatment with Antibodies

Function-blocking monoclonal antibodies can be injected intraperitoneally into mice, up to at least 48 h before sacrificing the animals and carrying out immunostaining. Previously, monoclonal antibodies against ICAM-1 (clone YN1/1.7.4), VCAM-1 (clone 6C7.1), and LYVE-1 (clones C1/8, B1/10 and mAb2125) have been injected and subsequently detected by whole-mount immunostaining, using goat anti-rat IgG conjugated with AlexaFluor®647 (Fig. 2) [21, 25, 26]. Rat IgG (R&D Systems) may be used for similar injections into mice in control groups.

3.3 Imaging

1. Ensure that samples are at room temperature before imaging, to allow condensation to evaporate.
2. Capture images by sequential scanning (rather than simultaneous scanning), to avoid bleed-through between channels, except for when there is no spectral overlap, for example, between Pacific Blue™ and AlexaFluor®647. Also, restricting the band of detection (in nm) for each channel will reduce bleed-through.
3. Low magnification (e.g., using a 10× objective) may be used to capture a large field of view, before using a higher magnification objective (e.g., 63×) to attain higher resolution (Fig. 3). Fine or highly discrete structures such as tight junctions may only be visible at higher magnification.
4. Capturing images in a *Z*-stack and exporting this as a three-dimensional rendered projection ensures that all vessels are in focus.

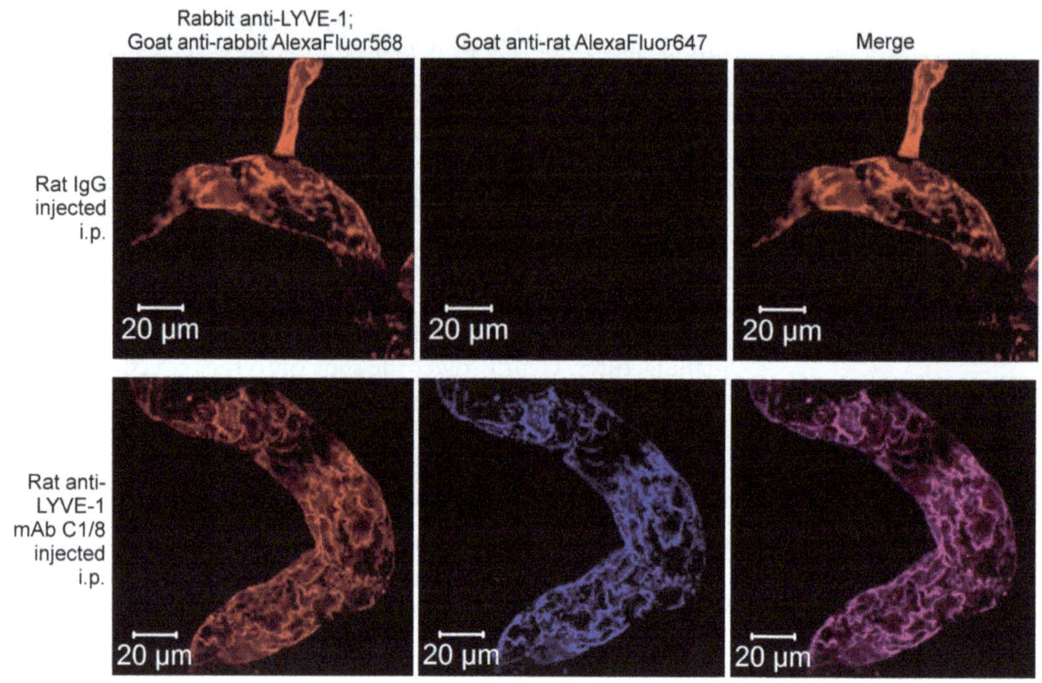

Fig. 2 Labeling of lymphatic vessels by preinjected mAbs. Intraperitoneal injections of either control rat IgG or anti-LYVE-1 (mAb clone C1/8), 0.5 mg, were carried out 48 h prior to 16 h PFA fixation (4 °C) and immunostaining with rabbit anti-LYVE-1 and goat anti-rabbit AlexaFluor®568, detecting bound C1/8 with goat anti-rat AlexaFluor®647. Note that labeling with pre-injected mAb is identical to the staining achieved with the polyclonal antibody. Images captured by Zeiss LSM 780 confocal microscope, using a 63×/1.4-NA oil immersion Plan-Apochromat (total magnification: 630×)

5. Raw data files must be retained. Exported .tif images may be analyzed in FIJI/ImageJ, free to download at https://imagej.net/Fiji. The contrast and brightness of individual channels may be optimized in either Photoshop (Adobe) or GIMP, free to download at https://www.gimp.org/.

4 Notes

1. Paraformaldehyde (PFA) is a white crystalline solid, formed from polymerization of methanal gas (formaldehyde) by heating. There are numerous hazards associated with this chemical: it is flammable, carcinogenic, irritant to the skin, eyes and respiratory tract, a reproductive toxin and also toxic by skin contact and inhalation. For safety, wear gloves, mask, personal protective clothing, and safety glasses. Also, work in a fume hood to avoid breathing formaldehyde fumes or powder. Dispose of waste PFA in line with local safety regulations.

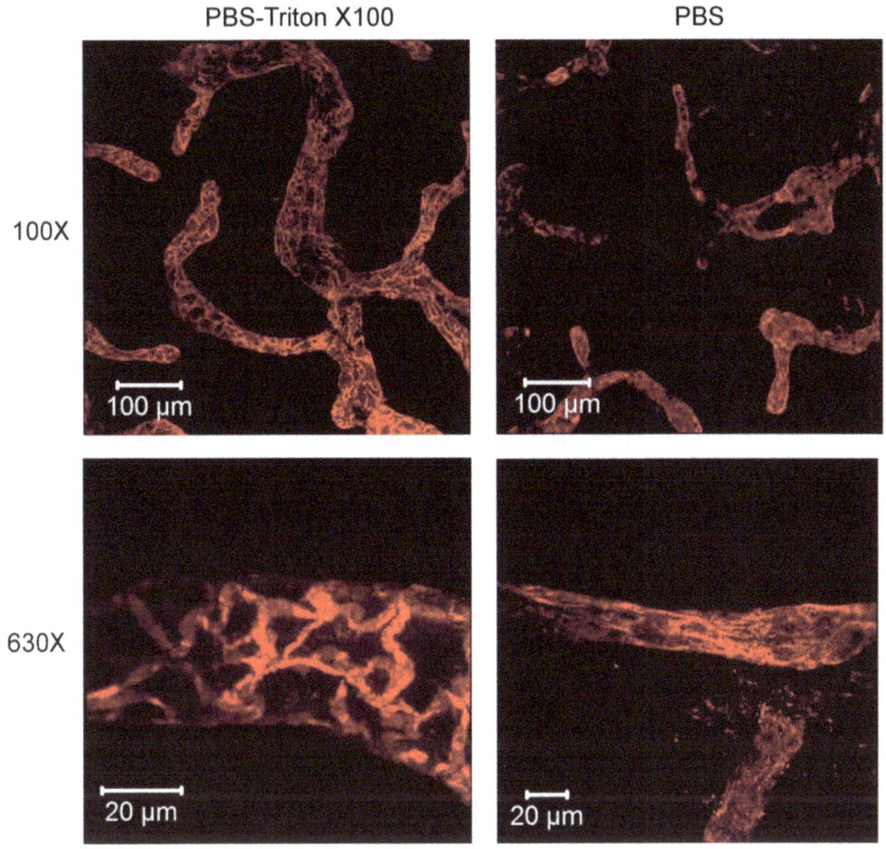

Fig. 3 Influence of permeabilization on LYVE-1 staining. Mouse dermis was fixed in PFA at room temperature for 2 h, then immunostained with rabbit anti-LYVE-1 and goat anti-rabbit AlexaFluor®568 in either the presence or absence of Triton X-100. Images captured by Zeiss LSM 780 confocal microscope, using a 10×/0.3 DIC M27 Plan-Apochromat (total magnification: 100×) and 63×/1.4-NA oil immersion Plan-Apochromat (total magnification: 630×). Note that expression of LYVE-1 at cell junctions is more apparent following permeabilization, suggesting that LYVE-1 is present in the overlapping endothelial flaps

2. Beware of using old stocks of PFA. Through the Cannizzaro reaction, paraformaldehyde polymers break down into methanol and formic acid, thus decreasing the concentration of the fixative and potentially destroying the tissue.
3. Permeabilization improves accessibility of antibodies into the tissue and enhances detection of proteins such as LYVE-1 that are found at the overlapping lymphatic endothelial cell junctions (Fig. 2). However, for assessing surface levels of antigen only, Triton X-100 must be omitted from the protocol.
4. Applying high concentrations of irrelevant proteins such as BSA fills up nonspecific binding sites and reduces nonspecific staining by primary antibodies. Blocking serum from the host species in which the secondary antibodies are raised (usually goat or donkey) may also be included.

5. VECTASHIELD Anti-fade Mounting Medium inhibits fading and photobleaching of fluorescent proteins and dyes, improves the refractive index of the sample, and prevents the tissue from drying out, which may lead to distortion.
6. Use commercial coverslip sealant rather than cosmetic nail varnish/polish, as the quality of the latter may vary and not harden sufficiently.
7. Fixation by PFA incurs numerous advantages, including stabilizing fine structures, securing target antigens in their native locations, preventing loss of cellular constituents, inactivating lysosomal enzymes, and thus preventing autolysis and inhibiting the growth of microbes.
8. PFA reacts with the side chains of basic amino acids, particularly lysine, purines, and thiols (cysteine), forming mono, dimethylol or hydrozyl methyl derivatives that are covalently bound to tissue [27]. Subsequent cross-linking occurs by the formation of -CH2- methylene bridges and involve other functional groups including amides, asparagines, and guanidine and tyrosine carbon rings. The minimum times required for permeation into the tissue and formation of covalent links are not known. The length of fixation should be determined according to individual antibodies, depending upon the sensitivity to fixation-induced changes to epitopes.
9. Secondary antibodies that are preadsorbed against other species will give better results, with less cross-reactivity.

Acknowledgments

This work was supported by the Medical Research Council Human Immunology Core Grant MC_UU_00008.

References

1. Johnson LA (2021) In sickness and in health: the immunological roles of the lymphatic system. Int J Mol Sci 22(9):4458. https://doi.org/10.3390/ijms22094458
2. Oliver G, Kipnis J, Randolph GJ, Harvey NL (2020) The lymphatic vasculature in the 21 (st) century: novel functional roles in homeostasis and disease. Cell 182(2):270–296. https://doi.org/10.1016/j.cell.2020.06.039
3. Petrova TV, Koh GY (2020) Biological functions of lymphatic vessels. Science 369:6500. https://doi.org/10.1126/science.aax4063
4. Baluk P, Fuxe J, Hashizume H, Romano T, Lashnits E, Butz S et al (2007) Functionally specialized junctions between endothelial cells of lymphatic vessels. J Exp Med 204(10):2349–2362. https://doi.org/10.1084/jem.20062596
5. Banerji S, Ni J, Wang SX, Clasper S, Su J, Tammi R et al (1999) LYVE-1, a new homologue of the CD44 glycoprotein, is a lymph-specific receptor for hyaluronan. J Cell Biol 144(4):789–801. https://doi.org/10.1083/jcb.144.4.789
6. Breiteneder-Geleff S, Soleiman A, Kowalski H, Horvat R, Amann G, Kriehuber E et al (1999) Angiosarcomas express mixed endothelial phenotypes of blood and lymphatic capillaries: podoplanin as a specific marker for lymphatic endothelium. Am J Pathol 154(2):385–394.

https://doi.org/10.1016/S0002-9440(10)65285-6

7. Hirakawa S, Hong YK, Harvey N, Schacht V, Matsuda K, Libermann T et al (2003) Identification of vascular lineage-specific genes by transcriptional profiling of isolated blood vascular and lymphatic endothelial cells. Am J Pathol 162(2):575–586. https://doi.org/10.1016/S0002-9440(10)63851-5
8. Kriehuber E, Breiteneder-Geleff S, Groeger M, Soleiman A, Schoppmann SF, Stingl G et al (2001) Isolation and characterization of dermal lymphatic and blood endothelial cells reveal stable and functionally specialized cell lineages. J Exp Med 194(6):797–808. https://doi.org/10.1084/jem.194.6.797
9. Wigle JT, Oliver G (1999) Prox1 function is required for the development of the murine lymphatic system. Cell 98(6):769–778. https://doi.org/10.1016/s0092-8674(00)81511-1
10. Kaipainen A, Korhonen J, Mustonen T, van Hinsbergh VW, Fang GH, Dumont D et al (1995) Expression of the fms-like tyrosine kinase 4 gene becomes restricted to lymphatic endothelium during development. Proc Natl Acad Sci U S A 92(8):3566–3570. https://doi.org/10.1073/pnas.92.8.3566
11. Lymboussaki A, Partanen TA, Olofsson B, Thomas-Crusells J, Fletcher CD, de Waal RM et al (1998) Expression of the vascular endothelial growth factor C receptor VEGFR-3 in lymphatic endothelium of the skin and in vascular tumors. Am J Pathol 153(2):395–403. https://doi.org/10.1016/S0002-9440(10)65583-6
12. Gunn MD, Tangemann K, Tam C, Cyster JG, Rosen SD, Williams LT (1998) A chemokine expressed in lymphoid high endothelial venules promotes the adhesion and chemotaxis of naive T lymphocytes. Proc Natl Acad Sci U S A 95(1):258–263. https://doi.org/10.1073/pnas.95.1.258
13. Pfeiffer F, Kumar V, Butz S, Vestweber D, Imhof BA, Stein JV et al (2008) Distinct molecular composition of blood and lymphatic vascular endothelial cell junctions establishes specific functional barriers within the peripheral lymph node. Eur J Immunol 38(8):2142–2155. https://doi.org/10.1002/eji.200838140
14. Bazigou E, Xie S, Chen C, Weston A, Miura N, Sorokin L et al (2009) Integrin-alpha9 is required for fibronectin matrix assembly during lymphatic valve morphogenesis. Dev Cell 17(2):175–186. https://doi.org/10.1016/j.devcel.2009.06.017
15. Talsma DT, Katta K, Boersema M, Adepu S, Naggi A, Torri G et al (2017) Increased migration of antigen presenting cells to newly-formed lymphatic vessels in transplanted kidneys by glycol-split heparin. PLoS One 12(6):e0180206. https://doi.org/10.1371/journal.pone.0180206
16. Lim HY, Lim SY, Tan CK, Thiam CH, Goh CC, Carbajo D et al (2018) Hyaluronan receptor LYVE-1-expressing macrophages maintain arterial tone through hyaluronan-mediated regulation of smooth muscle cell collagen. Immunity 49(2):326–41 e7. https://doi.org/10.1016/j.immuni.2018.06.008
17. Breiteneder-Geleff S, Matsui K, Soleiman A, Meraner P, Poczewski H, Kalt R et al (1997) Podoplanin, novel 43-kd membrane protein of glomerular epithelial cells, is down-regulated in puromycin nephrosis. Am J Pathol 151(4):1141–1152
18. Dobbs LG, Williams MC, Gonzalez R (1988) Monoclonal antibodies specific to apical surfaces of rat alveolar type I cells bind to surfaces of cultured, but not freshly isolated, type II cells. Biochim Biophys Acta 970(2):146–156. https://doi.org/10.1016/0167-4889(88)90173-5
19. Rishi AK, Joyce-Brady M, Fisher J, Dobbs LG, Floros J, VanderSpek J et al (1995) Cloning, characterization, and development expression of a rat lung alveolar type I cell gene in embryonic endodermal and neural derivatives. Dev Biol 167(1):294–306. https://doi.org/10.1006/dbio.1995.1024
20. Clasper S, Royston D, Baban D, Cao Y, Ewers S, Butz S et al (2008) A novel gene expression profile in lymphatics associated with tumor growth and nodal metastasis. Cancer Res 68(18):7293–7303. https://doi.org/10.1158/0008-5472.CAN-07-6506
21. Johnson LA, Clasper S, Holt AP, Lalor PF, Baban D, Jackson DG (2006) An inflammation-induced mechanism for leukocyte transmigration across lymphatic vessel endothelium. J Exp Med 203(12):2763–2777. https://doi.org/10.1084/jem.20051759
22. Johnson LA, Prevo R, Clasper S, Jackson DG (2007) Inflammation-induced uptake and degradation of the lymphatic endothelial hyaluronan receptor LYVE-1. J Biol Chem 282(46):33671–33680. https://doi.org/10.1074/jbc.M702889200
23. Vigl B, Aebischer D, Nitschke M, Iolyeva M, Rothlin T, Antsiferova O et al (2011) Tissue inflammation modulates gene expression of lymphatic endothelial cells and dendritic cell migration in a stimulus-dependent manner. Blood 118(1):205–215. https://doi.org/10.1182/blood-2010-12-326447

24. Baluk P, McDonald DM (2018) Imaging lymphatics in mouse lungs. Methods Mol Biol 1846:161–180. https://doi.org/10.1007/978-1-4939-8712-2_11
25. Johnson LA, Banerji S, Lawrance W, Gileadi U, Prota G, Holder KA et al (2017) Dendritic cells enter lymph vessels by hyaluronan-mediated docking to the endothelial receptor LYVE-1. Nat Immunol 18(7):762–770. https://doi.org/10.1038/ni.3750
26. Teoh D, Johnson LA, Hanke T, McMichael AJ, Jackson DG (2009) Blocking development of a CD8+ T cell response by targeting lymphatic recruitment of APC. J Immunol 182(4):2425–2431. https://doi.org/10.4049/jimmunol.0803661
27. Thavarajah R, Mudimbaimannar VK, Elizabeth J, Rao UK, Ranganathan K (2012) Chemical and physical basics of routine formaldehyde fixation. J Oral Maxillofac Pathol 16(3):400–405. https://doi.org/10.4103/0973-029X.102496

Chapter 8

Transmission Electron Microscopy of Endothelium

Zubair Ahmed Nizamundeen, Loubna ElSaboni, Grace Vetch, Jumana Karim, Chris R. Neal, Claire L. Allen, David O. Bates, and Kenton P. Arkill

Abstract

Transmission electron microscopy using resin sections still remains an exceptionally useful tool in evaluating cellular ultrastucture within tissue. For the endothelium the best method for maintaining such structure is perfusion fixation: fixing the tissue under physiological pressure. Here the focus is on a method of maintaining the vascular wall structure including the endothelial glycocalyx and extending this with tilt series tomography. Shown are typical histological sections from multiple capillary beds including brain, heart and retina using a lanthanide staining technique (LaDy GAGa) to highlight that the differences in normo-physiology are substantial.

It is hoped that users will find the notes useful in deciding which specific staining and imaging method would suit their needs so this technically challenging, and low throughput methodology, is used to its best effect.

Key words Endothelial glycocalyx, Retina capillary, Heart capillary, Brain capillary, Resin embedding

1 Introduction

Transmission electron microscopy (TEM) is exemplary at determining the sub-cellular scale structure. For capillaries and the endothelial cells this scale can be vital for understanding functional changes in pathologies or conditions such as angiogenesis, primarily due to the thicknesses of various layers such as the endothelial glycocalyx (eGX), the cytoplasm, the basement membrane (BM) and sub-BM cell regions (e.g., pericyte processes) all being <500 nm. Further, the organization of the structures gives rise to the tissue specific roles of the capillary walls. For general resin embedded TEM of endothelial cells there is no preparation difference to any other cell or tissue and here are some examples [1–3]. However, if the eGX layer (often shed during angiogenesis and likely an underappreciated contributor to generation of different

Andrew V. Benest (ed.), *Angiogenesis: Methods and Protocols*, Methods in Molecular Biology, vol. 2441,
https://doi.org/10.1007/978-1-0716-2059-5_8,

phenotypes [4]) of the endothelium is required then additional steps are required to preserve it, due to its delicate polysaccharide nature. Suitable stains include Alcian Blue 8GX [5, 6] and similar cupromeronic blue [7], Ruthenium Red [8, 9], Ferritin [10], Tannic Acid [11], and lanthanides including lanthanum [12, 13] terbium [14] and thorium [15], None of the standard stains prevent loss of the layer consistently. Cell culture options for stains are the same as tissue and examples include [16, 17].

To retain the structures in their physiological form, specialized tissue preparation is the only option, and to detect eGX stain to the endothelial surface, perfusion fixation must be applied. Additionally, perfusion fixation maintains general vascular architecture by maintaining correct pressures [18]. Here we describe our perfusion staining technique and include in the notes some hints and tips for different scenarios. Our technique uses a mix of lanthanum and dysprosium and is dubbed the "LaDy GAGa" method, with advantages of visualizing fine structure [19, 20].

2 Materials

2.1 Buffers

For ease, four buffers were made for perfusion fixation—*Buffers 0, 1, 2,* and *3*.

Buffer 0: 1 L of Mammalian ringer's buffer was made without $NaHCO_3$ to avoid lanthanide precipitation (154 mM NaCl, 5.64 mM KCl, 2.16 mM $CaCl_2$, 10 mM Glucose and 5.5 mM HEPES (Sigma)).

Buffer 1: To *Buffer 0*, 0.5% of $LaNO_3$.6H2O and 0.5% of $DyCl_3.6H_2O$ were added (*see* **Note 1**).

Buffer 2: To *Buffer 1*, 2.5% glutaraldehyde and 2% sucrose were added.

Buffer 3: To *Buffer 0*, 2.5% glutaraldehyde and 2% sucrose were added.

Each buffer was pH adjusted to 7.3 using 1 M NaOH and 1 M HCl.

A volume of 10 ml per mouse is used for of *Buffers 1* and *2*. *Buffer 3* is to start the EM embedding process so only the volume covering the samples is required (*see* **Note 2**).

2.2 Perfusion

For perfusion suggest 24-gauge needles.

Retort stand for perfusion with known pressure head, with a switching tap for perfusates.

2.3 Resin Embedding and Electron Microscopy

1. TEM coated slot grids (formvar or pioloform).
2. Enough 5 ml or 10 ml glass vials for each tissue from each mouse collected.

3. Also standard TEM resin embedding and sectioning chemicals and equipment including:
   (a) Resin (Epoxy e.g., Agar 512).
   (b) Cacodylate buffer.
   (c) Graded ethanol (50%, 70%, 90%, and 100% dry).
   (d) Ultrapure water H_2O 18.2 MΩ.
   (e) Propylene Oxide.
   (f) 1% osmium tetroxide
   (g) 1% aqueous uranyl acetate
   (h) Moulding blocks.
   (i) Diamond knife, Ultratome (e.g., Leica UC6), chloroform.

3 Methods

All procedures should be carried out under an active Home Office project licence protocol, or equivalent for the experimental location.

1. Inject mice i.p. with medetomidine hydrochloride (Sedastart, 1 mg/kg) and ketamine hydrochloride (Ketaset, 75 mg/kg). Ensure that the animal has reached a surgical plane of anesthesia by using a toe-pinch response. The animal must be unresponsive before continuing.
2. Using a pair of blunt scissors make a 5–6 cm lateral incision through the integument and abdominal wall from the lower abdomen.

   Carefully separate the liver from the diaphragm and continue the incision along the entire length of the rib cage to expose the pleural cavity.
3. Displacing the lungs, make a cut through the rib cage up to the collarbone. Make a similar cut on the contralateral side and gently lift away the sternum.

   Make a small incision to the posterior end of the left ventricle and pass a 24-gauge blunt needle through the cut ventricle into the ascending aorta, avoiding the aortic arch. Clamp the needle in place using a hemostat and to prevent leakage.

   Make an incision to the animal's right atrium to create a fluid outlet.

   Attach the buffer outlet port to the needle entering the left ventricle, taking care not to introduce any air bubbles. Adjust the angle of the needle as required to achieve maximum flow rate.

4. Perfuse the left side of the mouse heart at 60 mmHg (80 cm height water, use two with a tap to switch) pressure with perfusion *Buffer 1* to *flush*, followed by perfusion with *Buffer 2*, until the animal becomes rigid. This is normally 10 ml for each buffer per mouse. Clearing of the liver is an indicator of a good perfusion and fixation tremors should be observed within seconds.

   Isolate required tissue placing in *Buffer 3*, dissect to 1 mm^3. This can be used for short term (e.g., a week) storage at 4 °C until all samples ready for embedding.

5. Resin Embedding (*see* **Note 3**).
   (a) 1 mm^3 tissue samples were transferred to 0.1 M sodium cacodylate buffer for 5 min.
   (b) Fix in 1% OsO_4 for 1 h, followed by 3 washes in sodium cacodylate, and then 3 washes in ddH_2O (18.2 MΩ).
   (c) Stain overnight at 4 °C in 1% aqueous filtered uranyl acetate.
   (d) Wash twice using ddH_2O, and then dehydrated by incubating in 50%, 70%, 90%, 3 × 100% (water free) ethanol for 20 min each.
   (e) Transfer to 100% propylene oxide (must be glass vials) and use a rotator from here. TEM Resin infiltration can be performed done by incubating the samples in decreasing ratios of propylene oxide:resin (2:1, 1:1, 1:2) for 1 h each. The final 100% resin can be overnight at room temperature. If left overnight, follow with a further 1 h 100% resin step.
   (f) Transfer to appropriate mould for sectioning and polymerize for 48–72 h at 60 °C.
   (g) Use semi thin sections (500 nm to 1000 nm; UC6 Ultratome, Lieca) with toluidine blue to histological assessment. Trim to <1 mm a region of interest before ultrathin sectioning (70–100 nm). Section 300 nm sections for tilt series tomography if required. Use coated slot grids for imaging.

6. Transmission electron microscopy.
   (a) For 2D imaging, a 100 keV TEM (Tecnai T12 BioTwin, FEI) is suitable preferably with a biological objective to enhance contrast (*see* Fig. 1) (*see* **Note 4**).
   (b) 15 nm gold fiducials were added to 300 nm sections by drop casting. Remove excess fluid by gently blotting the solid copper part of the grid. Tilt series tomography is preferable with a 200 keV TEM (Tecnai G2 T20, FEI) high angle tilt holder. Micrographs were recorded with

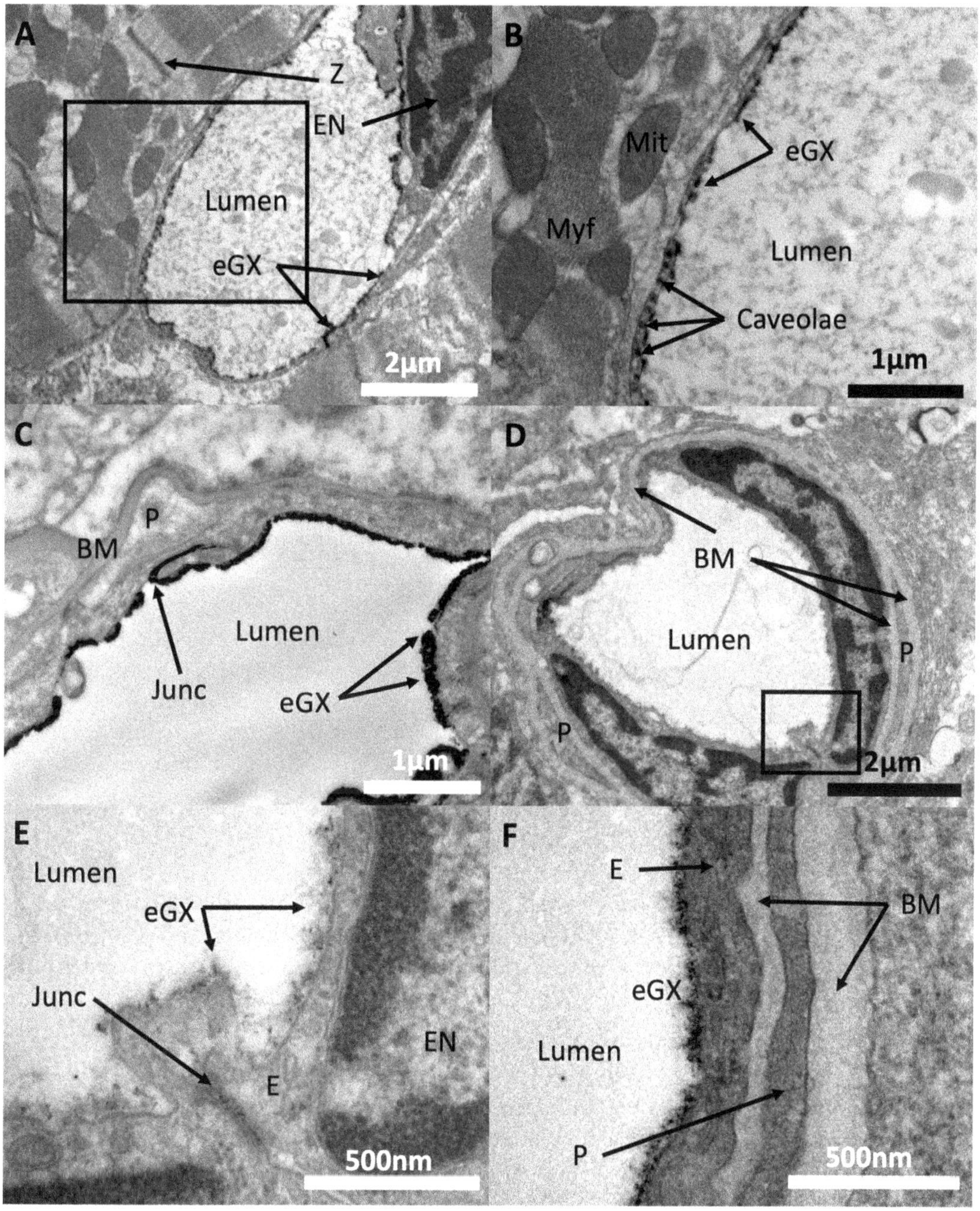

Fig. 1 Transmission Electron Micrographs of Mouse Heart, Brain, and Retinal Capillaries. (**a**) Heart Capillary (**b**) Higher magnification from box in (**a**). (**c**) Brain, including an intercellular junction. (**d**) Retinal Capillary. (**e**) High magnification of the boxed region in (**d**) and (**f**) highlighting the split basement membrane in retina. Labels: Endothelial Cell = E, Endothelial Cell Nucleus = EN, Endotheial Glycocalyx = eGX, Pericyte = P, Mitochrondria = Mit. Z-Line = Z, Myofibrils = Myf. Note these images have had their mean ± SD intensities corrected (Floated) for display purposes

the sections tilted by 1° ranging between ±60° at 0.70 nm/pixel (14kx magnification on this microscope). The exposure time was 2.0 s for images at 0° view, with an increase for each degree up to 3.2 s for ±60° views. This was to account for the increasing section thickness at higher viewing angles. Micrographs were recorded automatically and stored in a 16-bit MRC file stack. The tilted views were reconstructed into a single 3D density image stack using IMOD tomogram reconstruction software [21] (*see* Fig. 2) [20] *see* **Note 5**).

4 Notes

1. *Precipitation*

   Lanthanides tend to form a white carbonate precipitate. Limited air contact is therefore vital in preparation. We believe we have seen crystalized La in published works.

2. *Choice of Stain*

   Over time there have been many stains used for the eGX. These can be categorized into mordents that clump the eGX such as Alcian blue and small highly charged cations (Lanthanides). Our experience suggests that the mordents are good for eGX coverage, but the ions are better for structure where it remains. Note that we advise a 7.2 pH flush prior to a 6.6 pH buffer mix with Alcian blue for best results.

   (a) *Perfusion fixation*

   For rats suggest cuffing the tail if require tissue from the head. Pressure height should be appropriate, e.g., for rats 100 mmHg or human kidney use 100–120 mmHg. We are also becoming aware that some oncotic balance is beneficial [18, 19].

3. *Embedding options*

   We tend to perform uranyl acetate staining en-bloc, many people stain the sections afterwards. Larger tissue volumes require longer times for each step. We tend to not post stain sections with additional metals (lead citrate, etc.), this is best practice for tomography where these do not always penetrate the thicker sections evenly. Your preferred resin will likely make little or no difference, harder formulation resins (e.g., Durcupan™) tend to shrink less for tomography or other 3D techniques (e.g., Serial Block Face Scanning Electron Microscopy).

4. *Imaging tips*

   We tend to find in these samples the best resolution measured from the resin is around 2 nm, therefore Nyquist sampling would be 1 nm pixels. With most cameras with heavy

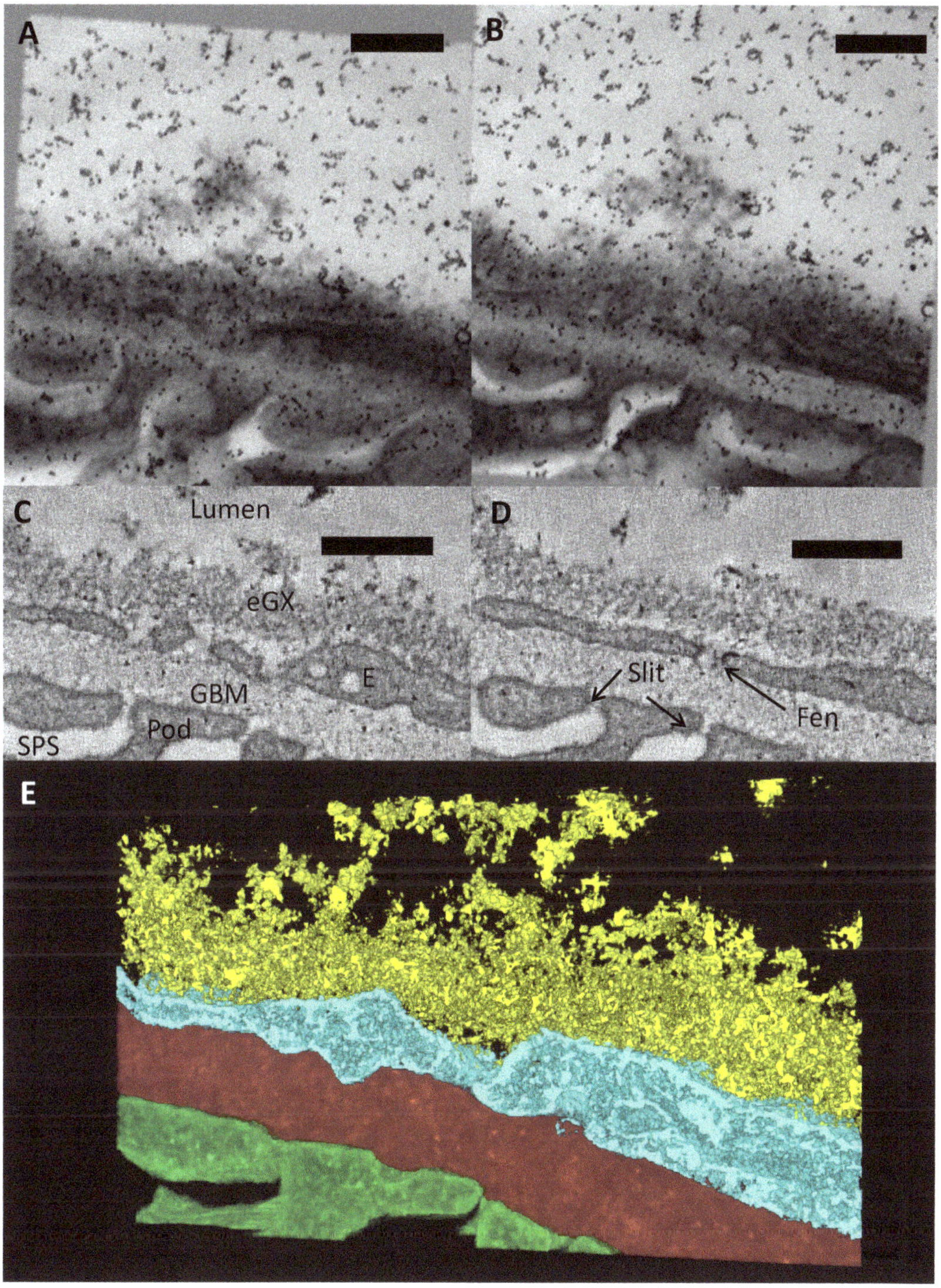

Fig. 2 Tomographic reconstruction of rat glomerular capillary glycocalyx using LaDy GAGa staining on a 300 nm thick section. Adapted from Arkill et al. 2014 [20] (**a**, **b**) raw transmission electron micrograph at $-30°$ and $30°$ respectively. Note 15 nm gold fiducial markers on surfaces. (**c**, **d**) reconstructed images of glycocalyx in cross-section 90 nm apart respectively. There is the appearance of a gap between the glycocalyx (eGX) and the endothelial cell (**e**). Small holes can be seen throughout eGX, these are gaps between the fibers. (**e**) The 3D projection (Yellow = eGX, Blue = Endothelial cell, Red = Glomerular Basement membrane, green = Podocyte) A-D: scale bar is 300 nm. Labels: E = endothelial cell, eGX = glycocalyx, Fen = Fenestration, GBM = glomerular basement membrane, SPS = subpodocyte space, Slit = Slit Diaphragm (the diaphragm is discernible), Pod = Podocyte (foot process)

metal staining there is little advantage with oversampling pixels when the reduction in field-of-view is considered. We therefore tend to use ≈0.75 nm pixels.

Make sure the microscope settings are saved either separately or as metadata in the image. It is recommended to keep images with the cameras natural bit depth rather than an 8-bit "snapshot." As section thickness can vary, quantification requiring density should have an internal control (e.g., tissue free area, etc.)

Take a section map (a montage is best) at the start, and for each vessel take an overview image. Bear in mind that the same "magnification range" has the same focus for TEMs we often also take the lowest magnification on the highest range used and as personal preference focus on approximately double the magnification (if still in the same magnification range) away from the intended position.

If section shrinkage is an issue for quantification, then we suggest the protocol in Mantel et al. 2012 [22] to back calculate the original thickness.

(a) Analysis tips

We use Fiji for most quantitative analysis [23]. It is useful to note the proportion of vessels perfused (by lack of blood cells). Consider that 2D analysis has a depth precision of the section thickness (80–100 nm) making measurements of objects subject to cut position. A tomogram as described here, after section shrinkage, gives a ≈6 nm section.

5. *Additional measurements*

We amended our project license to remove organs/biopsies prior to perfusion fixation for immunohistochemical techniques not possible with after glutaraldehyde fixation nor coated with lanthanides.

References

1. Eaton B, Leach L, Firth J (1993) Permeability of the fetal villous microvasculature in the isolated perfused term human placenta. J Physiol 463(1):141–155
2. Adamson RH et al (2004) Oncotic pressures opposing filtration across non-fenestrated rat microvessels. J Physiol 557(3):889–907
3. Oltean S et al (2015) Vascular endothelial growth factor-A165b is protective and restores endothelial glycocalyx in diabetic nephropathy. J Am Soc Nephrol 26(8):1889–1904
4. Brown M et al (1996) Appearance of the capillary endothelial glycocalyx in chronically stimulated rat skeletal muscles in relation to angiogenesis. Exp Physiol 81(6):1043–1046
5. van den Berg BM, Vink H, Spaan JAE (2003) The endothelial glycocalyx protects against myocardial edema. Circ Res 92(6):592–594
6. Betteridge KB et al (2017) Sialic acids regulate microvessel permeability, revealed by novel in vivo studies of endothelial glycocalyx structure and function. J Physiol 595(15):5015–5035
7. Meuwese MC et al (2010) Endothelial surface layer degradation by chronic hyaluronidase infusion induces proteinuria in apolipoprotein E-deficient mice. PLoS One 5(12)

8. Luft JH (1966) Fine structure of capillary and endocapillary layer as revealed by ruthenium red. Fed Proc 25(6P1):1773
9. Baldwin A, Winlove C (1984) Effects of perfusate composition on binding of ruthenium red and gold colloid to glycocalyx of rabbit aortic endothelium. J Histochem Cytochem 32(3):259–266
10. Loudon MF, Michel CC, White IF (1979) Labeling of vesicles in frog endothelial-cells with ferritin. J Physiol 296(NOV):97
11. Qvortrup K, Rostgaard J (1993) Ultrastructure of the endolymphatic duct in the rat - fixation and preservation. Acta Otolaryngol 113(6):731–740
12. Hjalmarsson C, Johansson BR, Haraldsson B (2004) Electron microscopic evaluation of the endothelial surface layer of glomerular capillaries. Microvasc Res 67(1):9–17
13. Chappell D et al (2009) The glycocalyx of the human umbilical vein endothelial cell an impressive structure ex vivo but not in culture. Circ Res 104(11):1313–1317
14. Wagner RC, Chen SC (1990) Ultrastructural distribution of terbium across capillary endothelium - detection by electron spectroscopic imaging and electron-energy loss spectroscopy. J Histochem Cytochem 38(2):275–282
15. Hegermann J et al (2015) Visualization of the glomerular endothelial glycocalyx by electron microscopy using cationic colloidal thorium dioxide. Histochem Cell Biol:1–11
16. Singh A et al (2007) Glomerular endothelial glycocalyx constitutes a barrier to protein permeability. J Am Soc Nephrol 18(11):2885–2893
17. Psefteli P-M et al (2021) Glycocalyx sialic acids regulate Nrf2-mediated signaling by fluid shear stress in human endothelial cells. Redox Biol 38:101816
18. Neal CR et al (2018) Novel haemodynamic structures in the human glomerulus. Am J Physiol Renal Physiol 315:F1370–F1384
19. Arkill KP et al (2012) 3D reconstruction of the glycocalyx structure in mammalian capillaries using electron tomography. Microcirculation 19(4):343–351
20. Arkill KP et al (2014) Resolution of the three dimensional structure of components of the glomerular filtration barrier. BMC Nephrol 15:24
21. Kremer JR, Mastronarde DN, McIntosh JR (1996) Computer visualization of three-dimensional image data using IMOD. J Struct Biol 116(1):71–76
22. Mantell JM et al (2012) Quantitative biological measurement in transmission electron tomography. In: Electron microscopy and analysis group conference 2011
23. Schindelin J et al (2012) Fiji: an open-source platform for biological-image analysis. Nat Methods 9(7):676–682

Chapter 9

Measurement of Revascularization in the Hind Limb After Experimental Ischemia in Mice

Sohni Ria Bhalla, Federica Riu, Maria J. C. Machado, and David O. Bates

Abstract

Peripheral vascular disease is a major cause of morbidity and mortality, and is a consequence of impaired blood flow to the limbs. This arises due to the inability of the tissue to develop sufficiently functional collateral vessel circulation to overcome occluded arteries, or microvascular impairment. The mouse hind limb model of hind limb ischemia can be used to investigate the impact of different treatment modalities, behavioral changes, or genetic knockout. Here we described the model in detail, providing examples of adverse events, and details of ex vivo analysis of blood vessel density.

Key words Ischemia, Peripheral vascular disease, Experimental models, VEGF, Exercise, Angiopoietin

1 Introduction

Cardiovascular disease (CVD) is a leading public health problem in the UK. Although death rates are declining, the number of patients diagnosed and living with the disease continue to rise [1]. It has a severe impact on the patient's day-to-day activities, functional capacity and reduces their overall quality of life. This places a large financial burden on the NHS and the economy. In fact, CVD account for 60% of total health care costs [2]. Globally, it accounts for one third of deaths [3].

The umbrella term, CVD, is used for a number of linked pathologies, including coronary artery disease (CAD), cerebral ischemia, peripheral arterial disease (PAD) and microvascular disease (MVD). These diseases often occur due to atherosclerosis, which is a slow progressive build-up of a plaque on the wall of the artery, which narrows the lumen diameter and reduces the blood supply to the associating tissue. An atheroma in the coronary artery reduces the blood supply to the myocardium resulting in CAD.

Andrew V. Benest (ed.), *Angiogenesis: Methods and Protocols*, Methods in Molecular Biology, vol. 2441,
https://doi.org/10.1007/978-1-0716-2059-5_9,

Frequently patients experience symptoms including angina or myocardial infarction (if blood flow is severely restricted) [4]. Cerebral ischemia affects the carotid artery in the brain. This can result in a stroke [5]. PAD (also known as peripheral vascular disease, PVD) affects the femoral or iliac artery in the lower limbs. In the majority of cases, patients remain asymptomatic. In mild cases, patients experience intermittent claudication, which includes cramp, pain and aches in the calves and lower limbs. Critical limb ischemia (CLI) is a severe manifestation of PAD which occurs from a complete blockage of the artery resulting in ulcers, gangrene or necrosis [6, 7]. MVD is an independent disease associated with small 100 μm vessels (including capillaries), which regulate vascular tone and blood delivery to the muscles and tissues throughout the body. Although it is commonly associated with retinopathy (eyes), nephropathy (kidneys), and neuropathy (feet) [8], it can amplify the onset of large vessel atherosclerotic diseases. In fact, it is strongly associated with amputations with 1 in 6 patients at risk. This is further amplified with PAD [9, 10].

The formation of collateral vessels is an essential adaptive mechanism that occurs after obstruction of an artery or loss of arterioles. It is necessary to compensate for the deficient blood supply to protect and preserve the tissue jeopardized by ischemia. Metabolic diseases (including high blood pressure, obesity, diabetes, and dyslipidemia) inversely correlate with the collateral flow grade, where a low grade correlates with fewer collaterals [11]. Similarly, patients with a high metabolic syndrome score had a lower Rentrop collateral grade, where 0 to 1 indicates little to no collateral blood flow [12]. A main contributing factor to poor collateral vessel development is diabetes [13]. This group of patients is 62% more likely to have poor collateral vessel development compared to non-diabetics [14]. In fact, this risk group are ten times more likely to require an amputation [15]. Similarly, obesity (BMI >30 kg/m^2) is another risk factor which results in a poor collateral circulation in ischemic heart disease [16]. Of the patients that are diabetic, 90% are obese suggesting a direct link between the two risk factors are contributing to CVD [17] Importantly, the content of vascular endothelial growth factor (VEGF-A) in the serum of diabetic patients is elevated. However, the monocyte function and migration potential is impaired in this group of patients, even when stimulated with VEGF-A [18]. Similarly, in another study, diabetic CAD patients had an increase in VEGF-A production in the monocytes compared to the control group [19]. Therefore, understanding why these patients are at a greater risk, but do not form collateral vessels, by overcoming this, could underpin the development of a radical and effective therapeutics to induce revascularization.

1.1 Collaterals in CVD

1.1.1 Angiogenesis

The formation of capillary microvessels from preexisting vasculature is known as angiogenesis, which plays a significant role in pathological and physiological conditions [20, 21]. There are two types of angiogenesis: non-sprouting (intussusceptive) and sprouting. Non-sprouting angiogenesis is the modification of preexisting vessels by forming an intravascular pillar. This divides the vessel, which subsequently branches to become two independent capillaries [22]. In comparison, sprouting angiogenesis occurs when existing capillaries acquire a tip cell phenotype resulting in the formation of a sprout on an existing vessel [20, 21]. This is directed toward the concentration gradient of angiogenic factors such as VEGF-A [23] and angiopoietin-1 (Ang-1) [24]. These processes are driven by hypoxia and mediated by the activation of the transcription factor, HIF-1α, which subsequently activates the gene expression of VEGF-A [25]. Recent evidence has identified that the non-canonical Wnt5a signaling pathway is involved in the angiogenic response in retinal myeloid cells [26], adipose tissue [27], and PAD patients [28].

1.1.2 Arteriolargenesis and Arteriogenesis

The vascular network consists of arterioles which connect to capillaries and are linked to the main arteries. However, the capillaries can develop into arterioles by a process called arteriolargenesis (also referred to as capillary arterialization). This is essential to regulate blood perfusion. It occurs when smooth muscle cells are recruited to the pre-existing, expanding capillary wall to develop into functional arterioles. This process was captured in the mesentery by Benest et al. in 2008 [29]. Using this model, it has been shown that a combination of Ang-1, VEGF-A and vasodilators (e.g., eNOS to produce nitric oxide or prazocin) induce smooth muscle cell recruitment and coverage to develop functional vessels [29].It was later discovered that inhibition of VEGF-A (with sFLT1) and/or Ang-1 (with sTie2) in the presence of eNOS alone inhibited recruitment of smooth muscle cells and arteriolargenesis [30]. This indicates that the three components are necessary to stimulate the process.

Arteriogenesis is an efficient adaptive mechanism. It is the remodelling and expansion of pre-existing collateral vessels to compensate for blood lost via other routes [31] [32]. The key stimulus in initiating arteriogenesis is physical mechanical forces—in particular shear stress. After an arterial occlusion, there is a change in blood flow resulting in increased blood pressure upstream, and increased fluid shear stress in the lumen of preexisting collateral vessels [33] [34]. The force acts on the endothelial cells on the vessel wall and activates the calcium dependent enzyme, eNOS, which converts l-arginine to l-citrulline and produces nitric oxide [35]. It diffuses into the smooth muscle cells, binds and stimulates the heme protein soluble guanylyl cyclase, which catalyzes the dephosphorylation of GTP to cyclic GMP [36]. Thus, the synthesis of nitric oxide by eNOS is necessary to increase the lumen diameter and blood flow recovery resulting in a vasodilatory effect [37]. In fact, arteriogenesis increases the blood flow by 10- to 20-fold and increases the vascular reserve to the associating tissue [38].

Understanding the mechanisms of revascularization and collateralization and the contribution of arteriogenesis and angiogenesis is critical for the development of new therapeutic approaches for cardiovascular disease. Rodent models of revascularization after ischemia of the femoral artery (hind limb ischemia) can be used to determine the impact of various interventions, including diet, exercise, viral transduction, gene knockout, or drug delivery. Here we describe the methodology for investigation of mechanisms of revascularization.

2 Materials

1. Block solution: 1%BSA + PBS.
2. Perfusion- fixation: 4% Paraformaldehyde in PBS.
3. Permeabilization solution: 0.1% Triton-X with PBS.

3 Methods

3.1 Hind Limb Ischemia Model: Surgery

All surgical procedures were performed on 10- to 12-week-old (unless stated otherwise) C57BLJ mice using sterile equipment and aseptic technique.

1. Anesthetize mice using 2–3% isoflurane (100% w/w, IsoFlo ®) in 100% oxygen at a rate of 2 L/min. To ensure mouse is completely anesthetized, withdrawal reflexes with toe pinching to confirm no movement before the mouse is moved to the imaging or surgical platform on a head pad (Harvard Appartatus).
2. Insert rectal probe (Harvard Apparatus) to maintain body temperature at 37 °C throughout surgery and imaging.
3. Mouse is positioned on its front with back paws facing upwards. The hind paws are imaged using the laser speckle imaging system (FLPI-2, Moors Instruments).
4. Analgesia is provided by a subcutaneous injection of buprenorphine (0.05 mg/kg).
5. Apply hair removal cream (Nair) to the surgical area of the left hind limb and sterilize the area with a chlorohexidine-based solution.
6. The left paw is extended, taped down and covered with a sterile surgical drape (only the surgical area is exposed).
7. Perform a 1 cm incision from the knee to the abdomen.

8. Separate the fat from the connective tissue to expose the femoral artery from the vein and nerve.
9. Pass a piece of 6-0 W812 Mersilk Suture (Ethicon) underneath the femoral artery and the vessel is tied off above the superficial epigastric artery. Pass a second piece of suture lower down the knee (near the sapheno-popliteal bifurcation).
10. Electro-coagulate the vessel between the two sutures using a cauterizer (Wuhan Spring, Scenery Medical Instruments Co Ltd., Model CHR III, Evergreen).
11. Close incision using surgical clay Adams clips (Scientific Labs) and a topical tissue adhesive, GLUture (Zoetitis).
12. After the surgery, place the mouse on the imaging platform and image the hind paws to confirm successful surgery and ischemia induction. A summary of the procedure is shown in Fig. 1.

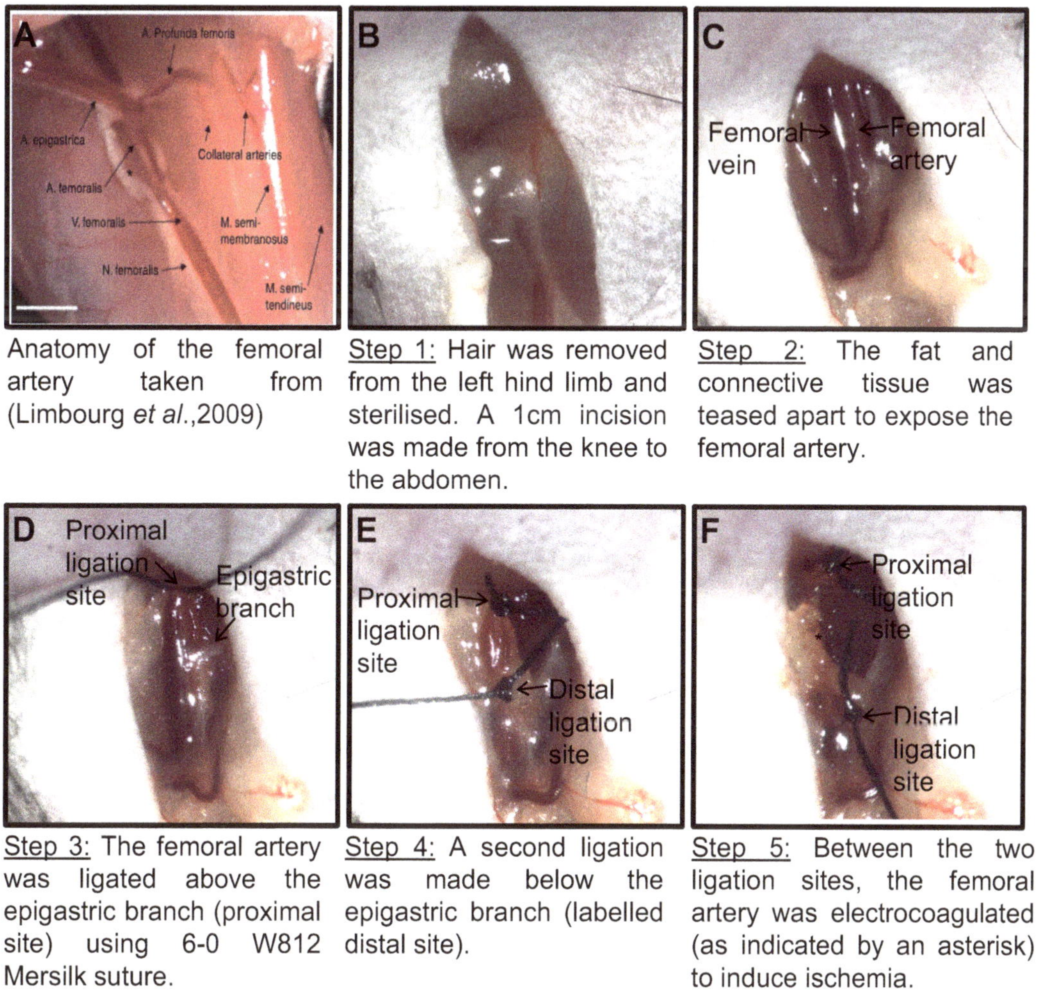

Fig. 1 Step-by-step guide to induce hind limb ischemia in mice (**a–f**). Panel A taken from (25)

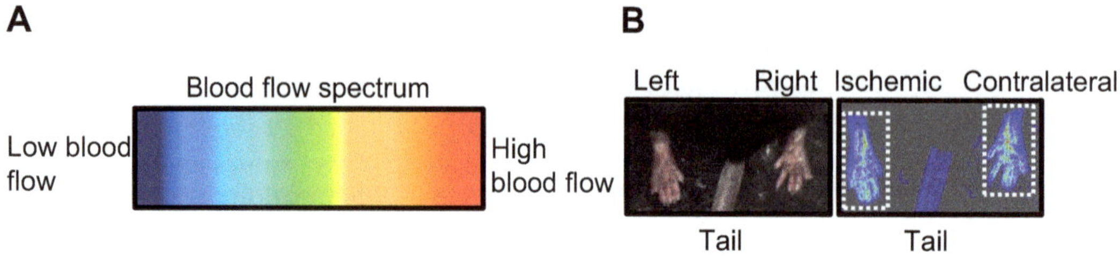

Fig. 2 Measuring the speckle intensity using the Laser Speckle Imaging System. (**a**) The blood flow spectrum highlights the levels of blood flow. (**b**) Example of the paws and laser speckle image highlighting the left (ischemic) and right (contralateral) paw and the tail. The white box around the paw indicates region of interest (ROI) used to calculate the speckle intensity

13. Allow animals to recover in a warmed, clean animal cage with mash. Monitor the mouse for a few hours after surgery to ensure there are no signs of pain or suffering. Continue to monitor the wound, paw and weights for the first seven days, and then weekly thereafter. The blood flow to the paws are imaged on post-operative days 3, 7, 14 and 21.

3.2 Measurement of Blood Flow Recovery

1. Anesthetize mice using 2–3% isoflurane (100% w/w, IsoFlo ®) in 100% oxygen at a rate of 2 L/min. To ensure mouse is completely anesthetized, withdrawal reflexes with toe pinching to confirm no movement.
2. The mouse is transferred to the imaging platform. Laser speckle images are taken using the Moor FLPI-2 full field laser perfusion imager.

   Note: The system relies on an infrared laser beam directed toward the paw. The backscattered light from the moving blood cells is detected as colored speckles which indicate the level on the spectrum as shown in Fig. 2a.
3. Speckle intensity is measured in the region of interest (ROI). Blood flow recovery can be calculated by using the ratio of the ischemic vs contralateral speckle intensity Fig. 2. The blood flow to the paws is imaged on post-operative days 3, 7, 14, and 21 and/or 28.

3.3 Adverse Effects

Of more than 150 mice that have undergone the hind limb ischemia procedure of the left femoral artery over the last 2 years, a small number of mice experience unexpected adverse effects, presumably due to the ligation at the proximal site being too high (shown in Fig. 3). Mice usually acquire a pale paw immediately after surgery indicating reduced blood flow, but in between day 1 to 3 post-surgery, 2 mice developed a swollen foot, 4 mice developed necrotic toes and 12 mice experienced tail de-gloving (early signs of tail necrosis). When this occurred, mice were culled before continuation of the study to prevent pain, distress, and suffering.

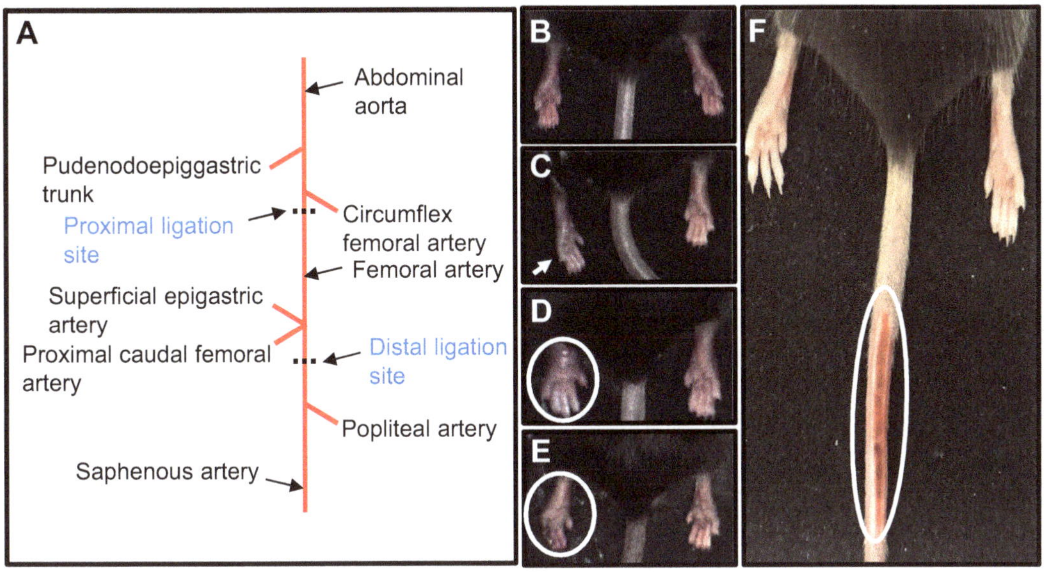

Fig. 3 Adverse effects that occurred possibly due to incorrect ligation position on the left femoral artery. (**a**) Anatomy of the left femoral artery highlighting the incorrect ligation site as shown black dotted line. The representative images highlight (**b**) Pre-op day 0: healthy paws and (**c**) Post-operative: day 0 discolored pale left paw (white arrow). The ischemic damage (highlighted by a white circle) developed within 3 days after the ligation: (**d**) Swollen paw (**e**) Necrotic toes and (**f**) Tail de-gloving (early sign of tail necrosis)

3.4 Immunofluorescence on Gastrocnemius Muscle

For immunofluorescence determination of vessel density, mice are administered an i.p. injection of 1 mg/kg medetomidine and 75 mg/kg ketamine. Once anesthesia is confirmed by lack of withdrawal reflexes, cardiac perfusion is performed on the animals with 25 mL of PBS followed by 25 mL of 4% paraformaldehyde (PFA)/PBS through the apex of the ventricle. The gastrocnemius muscle is dissected and either snap-frozen or post-fixed in 4% PFA overnight. The following day, the tissue is washed in PBS, cryo-protected in 30% sucrose and embedded in optimal temperature compound (OCT) then stored at −80 °C.

Multiple 5 μm sections are cut using a cryostat and mounted on superfrost plus (Thermoscientific) glass slides. All staining steps are carried out in a dark humidified StainTray slide staining system at room temperature, unless stated otherwise.

1. Fix sections in acetone for 5 min at −20 °C and left to dry at room temperature for 10 min.
2. Draw a circle around each section using a PAP pen. Wash section in PBS three times for 5 min.
3. Permeabilize each section in 0.1% Triton-X with PBS for 5 min. Repeat wash step.

4. Block section in 1% bovine serum albumin (BSA)/phosphate buffered saline (PBS) for 1 h.
5. Incubate slides overnight at 4 °C in primary antibody, IB_4 (1: 150) and α-smooth muscle actin-cy3 (1:300, Sigma-Aldrich) in block solution. A non-primary antibody slide is used as a negative control.
6. Wash sections and incubate with Alexa Fluor streptavidin 488 (1:500) and DAPI (1:1000) for 1 h at room temperature.
7. Repeat wash step. Cover slips are mounted with ProLong Gold Anti Fade Reagent (Cell signaling) and sealed with clear nail varnish the following day.
8. Images are taken at 10× magnification with a Confocal Microscope or Epi Fluorescence Microscope. For each gastrocnemius muscle (contralateral and ischemic), 4 sections are taken and 2 fields of view per section. For quantification, capillaries and arterioles are counted using FIJI and averaged per mm^2.

3.5 Statistical Analysis

All data is analyzed using Microsoft Excel and GraphPad Prism. Specific software's included Moors Instruments, Licor Image Studio Software (lite), and FIJI for image analysis. Power calculations are performed using G*Power. Data is presented as the mean ± standard error mean (SEM), unless stated otherwise. For blood flow measurements use a two-way ANOVA to compare groups. An N of 8 provides sufficient power for most studies.

References

1. Bhatnagar P et al (2016) Trends in the epidemiology of cardiovascular disease in the UK. Heart 102(24):1945–1952
2. Luengo-Fernandez R et al (2006) Cost of cardiovascular diseases in the United Kingdom. Heart 92(10):1384–1389
3. Wong ND (2014) Epidemiological studies of CHD and the evolution of preventive cardiology. Nat Rev Cardiol 11(5):276–289
4. Sanchis-Gomar F et al (2016) Epidemiology of coronary heart disease and acute coronary syndrome. Ann Transl Med 4(13):256
5. Zhao H et al (2013) Association of carotid atherosclerotic plaque features with acute ischemic stroke: a magnetic resonance imaging study. Eur J Radiol 82(9):e465–e470
6. Gornik HL, Beckman JA (2005) Cardiology patient page. Peripheral arterial disease. Circulation 111(13):e169–e172
7. Gardner AW, Afaq A (2008) Management of lower extremity peripheral arterial disease. J Cardiopulm Rehabil Prev 28(6):349–357
8. Beckman JA, Creager MA (2016) Vascular complications of diabetes. Circ Res 118(11):1771–1785
9. Beckman JA et al (2019) Microvascular disease, peripheral artery disease, and amputation. Circulation 140(6):449–458
10. Behroozian A, Beckman JA (2020) Microvascular disease increases amputation in patients with peripheral artery disease. Arterioscler Thromb Vasc Biol 40(3):534–540
11. Mouquet F et al (2009) Metabolic syndrome and collateral vessel formation in patients with documented occluded coronary arteries: association with hyperglycaemia, insulin-resistance, adiponectin and plasminogen activator inhibitor-1. Eur Heart J 30(7):840–849
12. Sasmaz H, Yilmaz MB (2009) Coronary collaterals in obese patients: impact of metabolic syndrome. Angiology 60(2):164–168
13. Abaci A et al (1999) Effect of diabetes mellitus on formation of coronary collateral vessels. Circulation 99(17):2239–2242

14. Hsu PC et al (2014) Coronary collateral circulation in patients of coronary ectasia with significant coronary artery disease. PLoS One 9(1):e87001
15. White CJ, Gray WA (2007) Endovascular therapies for peripheral arterial disease: an evidence-based review. Circulation 116(19):2203–2215
16. Yilmaz MB et al (2003) Obesity is associated with impaired coronary collateral vessel development. Int J Obes Relat Metab Disord 27(12):1541–1545
17. Bramante CT, Lee CJ, Gudzune KA (2017) Treatment of obesity in patients with diabetes. Diabetes Spectr 30(4):237–243
18. Waltenberger J, Lange J, Kranz A (2000) Vascular endothelial growth factor-A-induced chemotaxis of monocytes is attenuated in patients with diabetes mellitus: a potential predictor for the individual capacity to develop collaterals. Circulation 102(2):185–190
19. Panutsopulos D et al (2003) Peripheral monocytes from diabetic patients with coronary artery disease display increased bFGF and VEGF mRNA expression. J Transl Med 1(1):6
20. Folkman J, Klagsbrun M (1987) Angiogenic factors. Science 235(4787):442–447
21. Risau W (1997) Mechanisms of angiogenesis. Nature 386(6626):671–674
22. Burri PH, Djonov V (2002) Intussusceptive angiogenesis--the alternative to capillary sprouting. Mol Asp Med 23(6S):S1–S27
23. Connolly DT et al (1989) Tumor vascular permeability factor stimulates endothelial cell growth and angiogenesis. J Clin Investig 84(5):1470–1478
24. Koblizek TI et al (1998) Angiopoietin-1 induces sprouting angiogenesis in vitro. Curr Biol 8(9):529–532
25. Detmar M et al (1997) Hypoxia regulates the expression of vascular permeability factor/vascular endothelial growth factor (VPF/VEGF) and its receptors in human skin. J Investig Dermatol 108(3):263–268
26. Stefater JA 3rd et al (2011) Regulation of angiogenesis by a non-canonical Wnt-Flt1 pathway in myeloid cells. Nature 474(7352):511–515
27. Karki S et al (2017) WNT5A regulates adipose tissue angiogenesis via anti-angiogenic VEGFA165b in obese humans. Am J Physiol Heart Circ Physiol 313(1):H200–H206
28. Kikuchi R et al (2014) An antiangiogenic isoform of VEGF-A contributes to impaired vascularization in peripheral artery disease. Nat Med 20(12):1464–1471
29. Benest AV et al (2008) Arteriolar genesis and angiogenesis induced by endothelial nitric oxide synthase overexpression results in a mature vasculature. Arterioscler Thromb Vasc Biol 28(8):1462–1468
30. Stone OA et al (2017) Differential regulation of blood flow-induced neovascularization and mural cell recruitment by vascular endothelial growth factor and angiopoietin signalling. J Physiol 595(5):1575–1591
31. Ito WD et al (1997) Angiogenesis but not collateral growth is associated with ischemia after femoral artery occlusion. Am J Phys 273 (3 Pt 2):H1255–H1265
32. Scholz D, Cai WJ, Schaper W (2001) Arteriogenesis, a new concept of vascular adaptation in occlusive disease. Angiogenesis 4(4):247–257
33. Tuttle JL et al (2001) Shear level influences resistance artery remodeling: wall dimensions, cell density, and eNOS expression. Am J Physiol Heart Circ Physiol 281(3): H1380–H1389
34. Pipp F et al (2004) Elevated fluid shear stress enhances postocclusive collateral artery growth and gene expression in the pig hind limb. Arterioscler Thromb Vasc Biol 24(9):1664–1668
35. Marletta MA (1994) Nitric oxide synthase: aspects concerning structure and catalysis. Cell 78(6):927–930
36. Denninger JW, Marletta MA (1999) Guanylate cyclase and the .NO/cGMP signaling pathway. Biochim Biophys Acta 1411(2–3):334–350
37. Dai X, Faber JE (2010) Endothelial nitric oxide synthase deficiency causes collateral vessel rarefaction and impairs activation of a cell cycle gene network during arteriogenesis. Circ Res 106(12):1870–1881
38. Schaper W, Scholz D (2003) Factors regulating arteriogenesis. Arterioscler Thromb Vasc Biol 23(7):1143–1151

Chapter 10

Imaging Blood Vessels and Lymphatics in Mouse Trachea Wholemounts

Peter Baluk and Donald M. McDonald

Abstract

Changes in blood vessels and lymphatics in health and disease are easier to understand and interpret when studied microscopically in three dimensions. The mouse trachea is a simple, yet powerful, and versatile model system in which to achieve this. We describe practical immunohistochemical methods for fluorescence and confocal microscopy of wholemounts of the mouse trachea to achieve this purpose in which the entire vasculature can be visualized from the organ level to the cellular and subcellular level. Blood vessels and lymphatics have highly stereotyped vascular architectures that repeat in arcades between the tracheal cartilages. Arterioles, capillaries, and venules can be easily identified for the blood vessels, while the lymphatics consist of initial lymphatics and collecting lymphatics. Even small abnormalities in either blood vessels or lymphatics can be noticed and evaluated in three dimensions. We and others have used the mouse trachea for examining in situ angiogenesis and lymphangiogenesis, vascular development and regression, vessel patency, differences in transgenic mice, and pathological changes, such as increased vascular permeability induced by inflammatory mediators.

Key words Blood vessels, Lymphatic vessels, Endothelial cells, Angiogenesis, Lymphangiogenesis, Vascular regression, Immunohistochemistry, Confocal microscopy

1 Introduction

The increasing availability of transgenic mice, specific antibodies, and confocal microscopes has made wholemount preparations more attractive for studies of angiogenesis and lymphangiogenesis. To date, numerous tissues have been examined as wholemounts, including the retina, cornea, choroid, meninges, skin, intestine and mesentery but the attributes of the mouse trachea are not well known. Yet, wholemounts of the trachea yield vastly more three-dimensional information than histological sections. The purpose of this chapter is to present practical step-by-step immunohistochemical methods for preparing the mouse trachea for studies of blood and lymphatic vessels by fluorescence and confocal microscopy.

Andrew V. Benest (ed.), *Angiogenesis: Methods and Protocols*, Methods in Molecular Biology, vol. 2441,
https://doi.org/10.1007/978-1-0716-2059-5_10,

This model system has been used by our laboratory and others for examining angiogenesis [1] and lymphangiogenesis [2], many other features including vascular development, remodeling and regression [3], testing therapeutic drugs [4], blood vessel patency [5, 6], vessel differences in transgenic mice, and changes in pathological conditions, such as increased vascular permeability in response to inflammatory mediators [7, 8] or infectious organisms [2].

2 Materials

2.1 Mice for Studying Blood and Lymphatic Vessels in Mouse Trachea

1. Tracheal blood vessels and lymphatics have been studied in wild-type and in several transgenic strains of mice. Rat tracheas can also be processed in a similar way but are thicker and more difficult to make flat. We have used doxycycline-regulated double-transgenic mice where the expression of a growth factor, e.g., VEGF-A [1] or VEGF-C [1] or marker of interest is driven by a doxycycline-responsive promoter gene Scgb1a1 that is expressed in secretory Club cells found in the epithelium of trachea and large bronchioles (JAX strain).
2. Prospero homeobox 1-enhanced green fluorescence protein (Prox1-EGFP) transgenic mice have been particularly useful for studies of lymphatics and are available from the Mutant Mouse Resource Center at the University of California Davis as strain MMRC and from laboratories that use them [9], Although the endogenous green fluorescence of lymphatics in these mice fades rapidly after paraformaldehyde fixation, it can be amplified and made permanent by immunohistochemical staining for EGFP. Prox1-EGFP mice can be easily genotyped by the presence of green fluorescence of Prox1-EGFP in lens cells after a bright light is shined into their eyes.
3. Mice should be housed under pathogen-free conditions in filter-top cages to avoid accidental respiratory tract infections, which can have profound remodelling effects on airway blood vessels and lymphatics.

2.2 Anesthetics

Ketamine (87 mg/kg) plus xylazine (13 mg/kg) are injected intraperitoneally together in the same syringe (*see* **Note 1**).

2.3 Perfusion Apparatus

1. 1-L aspirator bottle with a hose outlet at the bottom (VWR/Kimax) is used for perfusion of fixative.
2. An aneroid sphygmomanometer (Medline) is used to generate a fixed pressure for perfusion of fixative from the aspirator bottle.

3. Silicone tubing and rubber bung are used to connect the sphygmomanometer and perfusion tubing to the aspirator bottle.
4. Male Luer-to-Luer adapters (Cole Palmer Cadence) attach the silicone perfusion tubing to an 18-gauge plastic catheter (Becton Dickinson), 30 mm in length that serves as the perfusion cannula.
5. A plastic tray similar to a restaurant dinner tray serves as a perfusion table that collects fixative during the perfusion.
6. The perfusion board is made from cork or a thick sheet of silicone or polystyrene foam (Styrofoam).
7. Surgical tape (e.g., 3 M) tape is wrapped around the front and hind paws of anesthetized mice and push-pins are used to fix the tape to the perfusion board.
8. A rubber band attached to front teeth and perfusion board is used to keep the neck and trachea straight during the perfusion.
9. A shadow-free surgical lamp is used for illumination during surgery, perfusion, and dissection.
10. 70% ethanol is used to moisten mouse fur before surgery and dissection.

2.4 Instruments for Surgery and Dissection

1. 140-mm long coarse scissors are used to open the chest.
2. 90-mm long fine scissors from Fine Science Tools #14060-09 are used to cut the heart and trachea.
3. Hemostat (locking clamp) with curved smooth tips is used to clamp tubing from perfusion bottle.
4. Hemostat (locking clamp) with straight serrated tips is used to clamp the rib cage after opening the chest wall.
5. Dumont 11-cm straight forceps from Fine Science Tools are used for dissection.
6. Coarse straight forceps (10 mm long with serrated tips from Fine Science Tools are used for surgery and dissection.
7. Scalpel handle and rounded or straight scalpel blades are used for surgery and dissection.
8. Fine scissors from Fine Science Tools are used for fine dissection.
9. A shadowless surgical light provides ideal illumination during the perfusion.
10. A binocular dissecting microscope is useful for fine dissection.

2.5 Fixatives, Buffers, Reagents, Incubation Solutions, Embedding Medium, Etc.

1. *Fixative: 1% paraformaldehyde in 1× phosphate-buffered saline (PBS)*

   Fixation is with 1% paraformaldehyde. See Making 1% paraformaldehyde.

2. *Calcium and magnesium-free PBS and PBS-Triton*

   The composition of 1× PBS is 0.2 g/L KH_2PO_4, 2.16 g/L Na2HPO4.7H20, 0.2 g/L KCl, 8.0 g/L NaCl. Store at room temperature. 1× PBS is conveniently prepared by diluting PBS 10× concentrate (GenClone) with water 1:10. Make PBS-Triton by adding 3 mL of Triton-X 100 (Fisher) to 1 L of 1× PBS. Stir with magnetic stirrer until dissolved.

3. *"Immunomix" antibody incubation medium*

   "Immunomix," consisting of 1× PBS containing 0.3% Triton-X 100, 0.2% bovine serum albumin (BSA, Sigma), and 0.1% sodium azide as an antibacterial agent (Sigma), is used for primary antibody incubations for immunohistochemistry. Can be stored at 4 °C for many months. When doing immunostaining using primary antibodies raised in goat or sheep, omit the 0.2% bovine serum albumin, as it can cross-react with the goat or sheep antibodies.

4. *"SuperDAPI" mounting medium*

   "SuperDAPI" consists of VECTASHIELD with DAPI mounting (Vector Labs) with additional DAPI (Sigma) at a final concentration of 50 μg/mL. *See* Subheading 3 on how to prepare (*see* **Note 2**).

5. *Coverslip sealing resin*

   Tissues are mounted on microscope slides in Vectashield with SuperDAPI and the coverslips are sealed with Cytoseal (Thermo-Fisher) or nail varnish.

6. *Sylgard silicone resin for making dishes to pin wholemounts*

   Sylgard 184 silicone elastomer resin is available as a kit (Fisher Scientific).

7. *Stainless steel pins for tracheal wholemounts*

   10-mm long headless stainless-steel pins ("Austerlitz Minutien" insect pins: 0.20 mm diameter, Fine Science Tools) are used to pin tracheal wholemounts to the layer of Sylgard in 6-well dishes (any brand). 5–6 tracheas per well in a 6-well dish can be stained simultaneously in a diluted antibody volume of 1–2 mL.

8. *Bradykinin for inducing leakage from tracheal blood vessels*

   Bradykinin acetate (Sigma) is more effective at inducing vascular leakage in the mouse trachea than substance P or histamine.

9. *Fluorescent microspheres for microscopic imaging of vascular leakage*

Table 1
Markers of endothelial cells in blood vessels and lymphatics and other cells in mouse trachea

Cell type	Marker	Host	Vendor	Catalog #	Comments
Vascular endothelial cells					
	PECAM-1 (CD31)	Armenian Hamster Rat	Thermo BioLegend	MA3105 102502	Clone 2H8 Clone MEC13.3 Weaker expression also on lymphatic EC
	VE-cadherin (CD144)	Rabbit Goat Rat	Millipore/ Sigma R&D Systems BD Biosciences	MABT886T AF1002 550548	Clone VIi37 Clone 11D4.1
	Endoglin (CD105)	Rat	eBioscience	14-1051	Clone MJ7/18
	Endomucin	Rat Goat	Invitrogen R&D Systems	14-5851-81 AF4666	Clone V.7C7 On venules, not arterioles
	Von Willebrand factor	Rabbit	Dako	0082	
	MECA-32 (PV-1)	Rat	BD Biosciences	550,563	Clone MECA-32
Endothelial glycocalyx	*Lycopersicon esculentum* lectin	Plant	Vector Laboratories	DL-1174-1	
Endothelial glycocalyx	*Griffonia simplicifolia* isolectin B4	Plant	Vector Laboratories	FL-1201-5	
Endothelial basement membrane	Collagen-IV	Rabbit	CosmoBio	LB-1403	Strong on blood vessels and collecting lymphatics
VEGF receptor	VEGFR-2	Goat	R&D Systems	AF644	Weaker expression also on lymphatic EC
Lumen of patent blood and lymphatic vessels	Podocalyxin	Goat	R&D Systems	AF1556	
High endothelial venules (HEV)	MECA-79	Rat (IgM)	BD Bioscience	553,863	Found in bronchoalveolar lymphoid tissue (BALT)

(continued)

Table 1
(continued)

Cell type	Marker	Host	Vendor	Catalog #	Comments
Lymphatic endothelial cells					
	EGFP in Prox1-EGFP mice	Chicken	Aves	GFP 1020	Overall best marker Strong signal. Marks cytoplasm and nucleus
	LYVE-1	Rabbit Rat Goat	AngioBio R&D Systems R&D Systems	11-034 MAB2125 AF2125	Good marker Also found on some macrophages
	Prox1	Rabbit Goat	AngioBio R&D Systems	11-002P AF2727	Nuclei of lymphatic EC and neuroendocrine cells
	VEGFR-3	Goat	R&D Systems	AF743	Weaker signal and higher background than Prox1
	CCL21	Goat	R&D Systems	AF469	In Golgi apparatus of lymphatic endothelial cells
	Neuropilin-2	Goat	R&D Systems	AF564	Endothelial cells of lymphatics and venules
	Podoplanin	Syrian hamster	AngioBio	11-033 Clone 8.1.1	Also expressed by some tracheal epithelial cells
Activity markers					
Dividing cells	Phospho-histone H3	Rabbit	Millipore	06-570	
Dividing cells	BrdU	Rat	Abcam	Ab567	Needs pre-injection of BrdU
Dying cells	Activated caspase-3	Rabbit	R&D Systems Cell Signaling	AF835 9579	Clone D3E9
Adhesion molecule	ICAM (CD54)	Rat	eBioscience	14-1061-82	Clone 429
Adhesion molecule	VCAM (CD106)	Rat	Millipore/ Sigma	MAB2627	Clone 6C7.1
Adhesion molecule	P-selectin	Rat	BD Biosciences	550,289	Clone RB40.34
mTOR activity	p-4E-BP1	Rabbit	Cell Signaling	2855	In lymphatic sprouts
mTOR activity	p-S6RP	Rabbit	Cell Signaling	4858	In lymphatic sprouts
Leakage marker	RCA1 lectin	Plant	Vector Laboratories	B-1085-5	Biotinylated

(continued)

Table 1 (continued)

Cell type	Marker	Host	Vendor	Catalog #	Comments
Other cell types					
Smooth muscle cells	α-smooth muscle actin	Mouse	Millipore/ Sigma	F3777 (FITC) C6198 (Cy3)	Also marks airway smooth muscle cells
Pericytes	NG2	Rabbit	Millipore/ Sigma	Ab5320	
Pan-epithelial cells	E-cadherin Pan-cytokeratin	Rat Rabbit	Thermo Dako	13-1900 Z0622	Clone ECCD-2
Club epithelial cells	CCSP	Rabbit	Millipore/ Sigma	07-623	
Ciliated epithelial cells	Acetylated tubulin	Mouse	Millipore/ Sigma	T7451	Clone 6-11B-1
T-lymphocytes	CD3e	Armenian hamster	BioLegend	100,302	Clone 145-2C11
B-lymphocytes	B220	Rat	BioLegend	103,202	Clone RA3-6B2
Neutrophils	S100A8 A100A9	Goat Rabbit	R&D Systems Abcam	AF3059 Ab105472	
Macrophages	Iba1 Mrc1/CD206 F4/80	Rabbit Rat Rat	WAKO BioRad BioLegend	019-19741 MCA2235GA 123102	Also known as Aifl Clone BM8 Better in wax sections
Dendritic cells	CD11c	Rat	eBioscience	14–0114-81	Clone N418 Better in wax sections

Bright red or green-fluorescent latex microspheres are available in a range of sizes from 20 to 1000 nm from several vendors. We use Fluospheres from Molecular Probes/Invitrogen (*see* **Note 3**).

Immunohistochemical reagents

1. Antibodies and other reagents used for immunohistochemical staining of lymphatics and other types of cells in mouse trachea are listed in Table 1. Combinations of antibodies that stain lymphatics together with other cell types can be very informative.
2. Primary antibodies, and fluorophore-conjugated secondary antibodies (Jackson ImmunoResearch), are typically used at a working concentration of 0.5–2.5 μg/mL (1:100–1:500 dilution) (*see* **Note 4**).

Pipettes and tips (10 μL, 200 μL and 1000 μL) for making antibody solutions.

Glass slides and coverslips (Fisher).

Fluorescence and/or confocal microscope.

It should have ×5, ×10, and ×20 dry objective lenses and ×40 and × 100 oil-immersion lenses and digital camera (*see* **Note 5**).

Micrometer slide with 1 mm scale (Ted Pella).

Microscope accessories:

Immersion oil for oil-immersion objective lenses (e.g., Zeiss Immersol 518 N (Fisher Scientific), lens paper (Fisher Scientific) and lens cleaning liquid (EM Sciences) for cleaning lenses after use.

Computer with image capture software, ImageJ (https://imagej.nih.gov/ij/ ImageJ and Photoshop (Adobe) for processing data.

3 Methods

3.1 Making 1% Paraformaldehyde Fixative

1. Formaldehyde vapor is toxic! Work in a fume hood or at ducted ventilation (*see* **Note 6**). To make 1 L of 1% paraformaldehyde (PFA) solution, add 10 g of powdered paraformaldehyde (Fisher) to a 1-L Pyrex flask with a magnetic stirring bar inside and a thermometer (be careful not to break it).
2. Add approximately one-third of the total final volume of distilled or de-ionized water. Stir with a magnetic stirrer. The result will be a cloudy white suspension.
3. Heat the suspension carefully to 55–60 °C on a hotplate with magnetic stirrer in a fume hood. Slowly and carefully add 1 N NaOH drop by drop until the suspension changes from cloudy white to clear. Only a few drops are needed. The suspension will suddenly turn almost clear but will contain some suspended fine white precipitate. If the temperature rises above 65 °C, discard the suspension and start over. Allow the suspension to cool and then filter it with coarse filter paper in a funnel to get a clear solution.
4. Add sufficient 10× PBS to achieve a final PBS concentration of 1× and allow the suspension to cool. When the solution reaches room-temperature adjust the pH to approximately 7.4 by adding HCl or NaOH drop by drop. Adjust the volume to the correct final value by adding distilled or de-ionized water.
5. Store 1% paraformaldehyde at room temperature in a tightly capped bottle for up to 2 weeks. Do not freeze. Discard if a precipitate is formed (*see* **Note 7**).

3.2 Preparation of Antibody Incubation Medium

Add 3 mL of Triton-X 100, 2 grams of BSA, and 10 mL of 10% sodium azide solution to a 1-L bottle and fill to 1 L with 1× PBS. Add a magnetic stirring bar and stir slowly to dissolve the BSA. If you attempt to dissolve the BSA quickly, it will clump and take significantly longer to completely dissolve (*see* **Note 8**).

3.3 Preparation of Sylgard Resin Dishes

Instructions for polymerizing Sygard184 silicone resin are included in the 1-lb kit. Wear gloves. Add 1 part catalyst by weight to 10 parts resin in a plastic beaker and mix gently, trying not to introduce too many air bubbles into the mixture. If available, use a desiccator bell jar connected to a vacuum source to remove remaining air bubbles. Pour the resin into shallow plastic 6-well dishes or similar containers to a depth of 5–10 mm. Allow the viscous liquid resin to polymerize overnight on a flat surface.

3.4 Preparation of SuperDAPI Mounting Medium

To prepare SuperDAPI, make DAPI stock solution by dissolving DAPI dilactate (Sigma) at 1 mg/mL in deionized water. Store this DAPI stock bottle in the dark at 4 °C. Next, take a fresh 10 mL dropper bottle of Vectashield mounting medium (H-1000). Pipette 500 μL of DAPI stock solution into an empty glass dropper bottle. Add Vectashield drop by drop and swirl to mix it. Repeat a few times to dilute the DAPI solution with the entire 10 mL volume of Vectashield and mix well. Label the bottle as SuperDAPI and keep it at 4 °C in the dark.

3.5 Fixation, Cardiac Perfusion, and Trachea Dissection

1. For studies of vascular permeability, inject bradykinin (10 mg/kg iv) and fluorescent microspheres (20–50 μL) mixed in the same 28-gauge insulin syringe via tail vein immediately before anesthesia and perfusion.
2. After anesthesia, position the mouse ventral side up on the dissection board. Pinch rear paws to ensure adequate anesthesia by lack of reflexes. Secure the mouse to the dissection board with tape around limbs and push pins. Hold the head flat with a rubber band placed between the teeth and two push-pins to keep the trachea straight.
3. Moisten the fur with 70% ethanol and cut and remove the skin to expose the ribs and ventral chest wall.
4. Using coarse scissors, cut the chest open to make a U-shaped flap. The base of the U should be at the level of the eighth rib and across the sternum. Avoid cutting the lungs by gently moving them out of the way of the scissors. Also avoid cutting the diaphragm to prevent the abdominal contents from getting in the way. The vertical portion of the U should end at the level of the fourth rib. Reflect the chest flap and clamp it with the serrated-tipped hemostat.
5. Hold the beating heart with coarse forceps and, using the fine scissors, make a shallow cut about 2 mm long at the tip of the

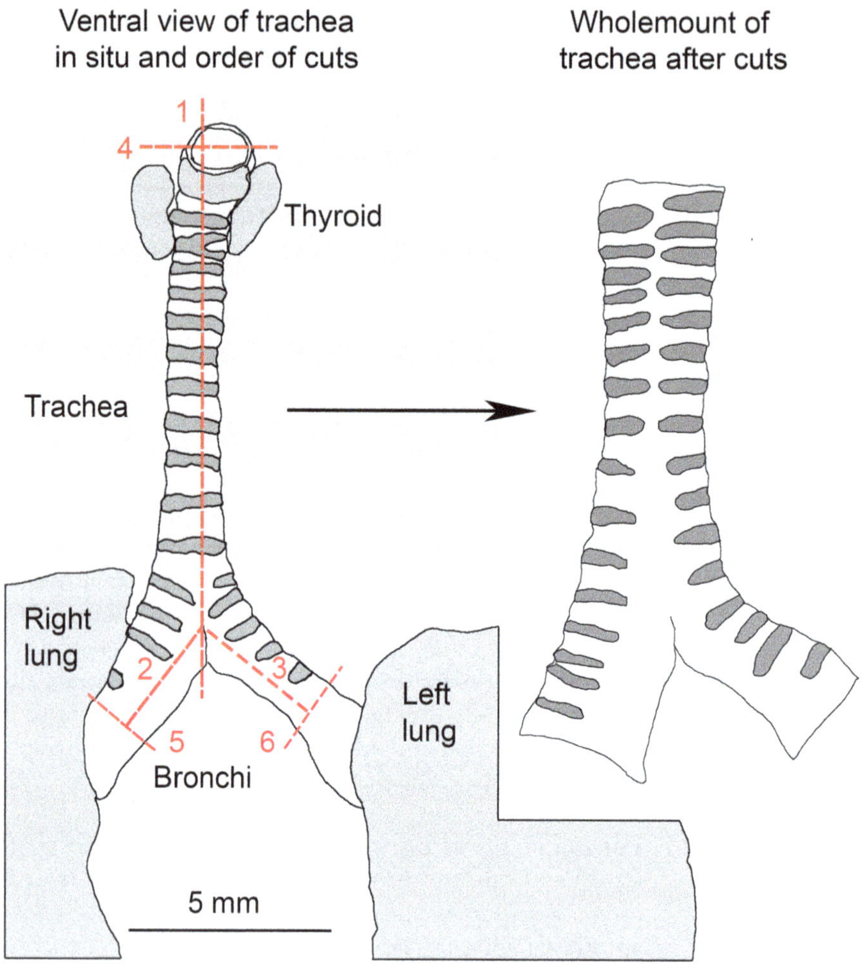

Fig. 1 Schematic diagram of mouse trachea in situ showing the order of cuts (dashed red lines) to make a wholemount preparation. 1: longitudinal cut along ventral aspect of trachea; 2 and 3: cuts along bronchi; 4: cut across larynx; 5 and 6: cuts across bronchi. The resulting wholemount is shown on the right

left ventricle. Gently tear open the left and right atria with fine forceps (don't use scissors) to create an exit route for blood and perfused fixative (*see* **Note 9**).

6. Insert the perfusion catheter into the aorta through the opening in the left ventricle and perfuse the systemic vasculature with fixative for 2 min at a pressure of 120 mmHg (*see* **Note 10**).
7. When the perfusate coming out from the right atrium is cleared of blood, finish the perfusion by clamping the perfusion line. Rinse away any blood remaining in the thoracic cavity with tap water from a hose or a squirt bottle.
8. Lift up the heart by the apex, cut the main vessels with scissors and remove the heart.

9. Using medium straight scissors cut the trachea in the order shown in Fig. 1. Cutting the trachea neatly while it is still stretched under tension in situ in the mouse is much easier than after it has been removed. Cut the trachea along the ventral midline with fine scissors to make a flat whole-mount. At this point keep the larynx and the bronchi attached to the trachea and use these regions as handles to manipulate the tissue with fine forceps or for securing with pins. Avoid damaging the cylindrical part of the trachea. The larynx and bronchi cannot be made into successful flat mounts, so they are cut off at the final step before mounting.
10. With fine forceps carefully remove the thyroid glands, recurrent laryngeal nerves at either side of the trachea and any major accumulations of adipose tissue on the adventitial surface. But don't worry too much about perfect cleaning, because the structures can be omitted from the final stack of confocal images.
11. Wash trachea with several changes of PBS for about 1 h at room temperature to thoroughly rinse out the fixative.

3.6 Incubation Procedure for Immunohistochemistry

1. Stretch the tracheas and pin them by the corners in the Sylgard dish, with up to 6 tracheas per well. Incubate with gentle rotation or agitation at room temperature. Mount the tracheas with a gap between them and the silicone layer so that incubation solutions can circulate and reach both sides of whole mounts and be readily exchanged when washing.
2. Rinse fixative from specimens with several washes of PBS. Time is not critical, provided all of the fixative is rinsed out; more changes, and longer is better. Usually, 6 × 10 min is enough.
3. Mouse tracheas can be processed in a small volume (1–2 mL) of incubation medium. Incubate tracheas in antibody incubation medium containing one or more primary antibodies AND 5–10% blocking serum from the host species in which the secondary antibodies are raised (usually goat or donkey). The blocking step with normal serum reduces nonspecific staining by the primary antibodies. Combinations of primary antibodies that are raised in different species can be used, e.g., rabbit + goat + rat with appropriately colored species-specific secondary antibodies. Incubate tissues at room temperature for at least 6 h (overnight is better) in a 6-well dish on a mechanical rotator.
4. After the primary antibody incubation step, wash the tracheas with several changes of PBS-Triton for several hours. Wash whole mounts with PBS-Triton for a minimum of 6 h or overnight at room temperature.
5. Incubate for 1–4 h or longer (overnight is best) for whole mounts in fluorescent secondary antibodies diluted 1:500 in PBS-Triton. Do not use serum in this step. Secondary

antibodies conjugated to Cy3 can be used for the red channel, conjugated to Alexa 488 for the green channel, and conjugated to Alexa 647 for the far-red channel of fluorescence or confocal microscopy.

6. Repeat the wash step with PBS/Triton.
7. After this penultimate washing step, fix the tissues again briefly for 2 min in 1% PFA. This important step helps to prevent fading and preserves the fluorescence by fixing the labeled antibodies. Wash briefly (3 × 2 min) with PBS to remove the fixative.
8. If you prefer, the whole process is easily adapted permanent specimens for viewing by transmitted light microscopy, by using HRP-labeled secondary antibodies or streptavidin/biotin procedures.
9. Fluorescent polystyrene microspheres can be used to mark sites of vascular leakage. *See* ref. 7 for dose, duration, etc.).
10. Transfer the trachea wholemounts into a small volume of SuperDAPI in a watch glass or depression slide (VWR). The tracheas will become clear and almost invisible as the SuperDAPI medium infiltrates them.
11. After the final washing step, immerse the tracheas in Vectashield + SuperDAPI for 2–5 min in a few drops of VectaShield in a depression slide or watch glass. This step gives time for the mounting medium to infiltrate into the tracheal whole mount, which as a result, becomes more transparent.
12. Mount the tracheal whole mounts in SuperDAPI medium on glass slides, using as little mounting medium as possible. For most purposes, it is best to mount the trachea with the epithelial side up, as the blood vessels are closer to this surface and will be closer to the objective lens. Cover the wholemount with coverslips. If you have used too much mounting medium and the coverslips are floating, absorb the excess by blotting with paper towels.
13. Apply gentle pressure with forceps to the outside of the coverslip and a longitudinal stretching motion to squish and elongate the trachea. This part requires some practice.
14. Carefully paint the edges of the coverslips with Cytoseal 60 (Fisher Scientific #831016) or nail varnish to seal them to the glass slides. This step permits any immersion oil to be wiped off the coverslip. Use a fume hood for this step, as Cytoseal 60 contains volatile and noxious solvents (toluene and butyl benzyl phthalate.) Allow the Cytoseal to harden for about 1 h in a fume hood at room temperature before viewing the slides on a microscope (*see* **Note 11**). Store slides in a slide box or plastic slide tray (VWR) in the dark in a refrigerator. Because of the final fixation, the fluorescence should last for up to 1 year.

3.7 Microscopy, Photography, and Tiling in Photoshop

1. Prior to observation with a fluorescence or confocal microscope, take the slides out of the refrigerator and let them come to room temperature for a few minutes so any condensation can evaporate.
2. Photograph the sections in appropriate color channels with camera exposure adjusted for brightness, contrast, and resolution. For conventional fluorescence microscopy, it is better to photograph red and green channels separately, rather than using dual band red/green fluorescence filters. Some bright fluorescence from one channel can bleed into the other channel. If this occurs, the bleed through image needs to be blackened in Photoshop (Window menu > Channels > select channel with bleed-through image > Edit > Fill > Black 100% > File > Save). Red and green channels can be pasted into R and G channels and separately adjusted and viewed in the Channels window (*see* **Note 12**).
3. Low magnification montages or panoramas of entire tracheas can be made by stitching together multiple photographs taken at a higher magnification and resolution. This "photo-tiling" can be done with Photoshop or other applications. For example, 10–20 images at 5× are usually sufficient to make a panorama of an entire section of mouse trachea. Images at 5× have a great depth of field, so almost all blood vessels and lymphatics are in focus. Photograph individual images so that they have about 30% horizontal and vertical overlap with adjacent images.
4. In Photoshop, open all of the individual images > File > Automate > Photomerge > select "Auto Layout" (if not already clicked) and "Blend Images Together" and click "Add Open Files". Click OK. After processing, Photoshop will assemble the images into a seamless whole. If there are gaps in the image, the original photos have not had enough overlap with their neighbors and need to be re-photographed. Sometimes individual stray images can be dragged manually into the correct position in the panorama. The panorama will be in multiple Photoshop layers and will have a large file size. When you are sure the panorama is correct, flatten the multiple merged images (Layers > Flatten Image > Save as JPG).
5. It is useful to take a photo of a calibrated micrometer scale slide to calculate the final magnification of the panorama. An easy and accurate way to put a scale bar on your final micrograph is to take a photo of the micrometer slide at the same magnification as your image. Put the photo of the micrometer slide into a new layer in PhotoShop, use it to make a scale bar of a suitable length, then delete it, leaving only the scale bar.

3.8 Pixel Intensity Analysis, Surface Plots, and Morphometry with ImageJ

1. Morphometric parameters such as the size, area density, fluorescence intensity, etc. of immunostained blood vessels and lymphatics in tracheal wholemounts can be measured with ImageJ) after image optimization in Photoshop.
2. Camera settings and Photoshop adjustments must be the same for all images in the batch being analyzed, because the measurement of area density is based on the percentage of pixels above a certain brightness threshold.
3. In Photoshop adjust the contrast and brightness of the color channel containing vessels of interest, such that the background where lymphatics are absent is almost black and the vessels are bright.
4. Launch ImageJ and open the image (File > Open > select the image of interest).
5. In the "Image" menu of ImageJ scroll down to "Color" then across to "Split Channels". Keep the color channel containing the vessels of interest and discard the others.
6. Surface plots in ImageJ are a useful way to compare the intensity of staining in images. Arrange the images side by side, open in ImageJ and choose the color channel to be analyzed as in **steps 4** and **5** above. In ImageJ Analyze menu select "Surface Plot". A topographic image with mountains and valleys representing fluorescence intensity should be produced. To make a color-coded heat map of the fluorescence intensities, in the ImageJ Image menu (third from left in ImageJ menu) select "Lookup Tables" at bottom and choose one that works for you, e.g., "16 colors". For an example, *see* Fig. 3c.
7. For additional analysis of your image, select the region of interest using the "Freehand selection" lasso tool (fourth from the left in the ImageJ icon menu).
8. In the "Analyze" menu of ImageJ scroll down to "Histogram" describing the number of pixels in the image with a brightness of 0 to 225. Click "Copy".
9. Paste the data into an Excel spreadsheet, and calculate the total number of pixels with intensity 0 to 225 in the region of interest bounded by the yellow line.
10. Also in Excel, calculate the total number of pixels above the chosen brightness threshold, and calculate the percentage they represent of the total number of pixels. The absolute value of the brightness threshold is arbitrary but must be less than the brightness of the structures identified as blood vessels or lymphatics and must be the same for all images being analyzed.
11. A plugin called Image-based tool for Counting Nuclei (ICTN) for ImageJ, which can be downloaded from https://imagej.

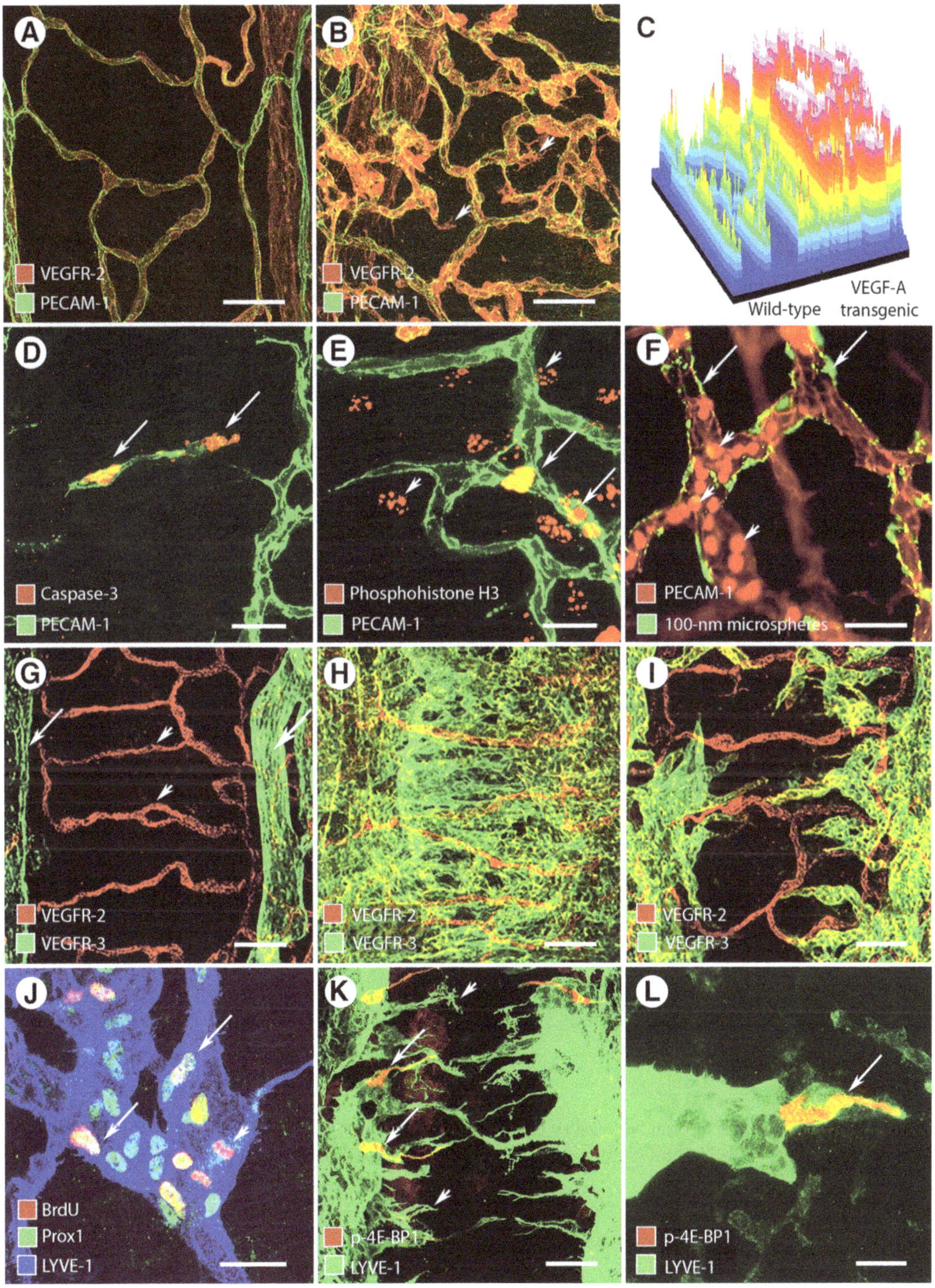

Fig. 3 (**a**, **b**) Capillaries over tracheal cartilages of transgenic mice at (**a**) baseline and (**b**) 7 days after doxycycline-induced overexpression of VEGF-A overexpression in VEGF-A transgenic mouse. Abundance of angiogenic vessels and increase in VEGFR-2 expression. Bar: 50 μm. (**c**) ImageJ surface plot showing increase in VEGFR-2 expression (fluorescence intensity) of A, B. (**d**, **e**) In mice, the tracheal vascular plexus is remodeled over the first week of neonatal life. (**d**) Endothelial cells first undergo apoptosis at P3, as shown by activated caspase-3 (arrows), and later (**e**) growth at P7 as shown by phosphohistone H3 (arrows).

nih.gov/ij/plugins/itcn.html, can be used for counting nuclei of Prox1-stained lymphatic endothelial cells.

12. Morphometric point counting can be used as an alternative to ImageJ for measuring area densities. For this approach, a rectangular test grid is overlaid on the image, and points of intersection are counted manually. A plugin to create such a grid in ImageJ can be accessed by clicking Analyze > scroll down to "Tools" > "Grid".
13. The whole process takes 3 days from mouse to image analysis. Day 1: perfusion washing, mounting in Sylgard dishes and primary antibody incubation overnight; Day 2: washing and secondary antibody incubation overnight; Day 3: washing, final fixation and mounting on glass slides, imaging and analysis. A low magnification confocal image of the architecture of mouse tracheal blood and lymphatic vessels is shown in Fig. 2. Examples of various useful immunohistochemical markers related to blood vessels and lymphatics are shown in Fig. 3.

4 Notes

1. Written records must be kept on the use of ketamine, as for other controlled substances, and made available for inspection by your institution's Environmental Health and Safety (EH&S) department.
2. Vectashield with DAPI mounting medium (Vector Labs product) does not contain enough DAPI to properly stain tracheal whole mounts. However, it does contain a useful proprietary anti-fade agent than prolongs fluorescence and minimizes bleaching.
3. A detailed product description of fluorescent microspheres from Molecular Probes/Invitrogen is available at:

   https://www.thermofisher.com/document-connect/document-connect.html?url=https%3A%2F%2Fassets.thermofisher.

Fig. 3 (continued) Arrowheads indicate other dividing cells. Bar: 20 μm. (**f**) Increased vascular permeability shown by extravasation of 100-nm green-fluorescent microspheres (arrows) from tracheal venules 4 min after bradykinin challenge (10 mg/kg iv). Numerous intravascular leukocytes adherent in venules (arrowheads) are also stained by the PECAM-1 antibody. Bar: 50 μm. (**g–i**) Tracheal lymphatics marked by VEGFR-3 staining at (**g**) baseline and (**hi**) after overexpression of VEGF-C in CCSP/VEGF-C transgenic mouse. (**h**) Doxycycline administered P0-P7, and (**i**) from P70-P77. Much stronger lymphangiogenesis in neonatal mouse. Arrows: lymphatics. Arrowheads: blood vessels. Bar 50 μm. (**j**) Tracheal lymphatic endothelial cell nuclei (arrows) labeled by BrdU staining 3 days after induction of VEGF-C in CCSP/VEGF-C transgenic mouse. Arrowhead marks mitotic cell. Bar: 20 μm. (**k, l**) Proliferating lymphatic endothelial cells marked by staining for p-4E-BP1 (arrows). Arrowheads mark lymphatic sprouts. Bar: 50 μm in K and 10 μm in L. (Images reprinted with permission; **a–c** from Ref. [1], **d**, **e** from Ref. [3], **f** from Ref. [7], **g–i** from Ref. [10], and **j–l** from Ref. [5])

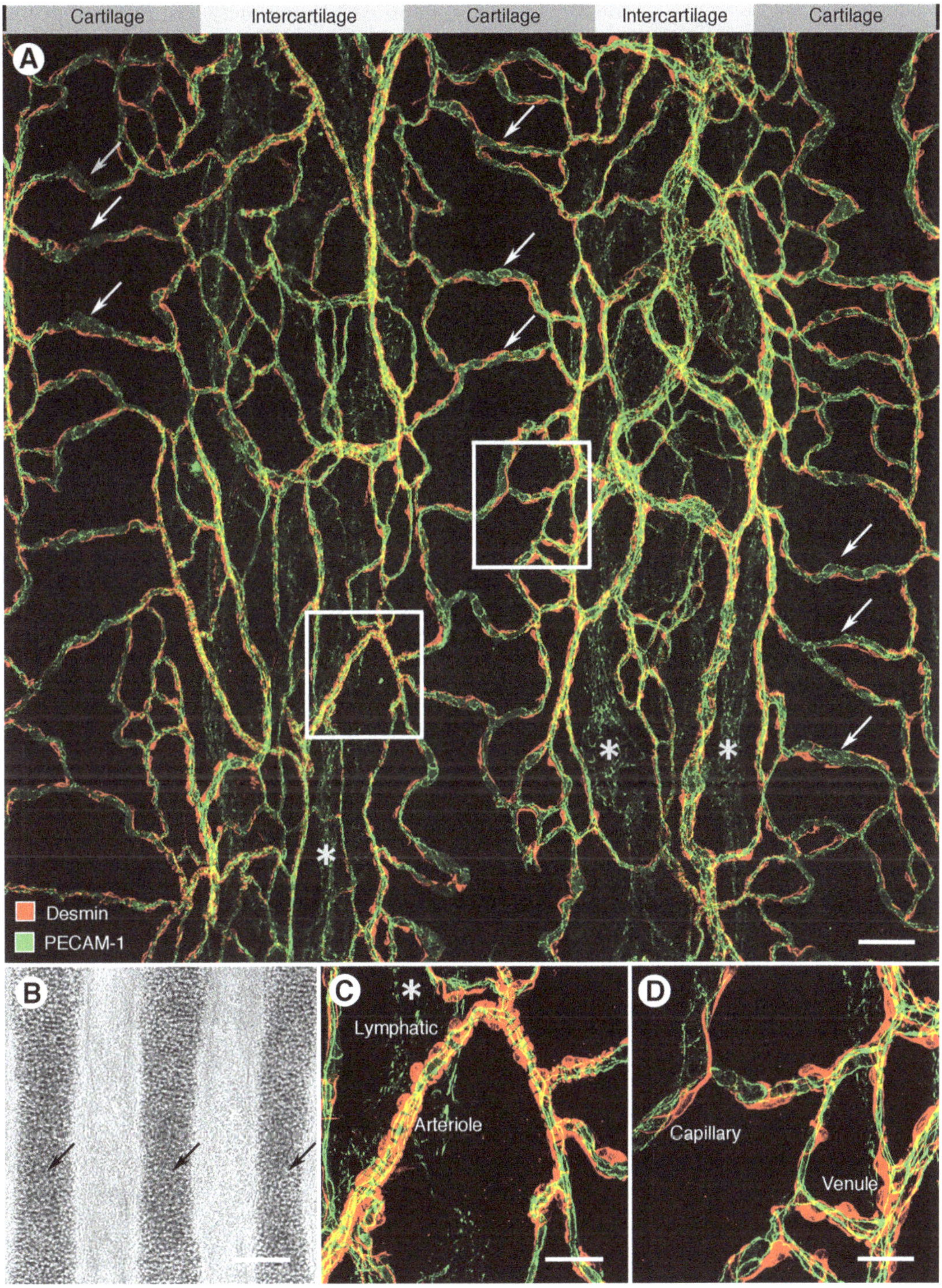

Fig. 2 Vasculature of adult mouse trachea. Whole mount preparation showing the vasculature of the adult trachea (endothelial cells, PECAM-1, green) with pericytes and smooth muscle cells (desmin, red) at P70. Long axis of the trachea oriented horizontally. (**a**) Confocal micrograph showing ladder-like arrangement of capillaries (arrows) that span three cartilage rings. Most arterioles, venules, and weakly stained lymphatic vessels (*) are confined to regions between cartilages. Bar: 50 μm. Boxed regions are enlarged in **c**, **d**. (**b**) Cartilages (arrows) of entire panel **a** are shown here in a brightfield image. (**c**) Arteriole with densely packed

com%2FTFS-Assets%2FLSG%2Fmanuals%2Fmp05000.pdf&title=RmxlblNwaGVyZXMgRmxlb3Jlc2NlbnQgTWljcm9zcGhlcmVz

4. Secondary antibodies (e.g., from Jackson ImmunoResearch Laboratories) that are pre-absorbed for other species are superior to unabsorbed ones. They are more expensive but result in lower background staining.

5. Any brand of microscope from a reliable maker is suitable. Low magnification lenses (×5, ×10, ×20) are useful for making panoramas and images with a great focal depth. For fluorescence microscopy, oil-immersion lenses ×40 and ×100 provide a brighter image with more resolution, but they require more care in their use and have a shallower focal depth. Ordinary paper towels or tissues can be used for wiping and cleaning glass slides and coverslips, but NOT for objective lenses. Carefully clean oil immersion lenses with purpose-made lens paper, which can scratch and damage the lens surface. Keep the microscope stage clean and dry; wipe up any spilled liquid with tissue paper.

6. The Occupational Safety Health Administration (OSHA) limit for formaldehyde in the workplace is 0.75 parts formaldehyde per million parts of air (0.75 ppm) measured as an 8-h time-weighted average, or a short-term exposure limit of 2 ppm during a 15-min period. Check your local EH&S regulations for conditions for performing perfusion with paraformaldehyde solution and take any necessary training courses.

7. Do not freeze aliquots of paraformaldehyde in solution, as the activity will be lost. Do not use formalin (37% paraformaldehyde +15% methanol) as the fixative. Local safety regulations may require bottles to have warning labels of "Carcinogen" or similar. Dispose of waste paraformaldehyde in accord with local safety regulations.

8. Sodium azide is toxic and should be handled with care. Read the safety data sheet before handling it and check with local regulations and authorities for policies governing its use and disposal. Wear gloves, mask, and protective clothing. For safety, sodium azide is best stored and handled as a solution in a breakproof bottle surrounded by a safety container. Make a 10% aqueous solution by carefully dissolving 10 g sodium azide powder in 100 mL of deionized water.

Fig. 2 (continued) circumferentially oriented smooth muscle cells and a weakly stained lymphatic (*). (**d**) Venule and capillaries with obliquely or longitudinally oriented pericytes stained for desmin. Scale bar in **d**: 50 μm in **a**, 150 μm in **b**, and 20 μm in **c**, **d**. (Image reprinted with permission from Ref. [3])

9. When creating a route for outflow of blood and fixative during the perfusion, use a toothed forceps and do not cut the heart atria with scissors, as there is a risk of accidental cutting of the aorta or other structures by mistake.
10. In our experience, 1% paraformaldehyde solution is just as good for fixing mouse tissues for immunohistochemistry with most antibodies and produces fewer fumes than 4%. Perfusion should be done in a well-ventilated place, e.g., a fume hood or perfusion board equipped with approved ventilation nearby to minimize the danger of breathing toxic formaldehyde fumes. For safety, wear gloves, mask, and personal protective clothing/equipment.
11. When using objective lenses with short working distances (20× and above), be careful to avoid bringing the lens into contact with Cytoseal that has not fully hardened. If that should accidentally happen, clean off the Cytoseal immediately with lens paper and lens cleaning fluid.
12. If mistakes are made in Photoshop, they can be reversed by using the History menu window or the Edit > Step Backwards command.

Acknowledgments

This work was supported in part by grants R01 HL127402, R01 HL059157, R01 HL143896, and P01 HL024136 from the NIH National Heart, Lung, and Blood Institute and 11CVD03 from the Leducq Foundation.

References

1. Baluk P, Lee CG, Link H, Ator E, Haskell A, Elias JA et al (2004) Regulated angiogenesis and vascular regression in mice overexpressing vascular endothelial growth factor in airways. Am J Pathol 165(4):1071–1085. https://doi.org/10.1016/s0002-9440(10)63369-x
2. Baluk P, Tammela T, Ator E, Lyubynska N, Achen MG, Hicklin DJ et al (2005) Pathogenesis of persistent lymphatic vessel hyperplasia in chronic airway inflammation. J Clin Invest 115(2):247–257. https://doi.org/10.1172/jci22037
3. Ni A, Lashnits E, Yao LC, Baluk P, McDonald DM (2010) Rapid remodeling of airway vascular architecture at birth. Dev Dyn 239(9):2354–2366. https://doi.org/10.1002/dvdy.22379
4. Baffert F, Le T, Sennino B, Thurston G, Kuo CJ, Hu-Lowe D et al (2006) Cellular changes in normal blood capillaries undergoing regression after inhibition of VEGF signaling. Am J Physiol Heart Circ Physiol 290(2):H547–H559. https://doi.org/10.1152/ajpheart.00616.2005
5. Baluk P, Yao LC, Flores JC, Choi D, Hong YK, McDonald DM (2017) Rapamycin reversal of VEGF-C-driven lymphatic anomalies in the respiratory tract. JCI Insight 2(16):e90103. https://doi.org/10.1172/jci.insight.90103
6. Jiang X, Tian W, Tu AB, Pasupneti S, Shuffle E, Dahms P et al (2019) Endothelial hypoxia-inducible factor-2α is required for the maintenance of airway microvasculature. Circulation 139(4):502–517. https://doi.org/10.1161/circulationaha.118.036157
7. Baffert F, Le T, Thurston G, McDonald DM (2006) Angiopoietin-1 decreases plasma leakage by reducing number and size of endothelial

gaps in venules. Am J Physiol Heart Circ Physiol 290(1):H107–H118. https://doi.org/10.1152/ajpheart.00542.2005

8. Sun Z, Li X, Massena S, Kutschera S, Padhan N, Gualandi L et al (2012) VEGFR2 induces c-Src signaling and vascular permeability in vivo via the adaptor protein TSAd. J Exp Med 209(7):1363–1377. https://doi.org/10.1084/jem.20111343
9. Choi I, Chung HK, Ramu S, Lee HN, Kim KE, Lee S et al (2011) Visualization of lymphatic vessels by Prox1-promoter directed GFP reporter in a bacterial artificial chromosome-based transgenic mouse. Blood 117(1):362–365. https://doi.org/10.1182/blood-2010-07-298562
10. Yao LC, Testini C, Tvorogov D, Anisimov A, Vargas SO, Baluk P et al (2014) Pulmonary lymphangiectasia resulting from vascular endothelial growth factor-C overexpression during a critical period. Circ Res 114(5):806–822. https://doi.org/10.1161/circresaha.114.303119

Chapter 11

Noninvasive Measurement of Retinal Microvascular Permeability During Loss of Endothelial Quiescence

Naseeb K. Malhi, David O. Bates, Kenton P. Arkill, and Claire L. Allen

Abstract

In the retina EC dysfunction and angiogenesis are driven by an altered microenvironment e.g., diabetes, leading to hypoxia and inflammation in the retinal layers, resulting in excessive vascular leakage and growth. The gold standard for measuring blood–retinal barrier permeability in response to disease and or therapy has been the gold standard Evans blue (EB) assay. However, this technique has limitations in vivo, including nonspecific tissue binding and toxicity. Here we describe a novel imaging methodology combining sodium fluorescein fundus angiography (FFA) with mathematical quantification allowing retinal permeability to be noninvasively and accurately measured at multiple time points in the same animal, minimizing animal use in line with the 3Rs framework. In addition, this technique is a nontoxic, high throughput, sensitive, and cost-effective alternative technique to the Evans blue assay. Moreover, this technique can be translated to other species.

Key words Retina, Endothelial Cells, Quiescence, Blood–retinal permeability, Fundus fluorescein angiography, Fick's Law, Quantitative, Sensitive, Nontoxic

1 Introduction

Selective permeable barriers are vital for organ function and homeostasis. Such barriers include the endothelial cell (EC) barrier of blood vessels, the blood–brain barrier (BBB) [1], the blood–retinal barrier (BRB) [2], the blood–spinal cord barrier [3], and the blood–placental barrier [4].

Blood vessels under normal conditions display a non-proliferative, non-motile, tightly controlled junctions, non-adherent surface, and quiescent phenotype. This provides the ideal surface for efficient and regulated cell fluid, and solute exchange between the blood and interstitium [5]. The EC maintains this quiescent phenotype through the activity of a combination of epigenetic [6], surface receptor and junctional receptor activity [5], metabolic [7], and of course transcriptional [8] control

Andrew V. Benest (ed.), *Angiogenesis: Methods and Protocols*, Methods in Molecular Biology, vol. 2441,
https://doi.org/10.1007/978-1-0716-2059-5_11,

of key genes. They mediate tissue perfusion [9], counteract thrombosis [10] and vascular inflammation [9], and maintain vasoregulation [11] and barrier function [12].

Selective transport of micromolecular substances across these fundamental barriers occurs under physiological conditions but they are relatively impermeable to macromolecules including albumin [13]. When diseases such as diabetic retinopathy [14], sepsis [15], capillary leak syndrome [16], or cancer [17] result in disruption of these barriers there is an increase in vascular permeability.

The EB assay is rapid, sensitive, and has become a widely used method for estimating barrier integrity and vascular permeability. EB dye has a high affinity for albumin [6] and has a poor ability to cross-selective permeable barriers under normal circumstances and remains predominantly within the blood circulation [18] and dyes such as EB may extravasate from the circulation into surrounding tissues. Accumulation of EB dye can be extracted from stained tissues and quantified by spectrophotometry [19, 20]. EB is extensively used to estimate changes in vascular permeability in several models including stroke [21], cerebral ischemia [19], skin [22, 23]), endothelial damage caused by trauma [24], and the breast cancer brain metastasis model [17]. EB has also been used to measure breakdown in the blood–spinal cord barrier [25, 26], and the BRB [14, 27].

However, this methodology has several significant assumptions associated with its use which can lead to oversimplification of the research findings, if unaccounted for. These include: (a) the contribution of advective flux of albumin (and dye bound albumin) to the total flux varying from minimal to highly significantly dependent on hemodynamic parameters rather than barrier properties (b) a substantive percent of free dye present in the animal following administration, resulting in flux being highly blood flow dependent rather than barrier dependent [1, 28] (c) lack of specific binding to albumin, (d) injection of dye dissolved in physiological solutions affects the structure of the dye [29], (e) EB binds albumin when it is in the tissues, (f) change in tissue clearance can account for the changes (g) problems with spectroscopic methods that have been used to estimate the amount of dye, and (h) *in vivo* toxicity [28, 30]. These issues have been highlighted many times and are extensively discussed in the literature.

This method describes a nontoxic and high-throughput alternative technique to the EB assay to measure retinal permeability in a diabetic model as an exemplar [31]. In combination with the Micron IV™ ophthalmoscope (Phoenix Technology Group Inc.), we have used FFA, and a novel post imaging analysis methodology described in this chapter. This technique can be carried out in less than 15 min, is minimally invasive and can be repeated in animals, resulting in a reduction of the number of animals required when compared to the EB assay, which takes ~2.5 h to carry out per animal and requires multiple cohorts of animals per time point. In addition, we have estimated that our novel methodology is significantly more efficient and cost-effective than the EB assay to perform.

Hence, reversal or prevention of EC dysfunction is a therapeutic objective and understanding how quiescent endothelial cells (QECs) protect themselves against dysfunction is an unmet medical need.

2 Materials

Prepare all solutions using ultrapure water (prepared by purifying deionized water, to attain a sensitivity of 18 MΩ at 25 °C) and analytical grade reagents. Prepare and store all reagents at room temperature (unless indicated otherwise). Diligently follow all waste disposal regulations when disposing of materials. We do not add sodium azide to reagents.

2.1 Induction of Type I Diabetes

- Adult male Brown Norway rats (250–300 g) (*see* **Note 1**).
- 0.1 M citric acid monohydrate: Dissolve 2.1 g in 100 mL of water and store at room temperature.
- 0.1 M trisodium citrate, dihydrate: Dissolve 2.9 g in 100 mL of water and store at room temperature.
- 0.1 M citrate buffer, pH 4.5: Combine 3 parts citric acid monohydrate and 2 parts 0.1 M trisodium citrate. Dihydrate and adjust the pH to 4.5 with 1 N sodium hydroxide solution. Filter through a 0.45 μm Corning filter and store at 4 °C (*see* **Note 2**).
- Streptozotocin (STZ): Weigh 100 mg of STZ and dissolve in 2 mL 0.1 M citrate buffer pH 4.5. Filter as previous, store on ice, and use within 30 min of preparation (*see* **Note 3**).
- Sterile sodium chloride solution (0.9%, w/v).
- Insulin syringes and needle (0.5 mL, 27-gauge, 0.5″) (*see* **Note 4**).
- Fifteen percent sucrose solution (w/v). Weigh out 75 g of sucrose and transfer to a 500 mL graduated cylinder containing 300 mL of water. Mix for approximately 10 min and make up to 500 mL with water. Dispense into a separate drinking bottle (*see* **Note 5**).

2.2 Insulin Supplementation

- IsoFlo® (100%, w/v Isoflurane) (*see* **Note 6**).
- Accu-Chek® blood glucose monitor and glucose strips.
- Animal fur clippers.
- Chlorhexidine solution (2%, w/v).

- 16-gauge needle.
- Trocar.
- Linplant sustained release insulin implants (LinShin Canada Inc.).

2.3 Sodium Fluorescein Fundus Angiography (FFA)

- Micron IV™ ophthalmoscope (Phoenix Technology Group Inc.).
- Imaging cradle with fitted heat mat (Phoenix Technology Group Inc.).
- Anesthetic induction: 0.75 mL ketamine hydrochloride (Ketaset®, 100 mg/mL stock); 0.5 mL medetomidine hydrochloride (Sedastart®, 1 mg/mL stock); 0.75 mL sterile water (*see* **Note 7**).
- 1 mL syringes and 24-gauge sterile needles.
- Phenylephrine hydrochloride eye drops (5%, w/v).
- Tropicamide eye drops (0.8%. w/v).
- Viscotears.
- Ten percent sodium fluorescein (Na-Fl) solution (w/v): Weigh 5 mg of Na-Fl salt and dissolve in 50 mL sodium chloride solution (0.9%, w/v). Filter through a 0.45 μm Corning filter, protect from light and store at 4 °C to avoid degradation.
- Anesthetic recovery: Add 0.6 mL atipamezole hydrochloride (Sedastop®, 5 mg/mL stock) to 2.4 mL sterile water and mix thoroughly.
- Heat mat (37 °C), a recovery cage with absorbent bedding, mash and a fresh water bottle (*see* **Note 8**).
- Discover imaging software (Phoenix Technology Group Inc.).

2.4 Eye Enucleation and Retinal Flat-Mount Dissection

- Fine curved forceps.
- Sterile phosphate buffered saline (PBS).
- Paraformaldehyde (PFA, 4%, w/v): In a class II chemical safety cabinet dissolve a 10 mL vial of 16% PFA in 30 mL sterile PBS. Store at 4 °C.
- Fine curved Vannas iridectomy scissors.
- A microsurgical incision knife.
- Fine straight Vannas scissors.
- 48-well tissue culture dish.

2.5 Retinal Vessel Parameter Immunofluorescence

- Humidified slide chamber.
- 4% PFA.

- Tween-20.
- Triton X-100.
- Sterile PBS.
- PBS containing 0.1% Tween-20 (PBST): Dissolve 50 μL of Tween-20 in 50 mL of PBS. Store at 4 °C.
- PBS containing 0.1% Triton X-100 (PBSTx): Dissolve 50 μL of Triton X-100 in 50 mL of PBS. Store at 4 °C.
- 1% BSA/PBSTx: Weigh out 0.5 g of bovine serum albumin (BSA) and add to 40 mL of PBS. Add 50 μL of Triton X-100 and mix thoroughly before making up to 50 mL with PBS. Store at 4 °C.
- Isolectin B_4 (lectin from *Bandeiraea simplicifolia* conjugated to biotin).
- Streptavidin Alexa Fluor® 488 secondary conjugate.
- Glass microscope slides.
- Glass circular coverslips 19 mm × 19 mm.
- Vectashield antifade mounting media.

3 Methods

Carry out all procedures at room temperature unless otherwise specified.

3.1 Induction of Type I Diabetes

- Weigh the Brown Norway rats and give them a single intraperitoneal (i.p.) injection of STZ 50 mg/kg. Administer an equivalent volume (μL) as per equivalent animal weight (g). Inject controls with an equal volume of saline (*see* **Note 9**).
- In addition to water provide a 15% sucrose solution in a separate drinking bottle to alleviate the initial hypoglycaemic spike following STZ induction (*see* **Note 10**). Monitor and record the volume of sucrose intake over a 72-h period.
- Monitor the animals closely for 3 days for weight fluctuations and overall health (*see* **Note 11**).
- On day 4 post STZ induction withdraw a blood sample from the tail vein using a 24-gauge sterile needle and syringe. Pre-warm the animal(s) in a warming cabinet for approximately 10 min (*see* **Note 12**) prior to securing in a restraint cone. Wash the tail with a 2% chlorhexidine solution to disinfect the area and visualize the lateral tail vein. The lateral tail vein is accessed approximately one-third along the length of the tail from the tail tip, moving toward the base for multiple samples. Make a small incision across the vein using a microsurgical blade and withdraw

approximately 50 μL of blood using a pipette and sterile tip (*see* **Note 13**). Stop the blood flow by applying pressure to the wound for approximately 30 s.

- Check the blood glucose levels by applying a drop of blood to the glucose strip inserted into the Accu-Chek blood glucose monitor (*see* **Note 14**). Animals are considered hyperglycemic with blood glucose levels >15 mmol/L.
- Anesthetize the animal with gaseous isoflurane (2–5%), in a prostrate position and shave the nape of the neck with fur clippers.
- Sterilize the shaved area with 2% chlorohexidine and pinch the skin between the thumb and index fingers and insert a 16-gauge needle to create a pocket approximately 2 cm along the neck midline.
- Divide the Linplant insulin implant into thirds using a sterile surgical blade.
- Immerse the trocar in 2% chlorhexidine solution and push through the pocket just created. Immerse the implant briefly in chlorhexidine and insert into the proximal end of the trocar. Pinch the skin over the inserted implant before withdrawing the trocar (*see* **Note 15**).
- Add a drop of chlorhexidine over the skin opening, recover and monitor the animals (*see* **Note 16**).

3.2 Sodium Fluorescein Fundus Angiography (FFA)

- Weigh the animal and anesthetize with a mix of ketamine and medetomidine hydrochloride as described in Subheading 2.3. Administer an equivalent volume (μL) as per equivalent animal weight (g).
- Topically apply one drop of 5% phenylephrine hydrochloride and 0.8% tropicamide to the left eye, to dilate the pupil. Transfer to the imaging cradle fitted with a heat mat, once anesthetized.
- Apply Viscotears to the cornea of both eyes to prevent corneal dehydration and remove any whiskers in the field of view using a cotton tip.
- Advance the Micron IV ophthalmoscope toward the cornea until a seal is formed between the lens and Viscotears and align the optic disc in the center of the image. Adjust the brightness and focus to achieve a crisp image of the main retinal vessels (*see* **Notes 17** and **18**).
- Once the eye is correctly aligned capture bright field fundus images of the retina and check for any ocular abnormalities (*see* **Note 19**).

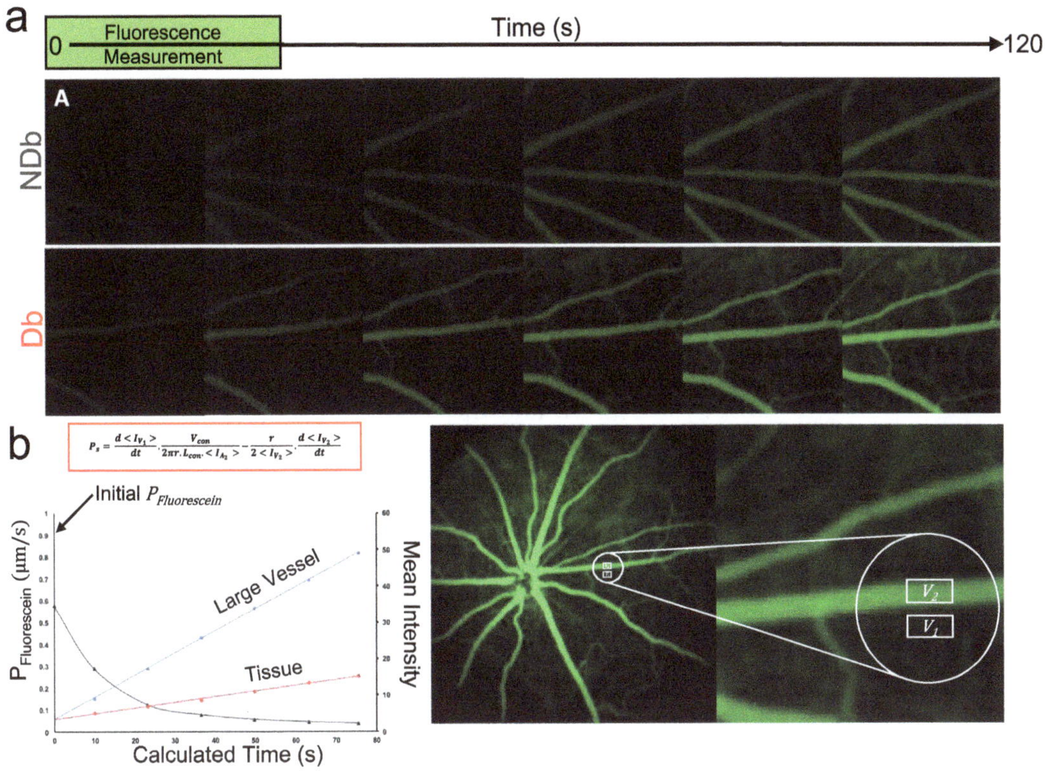

Fig. 1 In vivo fluorescein fundus angiography methodology. Sodium FFA was performed in non-(NDb) and diabetic (Db) Brown Norway rats (250–300 g) on days 0, 7, 14, 21, and 28. A 3 min (15 fps) video was captured (**a**). The mean intensity of Na-Fl in the retinal tissue ($< I_{V_1} >$) including unresolved capillaries and a main retinal vessel ($< I_{V_2} >$) were measured every 200 frames for 2 min. (**b**) Example of $< I_{V_1} >$ and $< I_{V_2} >$ over time used to determine initial fluorescein permeability ($P^{t=0}_{\text{Fluorescein}}$). (Figure adapted from Allen et al, 2020 [24])

- Select the filter set for FITC (green) imaging and record a 3 min video footage of the retina, at 15 frames per second, maximum gain.
- Administer a single 250 μL i.p. injection of 10% Na-Fl (w/v) by gently lifting the left leg and injecting into the left quadrant (*see* **Notes 20–22**).
- Stop the recording and save as an avi file format for further analysis.
- Recover the animal by administering an i.p. injection of 0.2 mL of atipamezole hydrochloride and place in a pre-warmed recovery cage containing fresh soft absorbent bedding, mash, and a water bottle (*see* **Note 23**).
- Repeat FFA on days 7, 14, 21, and 28 aligning the eye in the same position as captured on day 0, Fig. 1a (*see* **Note 24**). Figure 1 near here.

3.3 FFA Analysis

- Import the fluorescent angiogram avi files into Fiji software and open as a virtual stack.
- Using the regions of interest tool in Fiji to highlight an area within a main retinal vessel and a second nearby vessel termed the interstitium (including unresolved capillaries) using the rectangular selection tool.
- Plot an initial time course of the selected regions where there is detectable tissue fluorescence but no major vessel saturation, for analysis (*see* **Notes 25** and **26**).
- Save the above as comma separated variable (csv) text file and open in Fiji. To find the solute permeability (P_s) a macro (*see* Appendix 1) was used to fit the equation (*see* **Note 27**):

$$P_s = \frac{d < I_{V_1} >}{dt} \cdot \frac{V_{con}}{2\pi r.L_{con}. < I_{V_2} >} - \frac{r}{2 < I_{V_2} >} \cdot \frac{d < I_{V_2} >}{dt} \quad (1)$$

  where t is the time in seconds from the retinal bolus start, $< I_{V_1} >$ is the mean intensity in the interstitium (including unresolved capillaries), $< I_{V_2} >$ is the mean intensity of a major vessel, r is the mean exchange vessel radius. L_{con} and V_{con} are the length of vessels in a measured volume and that volume, respectively. r, L_{con} and V_{con} parameters were measured (*see* Subheading 3.4) from collected retinal tissue post euthanasia, from confocal volumes. Equation 1 is derived from Fick's law and is described in detail elsewhere [24].
- The output permeability decreases exponentially over time as the vascular and interstitial concentrations equilibrate (and this is not accounted for in Eq. 1). The macro extrapolates where the true $t = 0$ was, and the permeability at that point is taken (*see* **Note 28**) (Fig. 1b).

3.4 Vessel Parameter Analysis

- Post euthanasia, on day 28, enucleate both eyes with a pair of curved forceps and immediately fix them in 4% PFA (w/v), for 30 min. Once fixed transfer the eyes to sterile PBS.
- Remove the anterior of the eye by making an incision along the corneoscleral limbus and cutting around the circumference of the eye with a pair of curved Vannas scissors.
- Carefully remove the lens before excising the retina from the scleral/choroidal cup.
- Flat-mount the retina by making four incisions from the outer edge to the optic disc, to form a petal-like structure. Store the retinae in a 48-well tissue culture plate, in sterile PBS.

- Wash the retinae 3× with PBST, 5 min each time, on a rocking platform.
- Permeabilize with PBST× for 5 min at room temperature.
- Wash with PBST as previous.
- Block nonspecific binding sites with 1% BSA/PBSTx, for 1 h at room temperature.
- Stain the retinae with an endothelial cell specific marker, isolectin B_4 (lectin from *Bandeiraea simplicifolia* conjugated to biotin) diluted 1:400 in 1% BSA/PBSTx, overnight at 4 °C.
- Wash with PBST as previous.
- Apply secondary Alexa Fluor 488—streptavidin conjugate diluted 1:500 in 1% BSA/PBSTx, for 45 min at room temperature and protected from light.
- Wash with PBST as previous.
- Flat-mount the retina on a glass circular coverslip using a dissection microscope. Add a drop of Vectashield antifade mounting media to the retina and carefully apply a glass slide on top of the tissue without introducing air bubbles. Allow the mountant to dry and store slides at 4 °C and protected from direct light exposure.
- Image the retinae using a confocal microscope (Leica TCS SPE) and generate *z* stacks of the whole tissue thickness, including the superficial and deep plexuses (Fig. 2a).
- Import the *z* stacks into Fiji software [29] and manually measure vessel length per confocal volume by drawing along all vessels in the upper plexus, using the freehand line tool (*see* **Note 29**) (Fig. 2a, b).
- To determine retinal vessel radius, measure the diameter of all vessels per confocal volume using the freehand tool and divide the value by two to give the radius (Fig. 2c) (*see* **Note 29**).
- To determine the vascular retinal thickness, count the number of *z* stack positions between the upper and lower retinal plexus) (*see* **Note 29**). Figure 2a–d near here.

3.5 Statistical Analysis

- Import FFA data into Microsoft Excel (Microsoft Office Software) to generate the data files for analysis.
- Import the generated P_s values for Na-Fl into statistical software (e.g., Prism; GraphPad Software Inc., CA, U.S.A.) to generate graphs and perform statistical analyses.
- Use a one-way ANOVA with Bonferroni's correction to calculate vessel parameter significance between study cohorts.

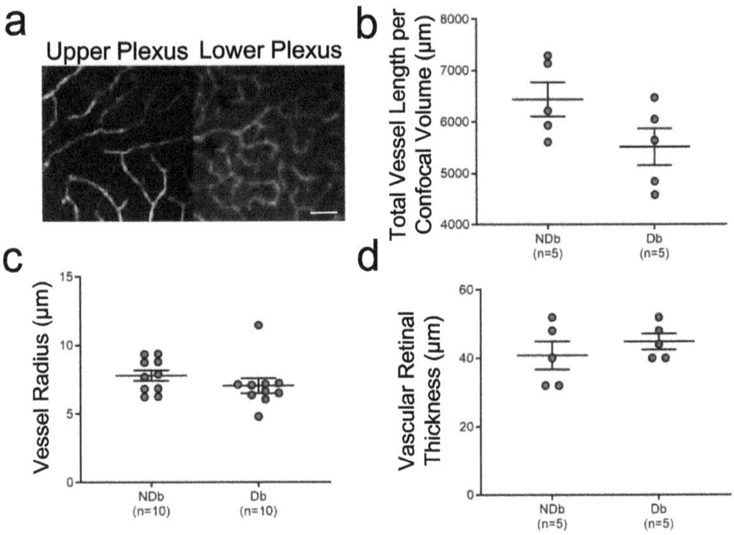

Fig. 2 Retinal vessel dimensions remain unchanged in nondiabetic versus diabetic Brown Norway rats. Average retinal vessel length per confocal volume, radius and vascular retinal thickness were measured in nondiabetic and diabetic cohorts prior to incorporating into the mathematical permeability equation. Eyes were enucleated on day 28, fixed in 4% PFA, hemisected and retinae stained with isolectin B4 (biotin conjugated from *Bandeiraea simplicifolia*, 1:400) followed by streptavidin Alexa Fluor 488 (1:500) (**a**). Total vessel length per confocal volume (**b**) and radius (**c**) measurements were calculated from non- and diabetic-retina, using Fiji software. (**d**) Vascular retinal thickness was measured from the upper to the lower plexus in the same animal cohorts. All data are shown as mean ± SEM. (Figure adapted from Allen et al, 2020 [24])

- To determine significance between initial calculated P_s values over a study period use a two-way ANOVA with Bonferroni's correction.
- Consider results statistically significant if $p < 0.05$ (*), $p < 0.01$ (**), $p < 0.001$ (***), $p < 0.0001$ (****). *See* Fig. 3 for exemplar results [31].

4 Notes

1. The described method can be used in rodents where there is pigment in the retinal layer (e.g., Brown Norway rats or C57bl/6j mice). This blocks most of the signal from the underlying choroid. The Micron IV ophthalmoscope is equipped with both lenses for mice and rats, and both have produced equally robust data using this method. Male rats are preferred because they achieve a higher average weight

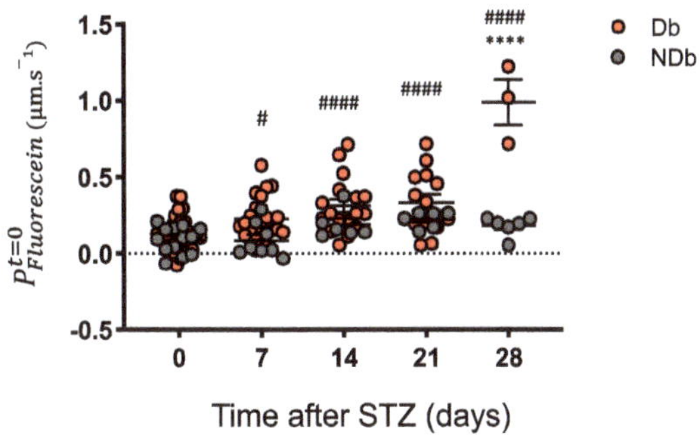

Fig. 3 Retinal sodium fluorescein permeability $P^{t=0}_{\text{Fluorescein}}$ significantly increases over 28 days in diabetic Brown Norway rats and remains unchanged in nondiabetic controls. Permeability measurements from 11 diabetic (Db, red closed circles) and 14 nondiabetic (NDb, grey closed circles) rats measured using corresponding group vessel spacing measurements. Bars show mean $\pm$ SEM. Statistical analysis was performed using a two-way ANOVA, with Bonferroni's post-hoc test, # $p < 0.05$, $^{\#\#\#\#}p < 0.001$ and $^{****}p < 0.0001$. $^{\#}$ denotes significance compared to day 0 in the Db group. *denotes significance between NDb vs Db per time point. (Figure adapted from Allen et al, 2020 [24])

compared to female rats and are more susceptible to the STZ toxicity without the protective oestrogen effect.

2. Prepare citrate buffer fresh each time.
3. Streptozotocin is a glucosamine-nitrosurea compound and alkylating agent that damages DNA and is particularly toxic to the insulin-producing beta cells of the pancreas. STZ can be used to induce insulin-dependent Type I diabetes mellitus in rodents. When weighing out STZ use a Class II chemical safety cabinet, double glove (nitrile), wear a lab coat, a mask and safety goggles. Wipe down all working surfaces with a 0.1% sodium hypochlorite solution and dispose of PPE via the correct route. Cover the bijoux containing STZ with foil to prevent light degradation and store on ice. Aliquots of STZ may be stored at −20 °C for later use. 100 mg of STZ will dose approximately 7–8 rats at 50 mg/kg.
4. Insulin syringes allow STZ solution to be accurately drawn and administered less invasively than standard needles and syringes.
5. If the animals are showing signs of dehydration (skin tenting) and hypoglycemia administer a 1 mL gavage of 15% sucrose solution daily for a 72-h period.
6. IsoFlo is a flammable, gaseous anesthetic. If a leak or spill occurs leave the room, ventilate the room, and allow the anesthetic to disperse.
7. Disadvantages of injectable anesthesia include lack of precision in controlling anesthetic depth, prolonged recovery time, and

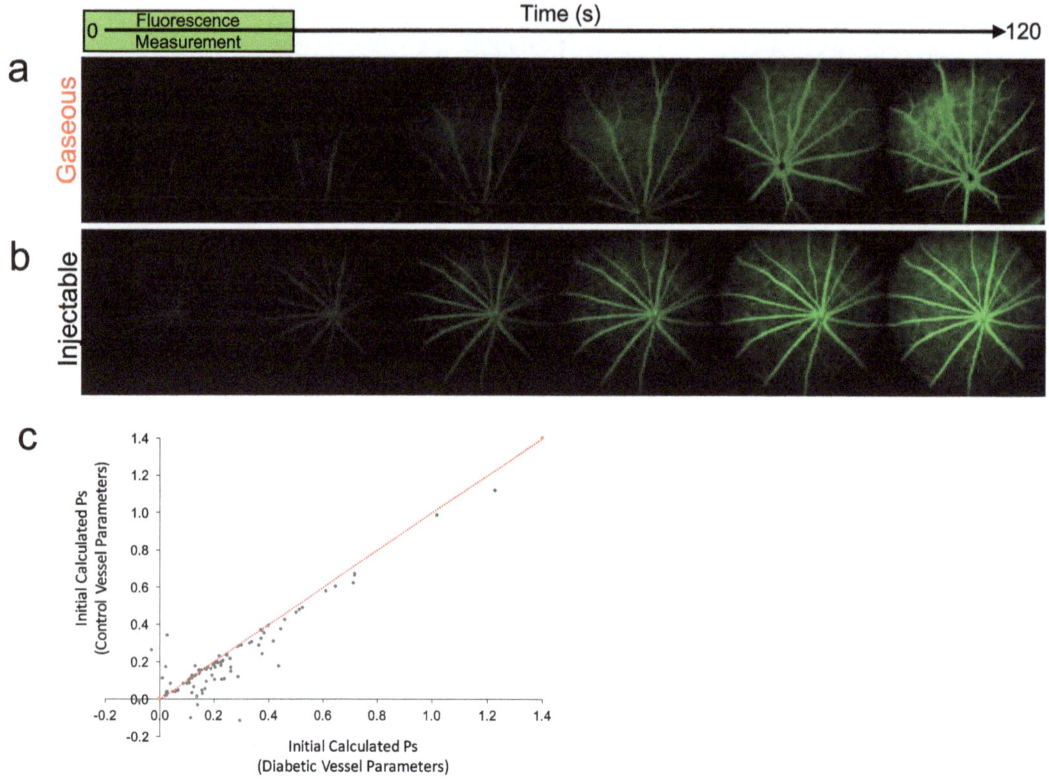

Fig. 4 Notes on in vivo fundus fluorescein angiography methods. Na-Fl FFA was performed in Brown Norway rats (250–300 g) using (**a**) gaseous (2–3% isoflurane) or (**b**) injectable (ketamine and medetomidine hydrochloride) anesthetic. A 3 min (15 fps) audio video interleave (avi) file was captured over time. A notable difference was observed in eye stability during image capture as denoted by the change in eyeball orientation and centering of the optic disc over time. (**c**) Differences in vascular parameters used. To calculate $P^{t=0}_{\text{Flourescein}}$ for an experimental group such as diabetics the measured vascular parameters can be drawn from the control or the diabetic group. (Figure adapted from Allen et al, 2020 [24])

physiological changes such as hypotension, hypercarbemia, and hypoxemia. Our previously published choice of anesthesia was gaseous (halothane/isoflurane) form. However, we regularly observed closing of the palpebra, eye movement, eye rolling, and therefore reverted to an injectable regimen to improve image quality output (Fig. 4).

8. Presoak food pellets in water for a minimum of 30 min to prepare mash. Pellets may also be soaked in 15% sucrose solution in the first 72 h following STZ induction.

9. Ensure that you are wearing appropriate PPE when handling STZ. Double glove (nitrile) wear an outer Tyvek, a mask and safety goggles. Wipe down all working surfaces with a 0.1% sodium hypochlorite solution and dispose of PPE via the correct route. Rodents will continue to excrete STA and its metabolites in their urine for approximately 72 h and should be

quarantined from other animals and staff. Cages must be decontaminated with a 0.1% sodium hypochlorite solution and all waste should be disposed of in purple top cytotoxic incineration bins.

10. During the first 24 h animals may show signs of hypoglycemia and provision of sucrose water during the induction period can reduce morbidity and mortality.
11. Monitor the welfare of the animals using an appropriate scoring system focusing on appearance, weight, natural behavioral characteristics, provoked behavior, and fluid volume parameters.
12. This can be stressful and can cause dehydration due to salivation, in addition to increasing metabolic rate, which may affect experimental data depending on the parameters observed. As a result, other routes such as saphenous vein sampling should be used where possible and where there is a need to take multiple samples.
13. The lateral tail vein is usually used and 50 μL to 0.2 mL of blood can be obtained per sample depending on the size of the animal and its health status. Illumination devices can be used to improve tail vein visualization, if necessary. To avoid bruising and damage to the tail, normally no more than two blood samples should be taken per session and in any one 24-h period. Alternate sides of the tail should be used, and successive needle punctures moved toward the tail base.
14. A minimum of 50 μL of blood is required to check the glucose levels. Insert the strip into the Accuchek monitor and directly insert the reading strip into a bolus of blood until adsorbed by the strip.
15. Each implant contains enough insulin to last for 60 days at the specified release rate.
16. There is no requirement for sutures or clips to close the wound.
17. Our microscope (Micron IV) is not the only dedicated retinal imager available. Of note is that the numerical aperture is limited by the eye both by distance and ocular lens. It is therefore unlikely that individual capillaries signal (5 μm diameter) could be satisfactorily resolved above noise.
18. Annotate the image with the animal ID and left or right eye, in the Discover imaging software.
19. There is a risk of development of cataracts with resultant blurring of posterior segment view.
20. When administering Na-Fl inject in the lower quadrant at a 30–40° angle toward the head, to avoid damage to the urinary

bladder, cecum, and other organs. Vary sides according to anesthetic dose administration.

21. The technique is limited to very high-density fluorescence (e.g., Na-Fl), other more sensitive microscopes with a monochrome high quantum efficiency camera would possibly be able to study bound dyes (e.g., FITC-Albumin), but with our system this was impossible. Our control groups, with our system, are at the limit of the permeability sensitively even with Na-Fl.
22. This method employs the use of i.p. injection to administer the dye (Na-Fl). Although the difference between i.p. and tail vein (intravenous, i.v.) injection were not fully studied it was noted that although the tail vein bolus was sharper (which would increase the permeability sensitivity), it was not steady and exhibited slight fluctuation. Further investigation was not pursued. Dye administration via the choroid artery would be the preferred route, but in our case, over a several week study, the increase in severity is not justified.
23. The animals will exhibit a green/yellowish hue to their eyes, extremities and continue to pass colored urine for approximately 24 h.
24. Due to the rapid uptake and delayed elimination of Na-Fl from the animal's circulatory system only one eye can be imaged per session.
25. Automation: For our studies so far, we have manually acquired the data from the images. Automated segmentation should be possible, e.g., by taking the main vessels as the high intensity and the tissue (capillaries and interstitium) as the low intensities. The difficulty, and why this was not pursued, was the eye movement with the gaseous anesthetic meant that our attempts with image alignment were not robust at the initial low intensity time points.
26. Phenomological Permeability (P_{Phen}) Analysis: It is possible to get significant data by dividing the tissue (and capillary) area mean intensity gradient by the large vessel mean intensity gradient:

$$P_{\text{Phen}} = \frac{< I_{V_1} >}{< I_{V_2} >}$$

The difference is not strictly the wall permeability, but the changes are likely related. The advantage is that it does not require analysis (and preparation workload) of the vessel parameters.

27. Curve Fitting: The macro must fit an exponential decay to the output. These can fail, we found the low permeability values were where it failed most. Our manual mean intensity measurement is once every 200 frames, in these cases more readings might help but it is fundamentally due to the interstitium filling

so slowly as to be hardly detectable above the capillaries. A sharp bolus would be needed (e.g., from carotid injection) for the lowest measurements.

28. Permeability output: This does not consider any change in pressure between the lumen and the interstitium. The pressure difference, one including oncotic and osmotic pressures, would be in the region of (1–5) cm.H_2O. A global change of pressure between groups could be checked for by (for example) a tail cuff but as this does not mean the pressures in the capillaries of interest have changed in the same manor, we did not take them as the information gained was outweighed by the technical challenges. There are clearly cases where this knowledge would be required.
29. Vascular parameters: The method used took the vascular parameters (radius and length) for each group throughout, i.e., the diabetic parameters were used for the diabetic calculation (Fig. 3). However, the measured parameters are only valid for the final measurement, and it is likely that the earlier measured parameters are more like the control group. In our case there is no overall experiment change or significant differences between the parameters and if we use the control group for all the parameters (Fig. 4). Using the control group vascular parameters throughout slightly underestimates the permeability for the diabetic group by approximately 10%, although much of this is due to the fluctuations for the low permeability values with the higher values having less difference. A full parameter analysis has not been performed so we did not pair the individual animals with their individual parameters, in part as the regions were not matched between instruments, and in part as to take enough regions in each animal when the group difference is low is a vast workload for unnecessary gain. There could be circumstances where this should be considered.

Acknowledgments

This work was supported by The Medical Research Council MR/P003214/1 (K.A); and MR/L01985X/1 (DOB); The National Eye Research Centre (DOB); and the Masonic Charitable Foundation (DOB).

Appendix

Macro used in Fiji. Data is expected in coma separated variable format with "Frame", "Interstitial Intensity", and "Vessel intensity" headings:

```
//------User Inputs from vascular parameters (opens as expected values in dialog)

meanR=7.8025;
lcon=6436.5;    // Measured length in microns
RetDepth=40.8; // distance from surface to top of RPE layer
AreaCon=134968; //xy area of confocal z stack
framerate=1;
//-----
Dialog.create("Questions");
Dialog.addNumber("Mean Vessel radius (um): ",meanR);
Dialog.addNumber("Length of Vessel in Confocal Volume (um): ",lcon);
Dialog.addNumber("Distance betwwen top and bottom capilaryplexus
(um): ",RetDepth);
Dialog.addNumber("Area of the Confocal Image: ",AreaCon);
Dialog.addNumber("Number   of   Frames/Second--leave as '1' input is time
(s): ",framerate);

Dialog.show();

meanR= Dialog.getNumber();

lcon= Dialog.getNumber();
```

```
RetDepth=Dialog.getNumber();
AreaCon=Dialog.getNumber();
framerate=Dialog.getNumber();
Vcon=AreaCon*RetDepth;
//------Chose a directory of .csv files
dir = getDirectory("Choose a Directory");
list = getFileList(dir);
ComTitles=newArray(list.length);
ComPs=newArray(list.length);
Name=newArray(list.length);
FolderResults=dir+"Folder Results.xls";
//setBatchMode(true);

showStatus("Running...");
for (j=0; j<list.length; j++) {
        Name[j]=list[j];
        path = dir+list[j];
                if(endsWith(path, ".csv")){
                        run("Results... ", "open=["+path+"]");
                        dotIndex = lastIndexOf(path, "."); //Finds the last dot
                        if (dotIndex!=-1){
                                path = substring(path, 0, dotIndex); // removes .csv
extension
                                NewPath=path+".xls";
```

```
        }
                dotIndex = lastIndexOf(Name[j], "."); //Finds the last dot
                if (dotIndex!=-1){
                        Name[j] = substring(list[j], 0, dotIndex); // removes .csv
extension
        }
//------ gets the array for calculation (if csv file has correct column headers)
Frame=newArray(nResults);
Time=newArray(nResults);
RealTime=newArray(nResults);
V1=newArray(nResults);
V2=newArray(nResults);
CalcV2=newArray(nResults);
Ps=newArray(nResults);
cPs=newArray(nResults);

for(i=0;i<nResults;i++){
    Frame[i]=getResult("Frame",i);
    Time[i]=Frame[i]/framerate;
    V1[i]=getResult("Interstitial Intensity",i);
    V2[i]=getResult("Vessel intensity",i);
}
//----linear fits to find t=0 timepoint and gradients
Fit.doFit("y = a*x+b",Time,V1);
```

```
gradV1=Fit.p(0);

interceptV1=Fit.p(1);

//Fit.plot;

Fit.doFit("y = a*x+b",Time,V2);

gradV2=Fit.p(0);

interceptV2=Fit.p(1);

//Fit.plot;

//print(gradV1+"  "+interceptV1);

RealZero=(interceptV2-interceptV1)/(gradV1-gradV2);

MissingI=Fit.f(RealZero);

//print(RealZero+"  "+MissingI);

for(i=0;i<Time.length;i++){

        CalcV2[i]=gradV2*Time[i]+interceptV2;

        cPs[i]=((gradV1*Vcon)/(Icon*2*PI*meanR*(CalcV2[i]-MissingI)))-
((meanR*gradV2)/(2*(CalcV2[i]-MissingI)));
}
//print(Ps[0]);

for(i=0;i<Time.length;i++){

        RealTime[i]=Time[i]-RealZero;
}
//---- Fits Ps to find Ps at t=0

Fit.doFit("Exponential with Offset",RealTime,cPs);

Fit.plot;

a=Fit.p(0);

c=Fit.p(2);
```

```
cPsT0=a+c;
//print(PsT0);
//----Results Table
run("Clear Results");
for(i=0;i<Time.length;i++){
	setResult("Frames",i,Frame[i]);
	setResult("Time",i,Time[i]);
	setResult("Real Time",i,RealTime[i]);
	setResult("Vessel-V2)",i,V2[i]);
	setResult("Tissue-V1)",i,V1[i]);
	setResult("CalcPs(um/s)",i,cPs[i]);

}
setResult("Initial Calculated Ps (um/s)",0,cPsT0);
ComTitles[j]=path;
ComPs[j]=cPsT0;
saveAs("Results", "NewPath");
}}//Closes main for and open(path) if loops

//---- making the results table
run("Clear Results");
for (i = 0; i <ComTitles.length; i++) {
	setResult("Label",i,Name[i]);
	setResult("Initial Ps(um/s)",i,ComPs[i]);
```

```
}

setResult("meanR",0,meanR);

setResult("lcon",0,lcon);

setResult("RetDepth",0,RetDepth);

setResult("AreaCon",0,AreaCon);

setResult("framerate",0,framerate);



saveAs("Results", "FolderResults");
```

References

1. Kaya M, Ahishali B (2011) The role of magnesium in edema and blood brain barrier disruption. In: Vink R, Nechifor M (eds) Magnesium in the central nervous system. University of Adelaide Press, Adelaide
2. Cunha-Vaz J, Bernardes R, Lobo C (2011) Blood-retinal barrier. Eur J Ophthalmol 21(6):S3–S9
3. Bartanusz V (2011) The blood-spinal cord barrier: morphology and clinical implications. Ann Neurol 70(2):194–206
4. Wang Y (2007) Endothelial barrier function in preeclampsia. Front Biosci 12:2412–2424
5. Marcu R, Choi YJ, Xue J, Fortin CL et al (2018) Human organ-specific endothelial cell heterogeneity. iScience 29(4):20–35
6. Ricard N, Bailly S, Guignabert C, Simons M (2021) The quiescent endothelium: signalling pathways regulating organ-specific endothelial normalcy. Nat Rev Cardiol 24:1–16
7. Kalucka J, Bierhansl L, Conchinha NV, Missiaen R et al (2018) Quiescent endothelial cells upregulate fatty acid β-oxidation for vasculoprotection via redox homeostasis. Cell Metab 28(6):881–894
8. Dryden NH, Sperone A, Martin-Almedina S et al (2012) The transcription factor Erg controls endothelial cell quiescence by repressing activity of nuclear factor (NF)-κB p65. J Biol Chem 287(15):12331–12342
9. Al-Soudi A, Kaaij MH, Tas SW (2017) Endothelial cells: from innocent bystanders to active participants in immune responses. Autoimmun Rev 16(9):951–962
10. Yau JW, Teoh H, Verma S (2015) Endothelial cell control of thrombosis. BMC Cardiovasc Disord 15:130
11. Konukoglu D, Uzun H (2017) Endothelial dysfunction and hypertension. Adv Exp Med 956:511–540
12. Komarova YA, Kruse K, Mehta D, Malik AB (2017) Protein interactions at endothelial junctions and signaling mechanisms regulating endothelial permeability. Circ Res 120(1): 179–206
13. Kozler P, Pokorny J (2003) Altered blood-brain barrier permeability and its effect on the distribution of Evans blue and sodium fluorescein in the rat brain applied by intracarotid injection. Physiol Res 52(5):607–614
14. Ved N (2017) Vascular endothelial growth factor-A165b ameliorates outer-retinal barrier and vascular dysfunction in the diabetic retina. Clin Sci (Lond) 131(12):1225–1243
15. Doi K (2008) Pre-existing renal disease promotes sepsis-induced acute kidney injury and worsens outcome. Kidney Int 74(8): 1017–1025
16. Zhang S, Wang S, Yao S (2004) Evidence for development of capillary leak syndrome associated with cardiopulmonary bypass in pediatric patients with the homozygous C4A null phenotype. Anesthesiology 100(6): 1387–1393
17. Do J (2014) Ex vivo Evans blue assessment of the blood brain barrier in three breast cancer brain metastasis models. Breast Cancer Res Treat 144(1):93–101
18. Wolman M (1981) Evaluation of the dye-protein tracers in pathophysiology of the blood-brain barrier. Acta Neuropathol 54(1): 55–61
19. Uyama O (1998) Quantitative evaluation of vascular permeability in the gerbil brain after

transient ischemia using Evans blue fluorescence. J Cereb Blood Flow Metab 8(2): 282–284
20. Bates DO (2010) Vascular endothelial growth factors and vascular permeability. Cardiovasc Res 87(2):262–271
21. Chen B, Fu BM (2004) An electrodiffusion-filtration model for effects of endothelial surface glycocalyx on microvessel permeability to macromolecules. J Biomech Eng 126(5): 614–624
22. Miles A, Miles E (1952) Vascular reactions to histamine, histamine-liberator and leukotaxine in the skin of guinea pigs. J Physiol 118: 228–257
23. Jackiewicz TA (1998) Evans blue dye modifies the ultrastructure of normal and regenerating arterial endothelium in rats. Microsurgery 18(1):47–54
24. Xu J (2001) Blockade of endogenous neurotrophic factors prevents the androgenic rescue of rat spinal motoneurons. J Neurosci 21(12): 4366–4372
25. Ved N (2018) Diabetes-induced microvascular complications at the level of the spinal cord: a contributing factor in diabetic neuropathic pain. J Physiol 596(16):3675–3693
26. Xu Q, Qaum T, Adamis A (2001) Sensitive blood-retinal barrier breakdown quantitation using Evans blue. Invest Ophthalmol Vis Sci 42:789–794
27. Moos T, Mollgard K (1993) Cerebrovascular permeability to azo dyes and plasma proteins in rodents of different ages. Neuropathol Appl Neurobiol 19(2):120–127
28. Hueper WC (1944) Reactions of the blood and organs of dogs after intravenous injections of solutions of methyl celluloses of graded molecular weights. Am J Pathol 20(4):37–71
29. Schindelin J (2012) Fiji: an open-source platform for biological-image analysis. Nat Methods 9(7):676–682
30. Huxley VH, Curry FE, Adamson RH (1987) Quantitative fluorescence microscopy on single capillaries: alpha-lactalbumin transport. Am J Phys 252(1):H188–H197
31. Allen CL (2020) Non-invasive measurement of retinal permeability in a diabetic rat model. Microcirculation 27(6):e12623

Chapter 12

An Ex Vivo Tissue Culture Method for Discovering Cell Dynamics Involved in Stromal Vascular Fraction Vasculogenesis Using the Mouse Mesentery

Dima Majbour, Ariana D. Suarez-Martinez, Nicholas A. Hodges, Arinola O. Lampejo, Banks M. Lomel, Elijah W. Rice, Hulan Shang, Adam J. Katz, and Walter L. Murfee

Abstract

Stromal vascular fraction (SVF), isolated from adipose tissue, identifies as a rich cell source comprised of endothelial cells, endothelial progenitor cells, pericytes, smooth muscle cells, fibroblasts, and immune cells. SVF represents a promising therapeutic heterogonous cell source for growing new blood microvessels due to its rich niche of cells. However, the spatiotemporal dynamics of SVF within living tissues remain largely unknown. The objective of this chapter is to describe a protocol for culturing SVF on mouse mesentery tissues in order to aid in the discovery of SVF dynamics and associated vessel growth over time. SVF was isolated from the inguinal adipose from adult mice and seeded onto mesentery tissues. Tissues were then cultured for up to 5 days and labeled with endothelial cell and pericyte markers. Representative results demonstrate the observation of SVF-derived vasculogenesis characterized by de novo vessel formation and subsequent vessel connection.

Key words Microcirculation, Stromal Vascular Fraction, Angiogenesis, Vasculogenesis, Tissue Culture, Mesentery

1 Introduction

Stromal Vascular Fraction (SVF) is a fat-derived heterogeneous cell source that contains endothelial cells, pericytes, smooth muscle cells, stem cells, and other accessory immune and stromal cells [1, 2]. Over the past decade, SVF has emerged as a promising vascularization cell therapy due to its ability to form new vascular networks [3–5]. In vitro studies have established the ability of SVF cells and in particular SVF-derived endothelial cells to undergo self-reassembly and the formation of tube-like structures, a process that is highly dependent on the extracellular matrix used as a substrate

Andrew V. Benest (ed.), *Angiogenesis: Methods and Protocols*, Methods in Molecular Biology, vol. 2441,
https://doi.org/10.1007/978-1-0716-2059-5_12,

[6, 7]. Furthermore, SVF implantation in vivo has been shown to result in the formation of functional microcirculation following implantation [8–11]. As such, transplantation of SVF represents a promising new cell-based therapy for the formation of new functional vasculature in the host environment, yet current SVF therapies have not reached their full potential due to the lack of understanding the dynamics of SVF-derived vessel formation over time. Previous studies suggest SVF can dramatically affect the presence of new vessels both in vivo and in vitro [1, 4, 12], however, the full understanding of the key players and how this process is taking place remains unclear. Discovery of SVF dynamics will guide new paradigms for research and design of enhanced tissue engineering models for therapeutics. There is a need for biomimetic models that bridge the gap between in vitro and in vivo in order to recapitulate a real tissue environment and observe the sequence of events.

The objective of this chapter is to describe a protocol for culturing stromal vascular fraction (SVF) on mouse mesentery tissues. This method enables pre-clinical observation of SVF de novo vessel formation in an intact tissue environment. Representative observations showcase the ability for SVF to form multicellular microvascular networks and the value of using the method for characterizing cell dynamics.

2 Materials

2.1 Animals

- C57Bl/6 adult (10–16 weeks) female mice.

2.2 Solutions and Reagents

- Type I Collagenase (Worthington Biochemical Corp).
- ACK Lysing Buffer, 1× (Thermo Fisher).
- Dulbecco's Phosphate-buffered saline with calcium and magnesium (DPBS) +/+.
- DPBS +/+ & 5% Fetal Bovine Serum (FBS).
- Optional: VybrantTM CM-DiI cell-labeling solution (V-22888 from Molecular Probes).
- Medium: Minimum Essential Medium (MEM), supplemented with 1% penicillin-streptomycin (Pen-Strep), and 10% fetal bovine serum (FBS).
- 0.9% Sodium chloride irrigation (saline).
- Hair removal cream.
- 70% ethanol.
- 70% isopropanol.
- Povidone-Iodine.

- 100% Methanol.
- Bovine Serum Albumin (BSA).
- Saponin.
- Glycerol.

2.3 Surgical Tools

- Autoclaved scissors.
- Autoclaved curved forceps.
- Autoclaved metal spatula.
- 100 mm sterile petri dish.
- 50 mL sterile conical tubes.
- 5 mL sterile syringe.
- 10 mL sterile syringe.
- 0.2 μm sterile syringe filters (Millipore).
- 70 μm sterile cytostrainer (ASI).
- 250 μm sterile cytostrainer.
- 1 mL sterile centrifuge tube.
- 6-well tissue culture sterile plate.

2.4 Other Materials

- Large benchtop centrifuge.
- Cell counter.
- Scale.
- Autoclave.
- Biosafety cabinet hood.
- Surgical stage made out of Styrofoam.
- Electric heating pad.
- Pre-cut blue surgical drape with an elliptical hole (1 by 3/4 in.) at the center.
- Absorbent bench underpad with waterproof barrier.
- Gauze sponge.
- Cotton-tip applicators.
- Cell-Crown inserts with 5 μm polycarbonate filter membrane.
- Microscope glass slides.

3 Methods

3.1 Preparation of Surgical Tools and Space

1. Place tools that need to be sterilized in pouches and autoclave (*see* **Note 1**).

2. In a biosafety cabinet prepare 30 mL of sterile saline and two petri dishes, one containing 10 mL of DPBS and the other containing 10 mL of MEM supplemented with 1% Pen-Strep.
3. Incubate the saline in a water bath set to 37 °C and place the petri dishes in an incubator.
4. Wipe down the designated surgical area with 70% ethanol.
5. Lay down and tape the corners of the absorbent pad.
6. Place the heating pad to the left of the surgical area and place a sterile blue drape on top of the heating pad.
7. Place a bowl with a Kimwipe at the top left of the surgical area to dispose of any waste.
8. Wrap and secure with tape a small absorbent pad onto a surgical stage made out of Styrofoam and place it on top of the surgical area (*see* **Note 2**).
9. Lay down a blue drape at the right side of the surgical area and place the sterile instruments that will be used during harvesting along with some sterile gauze on the drape.
10. Bring the conical tube containing sterile saline from the water bath to the surgical area and place it in a rack at the top right side of the surgical area.

3.2 Preparation of the Mouse for Harvesting Fat and Mesentery Tissues

1. Euthanize the mice via CO_2 asphyxiation followed by cervical dislocation (*see* **Note 3**).
2. Weigh the mouse.
3. On top of a paper towel, lay the mouse on its back.
4. Pour the hair removal cream on its abdomen and with a long cotton-tip applicator spread it all over its abdomen.
5. After 4.5 min, remove the hair removal cream and with a 3 × 3 in. cotton gauze, remove the extra cream.
6. Drip 70% isopropyl on the hairless area and wipe it with a gauze (*see* **Note 4**).
7. Sterilize the abdomen with iodine and repeat **steps 6** and **7** two more times.
8. Move the mouse on top of the surgical stage and tape down its limbs to prevent any movement while harvesting.

3.3 Harvesting Fat

1. Using sterile microforceps and a sterile scalpel blade (#10) make a superficial incision along the midline of the abdomen while avoiding damage to the underlying muscle.
2. Complete an upside-down Y-incision by cutting from the bottom of the midline incision in a diagonal to both femoral regions of the mouse.

3. Working from the corners of the Y-incision and firmly holding the edge of the skin with a sterile Castroviejo needle holder, carefully scrape the inguinal fat pad from the skin using the scalpel blade.
4. Repeat for the other side.
5. Harvest the additional fat from the inguinal region to the hip.
6. Gently remove the fat from the muscle by gently lifting the fat with sterile microforceps while cutting underneath with a pair of microscissors without slicing larger blood vessels.
7. Place the fat that was removed in a sterile 50 mL conical tube containing DPBS that has been warmed to 37 °C.

3.4 Stromal Vascular Fraction Isolation Fat Preparation

1. Using curved forceps, transfer the fat from the 50 mL conical tube to the lid of the 100 mm sterile petri dish.
2. Using sterile scissors, cut the fat finely into very small pieces.
3. Transfer cut fat into another sterile 50 mL conical tube with sterile spatula.
4. Weigh the fat to determine the appropriate amount of collagenase.

3.4.1 Digestion

1. Make 0.15% Collagenase Type I solution using warmed DPBS +/+.
   (a) Make 2 mL per 1 g of fat (*see* **Note 5**).
   (b) If <1 g, always make 5 mL so it can be safely shaken.
   (c) Make enough for first and second digestion.

   Ex: 12 mL of DPBS +/+ and 0.018 g (18 mg) of collagenase.
2. Vortex collagenase and DPBS +/+.
3. Using a sterile 10 mL syringe, draw up 5 mL of collagenase solution and attach 0.22 μm syringe filter for sterilization.
4. Push the collagenase solution through the filter into the tube containing the chopped fat.
5. Place the fat with collagenase in a temperature regulated shaker, 150 RPM, that is set to 37 °C for 45 min.

3.4.2 250 μm Filter

1. Centrifuge the tube at 600 × *g* for 10 min.
2. Using the sterile spatula, transfer the top layer of fat into a new 50 mL conical tube and begin the second digestion by following digestion **steps 3–5**.
3. Aspirate the leftover collagenase.
4. Resuspend the cell pellet with 5 mL of warmed DPBS +/+ & 5% FBS.

5. Transfer the cell suspension into a new, sterile 50 mL conical tube through a 250 μm cell strainer with funnel.
6. Add 1.5 mL of warmed DPBS +/+ & 5% FBS through the 250 μm cell strainer.

3.4.3 Lysing Buffer

1. Centrifuge the tube at 600 × *g* for 5 min.
2. Aspirate the leftover medium, without disturbing the cell pellet.
3. Using a sterile 5 mL syringe, draw up 5 mL of ACK lysing buffer and attach a 0.22 μm syringe filter.
4. Push the sterilized lysing buffer into the tube with the cell pellet.
5. Gently swirl the lysing buffer to detach cells from the bottom.
6. Wrap tube with foil to avoid light and incubate at room temperature for 3 min, very carefully shaking by hand. Too long of an exposure or shaking too roughly can lose 50–80% of cells.

3.4.4 70 μm Filter

1. Transfer lysing buffer with cells into a new, sterile 50 mL conical tube through a 70 μm cell strainer.
2. Add 5 mL of warmed DPBS +/+ & 5% FBS through the cell strainer to neutralize the lysing buffer.
3. Add 5 mL of warmed DPBS +/+ & 5% FBS to emptied conical tube, swirl and transfer to the new tube through cell strainer.
4. Centrifuge tube at 600 × *g* for 5 min.
5. Aspirate leftover medium, without disturbing cell pellet.

3.4.5 Cell Count

1. Gently resuspend the cell pellet with 500 μL of DPBS +/+ & 5% FBS and then add another 500 μL.
2. Add 10 μL of cell suspension in a 1 mL microcentrifuge tube.
3. Take out the microcentrifuge tube from hood and add 10 μL of trypan blue and mix thoroughly.
4. Add 10 μL to each side of counting slide for the cell counter.
5. Use cell counter to obtain yield and viability of cells (*see* **Note 6**).

3.5 Harvesting Mesentery Tissue

1. Pinch the lower abdomen with a slightly curved microforceps and pull up to cut the muscle with scissors to expose the organs.
2. Place the pre-cut blue drape on top of the exposed organs.
3. Using two sterile microforceps, gently pinch the small intestine to spread it so the mesentery windows are taut on top of the blue drape.
4. Drip some warmed saline on top of the mesenteries.

5. With curved microforceps, pinch the fat at the edge of the mesenteries and slightly pull to begin cutting the mesenteric windows with microscissors (*see* **Notes** 7 and **8**).
6. Cut 2–4 connected windows at one time.
7. Rinse the excised tissues with warmed DPBS +/+ in the 60 mm petri dish and then transfer the tissue to warmed MEM + 1% PS in another 60 mm petri dish.
8. Cut all the mesenteric windows.
9. Once done harvesting all the mesentery tissues, place the petri dish with tissues in the incubator set at 37 °C, 5% CO_2, and 95% humidity.

3.6 SVF Transplantation

1. In a biosafety cabinet, take one cell-crown out and tighten the membrane by slightly pressing on the edges.
2. With a sterile microforceps, take out one of the mesenteric window groups and gently blot it on the sterile blue drape before placing it on the membrane in order to remove excess fluid.
3. Spread out the tissue over the membrane by using the straight microforceps to nudge the windows apart while the flat part of the angled forceps holds the opposite edge of the peripheral fat pad. Spread out the windows so the tissue is fully exposed and flattened (Fig. 1).
4. Transfer 50 μL of the SVF solution (*see* **Note 9**) consisting of MEM + 10% FBS with 1×10^7 cells/mL (*see* **Note 10**) to each group of mesenteric tissues.
5. Incubate for 20 min to allow for the SVF to attach to the tissues.
6. Add 1 mL of medium to the well and then carefully place the insert in the well by inverting it and gently sandwiching the tissue between the insert and the bottom of the well at an angle to avoid capturing air between the tissue and the well.
7. Add an additional 3 mL medium to the sides and center of the well.
8. Repeat **steps 1–5** for all tissues.
9. Transfer the plate to an incubator set at stand culture conditions (*see* **Note 11**).

3.7 Immunohistochemistry

1. Spread tissues onto microscope slides, carefully cutting away the fat border of each tissue (*see* **Note 12**).
2. Fix the tissues in 100% methanol for 30 min at −20 °C.
3. Perform 3, 10-min washes with PBS 1% Saponin.
4. With a Kimwipe, carefully tap dry each slide.

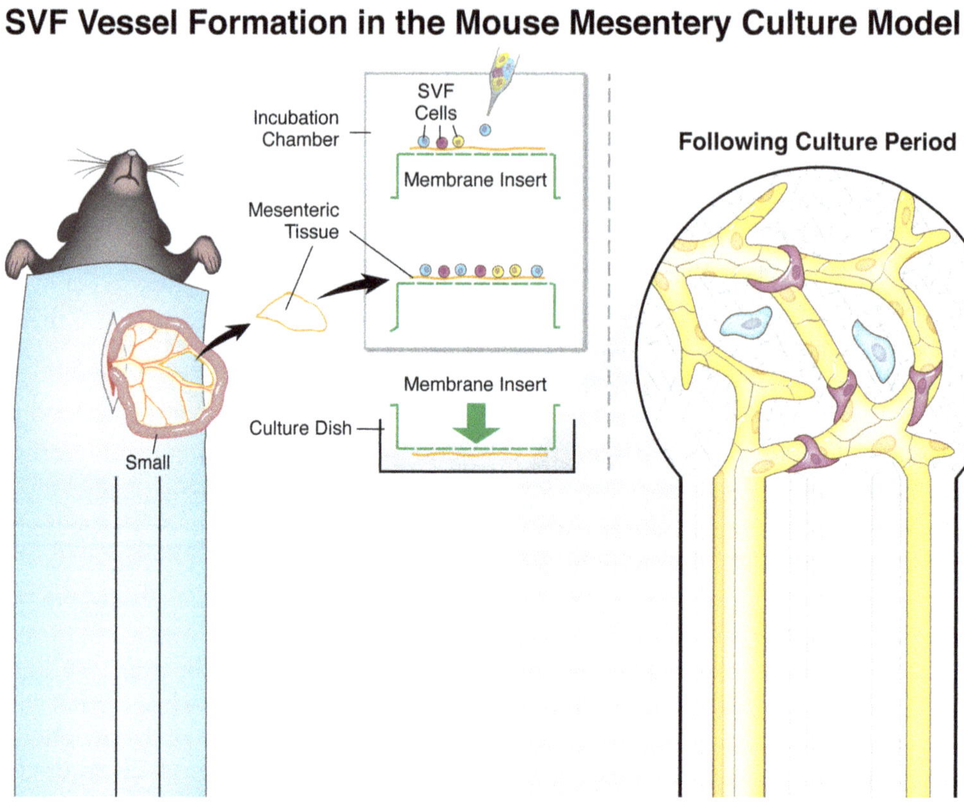

Fig. 1 Overview of SVF culture on Mouse Mesentery. Mesentery tissues are excised, rinsed, and spread across a membrane insert. SVF cell solution is deposited onto the mesenteric tissue and then the tissue is inverted and sandwiched between the insert and the well bottom. SVF in culture has the ability to form de novo vessels and multicellular microvascular networks supported by other mural cells such as pericytes. Printed with permission from Anita Impagliazzo

5. Using a Kimwipe or sterile cotton-tip applicator, carefully blot dry the slide area around the tissue.
6. Using a hydrophobic marker, circumscribe each tissue.
7. Pipet primary antibody solution onto each tissue (approximately 100 μL) and incubate in the dark at room temperature for 1 h (*see* **Note 13**).
8. Turn the slide on its side and tap off the antibody solution.
9. Using a dropper to completely cover the tissues perform 3, 10-min washes with PBS 1% Saponin.
10. Add 100 μL of secondary antibody solution onto each tissue and incubate in the dark at room temperature for 1 h (*see* **Note 14**).
11. Turn the slide on its side and tap off the antibody solution.
12. Using a dropper to completely cover the tissues, perform 3, 10-min washes with PBS 1% Saponin.

13. Remove the hydrophobic pen markings and use a disposable pipette to deposit a few drops of PBS:Glycerol solution with a 1:1 ratio onto each tissue.
14. Use a glass slide cover to then seal the tissues, minimizing air bubbles.
15. Seal the slides with nail polish or alternative sealant.
16. After sufficient time to allow for the nail polish to dry, the tissues can be imaged or stored long-term in a freezer (−20 °C). For short term storage, slides with tissues can be stored in a refrigerator (4 °C).

3.8 Imaging

1. As the mesenteric tissues are approximately 20–40 μm thin [13], tissues can be imaged before immunostaining with a standard epifluorescent microscope similar to cells in typical cell culture.
2. Images can be taken at different intervals, commonly ranging from every hour to everyday.
3. As the imaging interval increases the temperature and CO_2 will need to be maintained using an incubation chamber.

3.9 Representative Observations

SVF cells seeded onto avascular mouse mesentery tissues were able to form multicellular vessels by day 3 (Fig. 2). Vessel structures contained PECAM-positive endothelial cells and NG2 positive pericytes characteristically wrapped along the vessel structures. Immediately after transplantation and the start of the culture on day 0, no vessel structures were observed. Representative observations at day 3 and day 5 support an increase in vessel connectivity (Fig. 3). In order to identify novel cell structures during the vessel and network formation processes, specific vessel characteristics can be compared across different time points. As examples, Fig. 3 shows the number of PECAM-positive "blebs" per vascular area and the number of PECAM-positive vascular networks per vascular length decrease over time. "Blebs" are defined as irregular shaped PECAM-positive structures and seem to be associated with the early vessel segment formation. The decrease in networks, which are defined as vessel structures with at least one node or branch point, indicates an increase vessel segment connectivity over time. The value of the described method can be further exemplified by its use with tissues or SVF harvested from transgenic mouse strains. Figure 4 demonstrates the feasibility of tracking GFP-SVF populations in real time. In the example image sequence, individual GFP-cell clusters can be observed to connect.

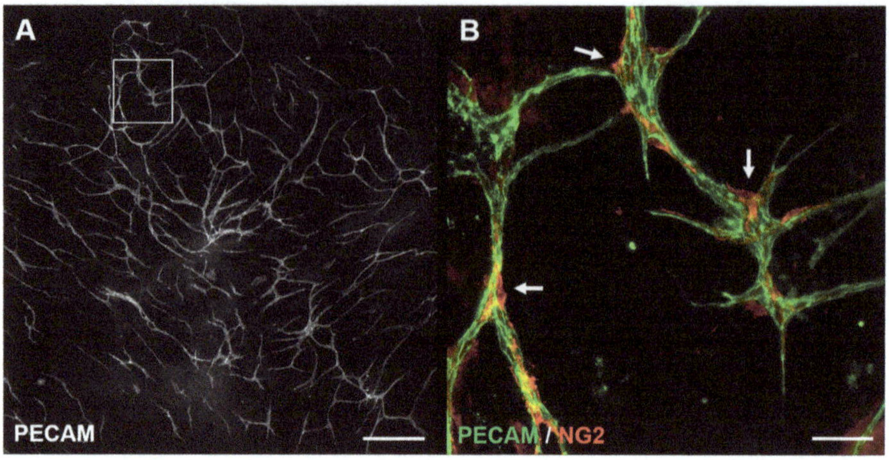

Fig. 2 Formation and characterization of SVF-derived networks. (**a**) Representative image of an SVF microvascular network containing PECAM-positive endothelial cells. Image taken at 4× magnification on Day 5 of culture. Scale bar = 500 μm. (**b**) The vessels outlined by the white box in panel (**a**) have been magnified to 30×. Vessel structure shows NG2-positive pericytes wrapping around the PECAM-positive vessels, indicated by white arrows. Scale bar = 70 μm

4 Notes

1. It is recommended that tools are autoclaved in multiple pouches according to method section.
2. The flat side of an empty Styrofoam tube holder can sufficiently act as a stage.
3. All animal procedures should be approved by an institutional animal care and use committee (IACUC) or its equivalent. Animal experiments described in the current protocol were approved by the University of Florida's Institutional Animal and Care Use Committee. The mice were euthanized in a CO_2 chamber by asphyxiation (10–30% of chamber volume/minute) followed by cervical dislocation.
4. Wipe from the center to the periphery of the shaved area in order to reduce contamination.
5. At least 2 grams of fat are needed, requiring 2–4 mice per gram of fat.
6. Viability of cell count must be ≥70%.
7. Leave a fat border around the tissues to provide cushion for the tissues when they are sandwiched between the membrane insert and the bottom of the well plate.
8. Starting where the forceps are pinching the tissue, cut evenly around the tissue to prevent one side from becoming too taut and tearing.

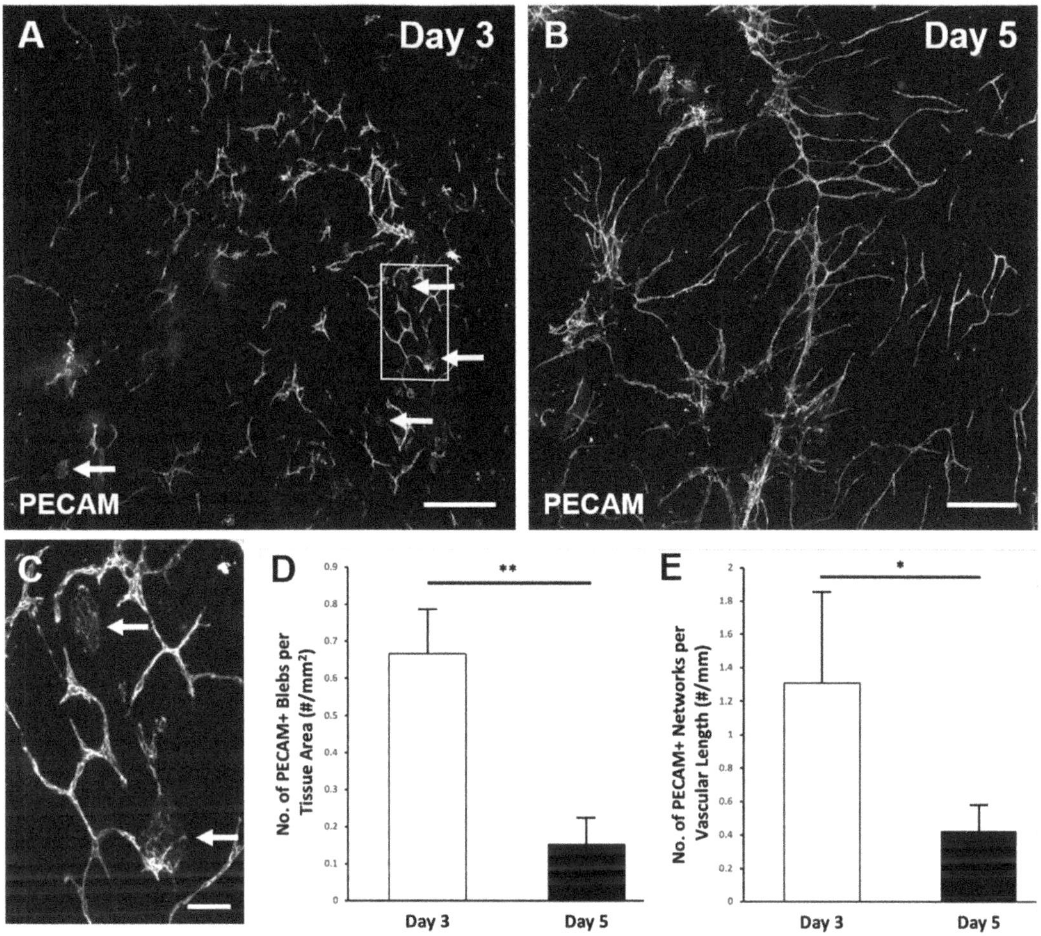

Fig. 3 Comparison of SVF networks and structures. (**a**) Vessels form SVF network imaged at 4× magnification on Day 3 of culture. Networks are small and fragmented. Irregularly shaped, PECAM-positive clumps of endothelial cells termed "blebs," indicated by white arrows, can also be seen at this stage in network. Scale bar = 500 μm. (**b**) Representative images of SVF microvascular network imaged at 4× magnification on Day 5 of culture. The presence of blebs is notably decreased, and networks appear to exhibit increased connectivity. Scale bar = 500 μm. (**c**) The blebs outlined by the white box in panel (**a**) have been magnified. Scale bar = 100 μm. (**d**) Changes in the number of blebs from Day 3 to Day 5 in SVF tissues was quantified per tissue area. (**e**) Changes in the number of networks from Day 3 to Day 5 in SVF tissues was quantified per tissue area. For the representative quantitative (**d**, **e**), 5 tissues harvested from 2 mice were imaged and analyzed per group. Statistical comparisons were made using a two-tailed unpaired Student's t-test assuming equal variance. Data represent averages of three independent analyses and are shown as averages +/− S.D. ** represents a significant difference between the two time points where $p < 0.0001$. * represents a significant difference between the two time points where $p < 0.01$

9. Depending on how large the group of mesenteric tissues is and the thickness of the surrounding fat, the total volume of SVF solution transferred to each group will vary. The range that we typically use was 20–50 μL. SVF solution must be contained

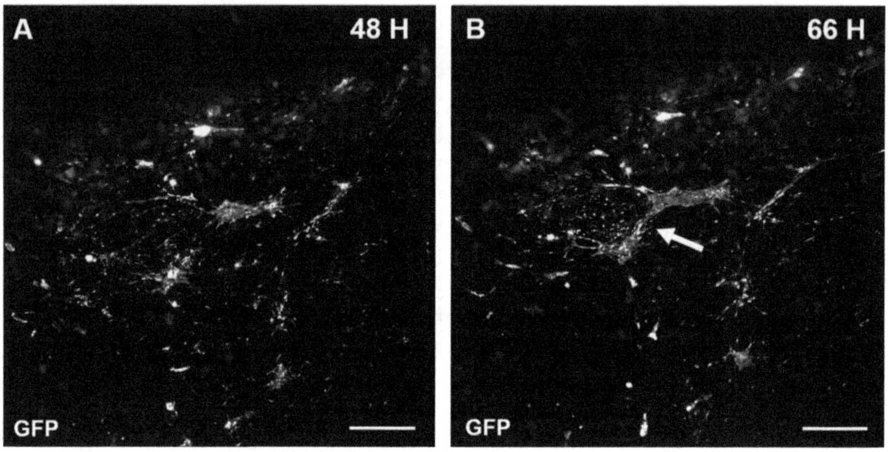

Fig. 4 Time-lapse Imaging of SVF Connectivity. Time-lapse imaging allows for the real time tracking of unique cells populations. SVF was harvested from GFP-transgenic mice and transplanted onto wild type mouse mesentery tissues. (**a**) GFP-positive, SVF-derived cell clusters imaged at 48 h. Scale bar = 500 μm. (**b**) Image of the same cell cluster imaged at 66 h. The individual cell clusters have formed a connection, indicated by the white arrow, with neighboring cell clusters and forming what appears to be the beginnings of a branched vessel segment. Scale bar = 500 μm

within the mesenteric windows as excess liquid can spill over the surrounding fat and lose SVF.

10. The exact cell concentration on each group of tissue varied due to the range of volume transferred to each group of mesenteries. Different cell concentration might be necessary depending on the type of cell being cultured as there is a critical concentration that must be reached before a robust vasculogenic response can take effect. Obtaining optimal cell concentration might need to be discovered through experimental trials.
11. Standard conditions: 5% CO_2 at 37 °C, and 95% humidity. Media was changed every 2 days. For the representative results shown in this chapter, media was changed daily. In our experience, media can be changed daily or every other day.
12. Whenever spreading the group of mesentery tissues on a glass slide, carefully cut out most of the outer fat with microscissors without disturbing the thin connective tissue where the cells are. After fully spreading the tissues and letting the thin connective tissue and rest of the fat dry, carefully cut out the rest of the fat with a scalpel blade (#10). Only cut out the fat if the thin connective tissue will not be affected (i.e., ripped).
13. For the representative results show in this chapter, the primary antibody solution used consisted of 5% normal goat serum (NGS), 1:200 biotinylated rat anti-mouse PECAM (BD Pharmigen; San Diego, CA) and 1:100 rabbit anti-

mouse NG2 (MilliporeSigma; Burlington, MA) diluted in a solution of PBS containing 1% Saponin and 2% Bovine Serum Albumin (BSA).

14. For the representative results show in this chapter, the secondary antibody solution used consisted of 5% NGS, 1:500 streptavidin secondary (Strep-CY2) (JacksonImmuno Research; West Grove, PA), 1:100 goat anti-rabbit secondary (GAR-CY3) (JacksonImmuno Research; West Grove, PA) diluted in a solution of PBS containing 1% Saponin and 2% BSA.

Acknowledgments

This work was supported by NIH grant R01AG049821 and funding from the University of Florida's Division of Research Program Development.

References

1. Jones VM, Suarez-Martinez AD, Hodges NA, Murfee WL, Llull R, Katz AJ (2021) A clinical perspective on adipose-derived cell therapy for enhancing microvascular health and function: implications and applications for reconstructive surgery. Microcirculation 28(3):e12672. https://doi.org/10.1111/micc.12672
2. Brown JC, Shang H, Li Y, Yang N, Patel N, Katz AJ (2017) Isolation of adipose-derived stromal vascular fraction cells using a novel point-of-care device: cell characterization and review of the literature. Tissue Eng Part C Methods 23(3):125–135. https://doi.org/10.1089/ten.TEC.2016.0377
3. Leblanc AJ, Nguyen QT, Touroo JS, Aird AL, Chang RC, Ng CK et al (2013) Adipose-derived cell construct stabilizes heart function and increases microvascular perfusion in an established infarct. Stem Cells Transl Med 2(11):896–905. https://doi.org/10.5966/sctm.2013-0046
4. Nunes SS, Maijub JG, Krishnan L, Ramakrishnan VM, Clayton LR, Williams SK et al (2013) Generation of a functional liver tissue mimic using adipose stromal vascular fraction cell-derived vasculatures. Sci Rep 3:2141. https://doi.org/10.1038/srep02141
5. Maijub JG, Boyd NL, Dale JR, Hoying JB, Morris ME, Williams SK (2015) Concentration-dependent vascularization of adipose stromal vascular fraction cells. Cell Transplant 24(10):2029–2039. https://doi.org/10.3727/096368914X685401
6. Muller S, Ader I, Creff J, Lemenager H, Achard P, Casteilla L et al (2019) Human adipose stromal-vascular fraction self-organizes to form vascularized adipose tissue in 3D cultures. Sci Rep 9(1):7250. https://doi.org/10.1038/s41598-019-43624-6
7. Madri JA, Williams SK (1983) Capillary endothelial cell cultures: phenotypic modulation by matrix components. J Cell Biol 97(1):153–165. https://doi.org/10.1083/jcb.97.1.153
8. Koh YJ, Koh BI, Kim H, Joo HJ, Jin HK, Jeon J et al (2011) Stromal vascular fraction from adipose tissue forms profound vascular network through the dynamic reassembly of blood endothelial cells. Arterioscler Thromb Vasc Biol 31(5):1141–1150. https://doi.org/10.1161/ATVBAHA.110.218206
9. Strassburg S, Nienhueser H, Bjorn Stark G, Finkenzeller G, Torio-Padron N (2016) Co-culture of adipose-derived stem cells and endothelial cells in fibrin induces angiogenesis and vasculogenesis in a chorioallantoic membrane model. J Tissue Eng Regen Med 10(6):496–506. https://doi.org/10.1002/term.1769
10. Karagergou E, Dionyssopoulos A, Karayannopoulou M, Psalla D, Theodoridis A, Demiri E et al (2018) Adipose-derived stromal vascular fraction aids epithelialisation and angiogenesis in an animal model. J Wound Care 27(10):637–644.

https://doi.org/10.12968/jowc.2018.27.10.637

11. Leblanc AJ, Touroo JS, Hoying JB, Williams SK (2012) Adipose stromal vascular fraction cell construct sustains coronary microvascular function after acute myocardial infarction. Am J Physiol Heart Circ Physiol 302(4): H973–H982. https://doi.org/10.1152/ajpheart.00735.2011
12. Zakhari JS, Zabonick J, Gettler B, Williams SK (2018) Vasculogenic and angiogenic potential of adipose stromal vascular fraction cell populations in vitro. In Vitro Cell Dev Biol Anim 54(1):32–40. https://doi.org/10.1007/s11626-017-0213-7
13. Norrby K (2006) In vivo models of angiogenesis. J Cell Mol Med 10(3):588–612. https://doi.org/10.1111/j.1582-4934.2006.tb00423.x

Chapter 13

Quantitative Three-Dimensional Analysis of the Lymphatic Vasculature in the Postnatal Mouse Heart

Konstantinos Klaourakis, Paul R. Riley, and Joaquim Miguel Vieira

Abstract

The development and maturation of the lymphatic vasculature are essential for organ function with disruption leading to severe phenotypes. For example, malfunction of cardiac lymphatics results in myocardial oedema, persistent inflammation and reduced cardiac output. Thus, it is important to study the process of cardiac lymphatic formation and growth from the early stages of fetal development to adulthood. In the murine heart the lymphatics continue to develop and expand postnatally with extensive growth and patterning occurring up to at least 2 weeks after birth. Here, we describe a protocol for whole-mount, multi-view imaging and quantification of lymphatic vessel parameters, including vessel junction number (i.e., branching density), vessel length, and number of vessel end points in the murine postnatal heart. This protocol is based on the use of reliable antibodies against key markers of lymphatic endothelial cells (LECs), specifically the glycoprotein lymphatic vessel endothelial hyaluronan receptor 1 (LYVE1), the vascular endothelial growth factor receptor 3 (VEGFR3; also known as Fms-related receptor tyrosine kinase 4, FLT4), the mucin-type protein podoplanin (PDPN), and the co-receptor neuropilin 2 (NRP2). For imaging and quantitative analysis of the sub-epicardial network in neonatal hearts, VEGFR3 was selected given its exclusive expression in the lymphatic endothelium. In addition to LECs, LYVE1 expression was detected in tissue-resident macrophages, PDPN in the epicardium, and NRP2 in the autonomic nervous system of the heart. Overall, we characterized the expression patterns of commonly used lymphatic markers in the context of the neonatal heart and provide an image analysis pipeline that can be adapted to study other organs and systems (e.g., blood vasculature and nerve system).

Key words Heart, Lymphangiogenesis, LYVE1, PDPN, NRP2, VEGFR3, Light-sheet microscopy, AngioTool

1 Introduction

The lymphatic vasculature is responsible for maintaining interstitial fluid balance and regulating the immune response by transporting fluids and immune cells from tissues to regional lymph nodes and back to the blood circulation. In the heart, structural and/or functional abnormalities of lymphatics lead to increased myocardial fluids, myocardial oedema and reduced cardiac output [1, 2]. Over

Andrew V. Benest (ed.), *Angiogenesis: Methods and Protocols*, Methods in Molecular Biology, vol. 2441,
https://doi.org/10.1007/978-1-0716-2059-5_13,

the past 5 years, there has been a growing interest in the study of cardiac lymphatics during development and in disease [3].

During mouse embryogenesis, the first cardiac LECs emerge from the extra-cardiac region near the outflow tract on the ventral side of the heart and from the sinus venosus on the dorsal side of the heart at embryonic day (E) 12.5 [4]. Lymphatic vessels then expand along the base to apex axis, branching laterally to fully cover both the dorsal and the ventral aspects of the growing heart [4]. Of significance, tissue-resident macrophages act as important regulators of the latter process via direct cell-cell interactions in a mechanism potentially dependent on macrophage hyaluronan [5]. Moreover, while the initial lymphatics in the fetal heart are restricted to the sub-epicardial compartment, in the postnatal heart, lymphatics grow deeper into the underlying myocardium, without reaching the endocardium [6, 7]. Fate-mapping studies using genetic drivers and whole-heart imaging have identified multiple sources for cardiac LECs, such as the *Pax3*-positive paraxial mesoderm-derived endothelium of the cardinal vein (*Tie2*-positive) [8], the *Vav1+*, *Pdgfrb+*, and *Csf1r+* hemogenic endothelium of the yolk sac [4], and more recently, the Isl1-positive second heart field, contributing lymphatic endothelium specifically to the ventral aspect of the heart [9, 10]. In addition to cell and molecular mechanisms regulating their development, the function of cardiac lymphatics in heart disease has also become the focus of active research in recent years. In particular, cardiac lymphatics have been shown to respond to myocardial infarction (MI) by activating a lymphangiogenic program, resulting in expansion of the network into the area of risk, reduction of interstitial oedema and clearance of immune cells [4, 11, 12]. Importantly, augmentation of this initial endogenous response by administration of lymphangiogenic agents (e.g., recombinant VEGF-C) translated into a significant improvement of the cardiac function following MI, as a result of enhanced trafficking of macrophages to regional lymph nodes, i.e. increased immune cell clearance [13]. Thus, these findings identified the cardiac lymphatic system as an important therapeutic axis to modulate inflammation post-injury [4, 11–17].

Through our understanding of the developmental program of LECs several genes have been proposed to serve as specific markers for lymphatics, such as *Lyve1*, *Flt4*, *Prox1*, *Pdpn*, and *Nrp2* [3]. However, given the emerging recognition of heterogeneity across the lymphatic vascular system, organ-specific differences in lymphatic vessel formation and function, and the existence of distinct origins for subpopulations of LECs, it is important to examine the specificity of these markers in the context of the organ of interest. Here, we provide a protocol for whole-heart staining and 3D-imaging of neonatal hearts, we discuss the limitation and benefits of each protein as a cardiac lymphatic marker, and describe a method to quantitatively assess the cardiac lymphatic vasculature complexity (e.g., length, branching density) that could be used in future studies characterizing the role of candidate genes during lymphatic development or the response of the vasculature to injury/disease and/or treatments of choice.

2 Materials

2.1 General

1. Phosphate-buffered saline (PBS; e.g., ThermoFisher).
2. 2-mL round-bottom tubes (e.g., BRAND).
3. 4% (w/v) Paraformaldehyde (PFA), pH 7.4 (e.g., Fisher Scientific).

2.2 Tissue Dissection

1. Dumont tweezers.
2. Dissecting scissors.
3. Petri dish.

2.3 Immunostaining

1. PBS containing 1% (v/v) Triton X-100 (e.g., Sigma-Aldrich).
2. Blocking solution: PBS containing 0.2% (v/v) Triton X-100, 2% (w/v) bovine serum albumin (BSA, e.g., Sigma-Aldrich) and 10% (v/v) serum (e.g., Sigma-Aldrich). Note that the selection of the serum solution is dependent on the species in which the secondary antibodies are raised.
3. Primary antibodies (Abs) against LECs (*see* **Note 1**):
   (a) Rabbit anti-mouse LYVE1 polyclonal Ab, AngioBio, dilution 1:400.
   (b) Goat anti-mouse VEGFR3 polyclonal Ab, R&D systems, dilution 1:50.
   (c) Hamster anti-mouse PDPN monoclonal Ab, Fitzgerald, dilution 1:200.
   (d) Goat anti-mouse NRP2 polyclonal Ab, R&D systems, dilution 1:200.
4. Secondary Abs:
   (a) Alexa Fluor 488 donkey α-rabbit IgG (ThermoFisher, dilution 1:400).
   (b) Alexa Fluor 488 goat α-hamster IgG (ThermoFisher, dilution 1:400).
   (c) Alexa Fluor 647 donkey α-goat IgG (ThermoFisher, dilution 1:400).

2.4 Imaging

1. Microwave oven.
2. Glass flask (*see* **Note 2**).

3. PBS containing 1.5% (w/v) low melting point agarose powder (e.g., Sigma-Aldrich).
4. Scissors.
5. 1 mL syringe (e.g., Terumo Disposable Syringes).
6. Light-sheet microscope (e.g., Zeiss Z.1).
7. Zen (black edition) software.

2.5 Computational Analysis

1. Image processing software (e.g., ImageJ [18]/FIJI [19]).
2. AngioTool software [20].
3. GraphPad Prism software.

3 Methods

3.1 Tissue Dissection

Sacrifice the neonatal mouse according to the relevant local ethical guidelines for animal care and experimentation (*see* **Note 3**). Harvest the heart, using Dumont tweezers and dissecting scissors, and transfer it into a 2-mL round-bottom tube.

3.2 Immunostaining

1. Fix the hearts in 4% PFA for 6 h or overnight at 4 °C, with constant rolling.
2. Wash off the PFA with PBS (twice at room temperature, RT).
3. Permeabilize the hearts in PBS containing 1% (v/v) Triton X-100 for 1 h at RT, with constant rolling.
4. Remove the permeabilization agent and add the blocking solution. Incubate samples overnight at 4 °C, with constant rolling.
5. Remove the blocking solution and incubate the hearts in primary antibodies diluted in blocking solution for 24–48 h at 4 °C, with constant rolling (*see* **Note 4**). Wash the hearts in PBS five to ten times for about 15 min each.
6. Incubate the hearts in secondary antibodies diluted in PBS overnight at 4 °C, with constant rolling (*see* **Note 5**). Note that fluorophore-conjugated antibodies are photosensitive, thus samples need to be protected from light.
7. Wash samples in PBS five to ten times for 15 min each, with constant rolling.
8. Post-fix the hearts in 4% PFA for 30–60 min at room temperature to cross-link antibody-antigen complexes.
9. It is recommended to proceed with imaging as soon as possible, however stained samples can be stored for up to 1–2 weeks in PBS at 4 °C.

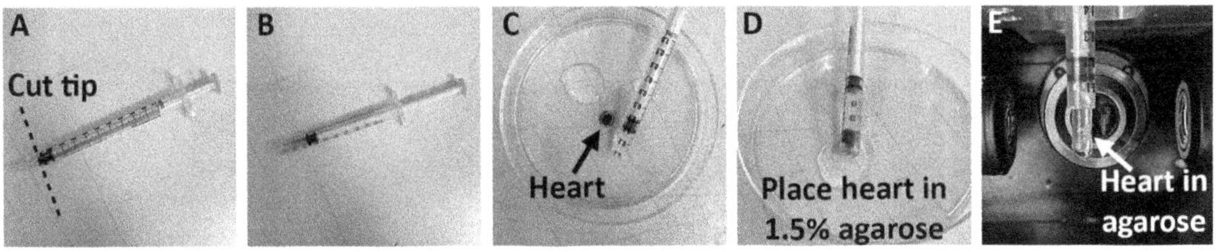

Fig. 1 Preparing the heart for mounting and imaging with a Zeiss Z.1 light-sheet microscope. (**a**) The heart is mounted in a 1-mL syringe. (**b**) Use a pair of scissors to cut off the tip of the syringe. (**c**) Prepare 1.5% (w/v) low melting point agarose and let it cool down until it is below 60 °C, but still liquid. (**d**) Place a small amount of the cooled down liquid agarose in the syringe and aspirate the heart. After a few minutes, the agarose should have fully polymerized. (**e**) Place the syringe containing the mounted the heart in the Z.1 light-sheet sample chamber and push out the agarose-containing heart. The heart is now ready to be imaged

3.3 Imaging

1. Add PBS to the glass flask (e.g., 50 mL).
2. Add the correct amount of low melting point agarose to prepare a 1.5% (w/v) solution into the flask while rapidly stirring to avoid clumping (e.g., 750 mg in 50 mL of PBS).
3. Heat the flask in a microwave oven to melt the agarose until it starts boiling. Ensure that all agarose particles dissolve completely.
4. Remove the flask from the microwave oven, and gently swirl it to mix the agarose solution (*see* **Note 6**).
5. Cut the tip of a 1-mL syringe to create a cylinder (Fig. 1a, b).
6. Wait until the agarose solution reaches below 60 °C, then take a small volume using the syringe and place one heart inside the agarose/syringe (Fig. 1c, d).
7. Wait 2 min until the agarose inside the syringe fully polymerizes. The heart can now be pushed out and imaged using the Zeiss Z.1 light-sheet (*see* **Note 7**, Fig. 1e).
8. On Zen software, apply Z-stack (optimal interval), tiling (two vertical), and multi-view (every 90°) functions to acquire images of the whole specimen. Imaging each heart with a light-sheet should take approximately 5 min.
9. The selection of the most suitable marker for the analysis of cardiac lymphatics is based on its specificity at the developmental stage of choice, as well as the aim of the study (Fig. 2). For postnatal hearts:
   (a) PDPN localizes in lymphatics and epicardium (Fig. 2a–e);
   (b) NRP2 localizes in lymphatics and axon tracts of the autonomic nervous system (Fig. 2f–j);
   (c) VEGFR3 is expressed in lymphatics (Fig. 2k–o);
   (d) LYVE1 is co-expressed in lymphatics and tissue-resident macrophages (Fig. 2p–t).

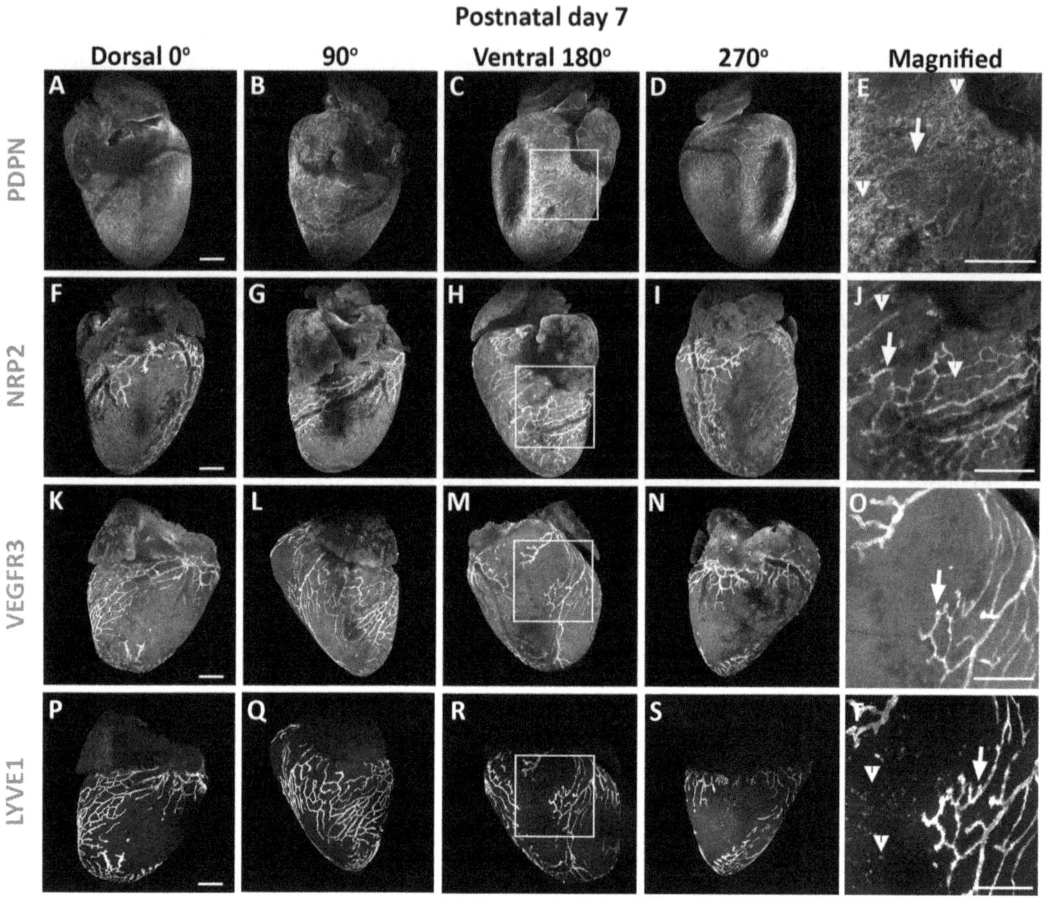

Fig. 2 Multi-view light-sheet imaging of murine hearts at postnatal day 7 using different lymphatic markers. (**a–e**) The anti-PDPN antibody labels the epicardium (white arrowheads at **e**), making challenging to distinguish the sub-epicardial lymphatic vessels (white arrow at **e**). (**f–j**) NRP2 is expressed in axons (white arrowheads at **j**) and lymphatics (white arrow at **j**). (**k–o**) VEGFR3 is specific for cardiac lymphatics (white arrow at **o**). (**p–t**) LYVE1 is expressed in tissue-resident macrophages (white arrowheads at **t**), as well as cardiac lymphatics (white arrow at **t**). Scale bars 0.5 mm

3.4 Computational Analysis

1. Open image using an image analysis software like ImageJ/Fiji.
2. Note down the scale of the image in pixel per mm.
3. Convert the image to 8-bit and save as TIFF, PNG, or JPEG.
4. Open AngioTool and import image (Fig. 3a).
5. In AngioTool "Settings" tab, input the correct scale in pixel per mm (*see* **Note 8**; Fig. 3b).
6. In AngioTool "Analysis" tab, define the diameter and intensity of lymphatics, making sure the outline overlay in the preview window is correct (Fig. 3a, c).

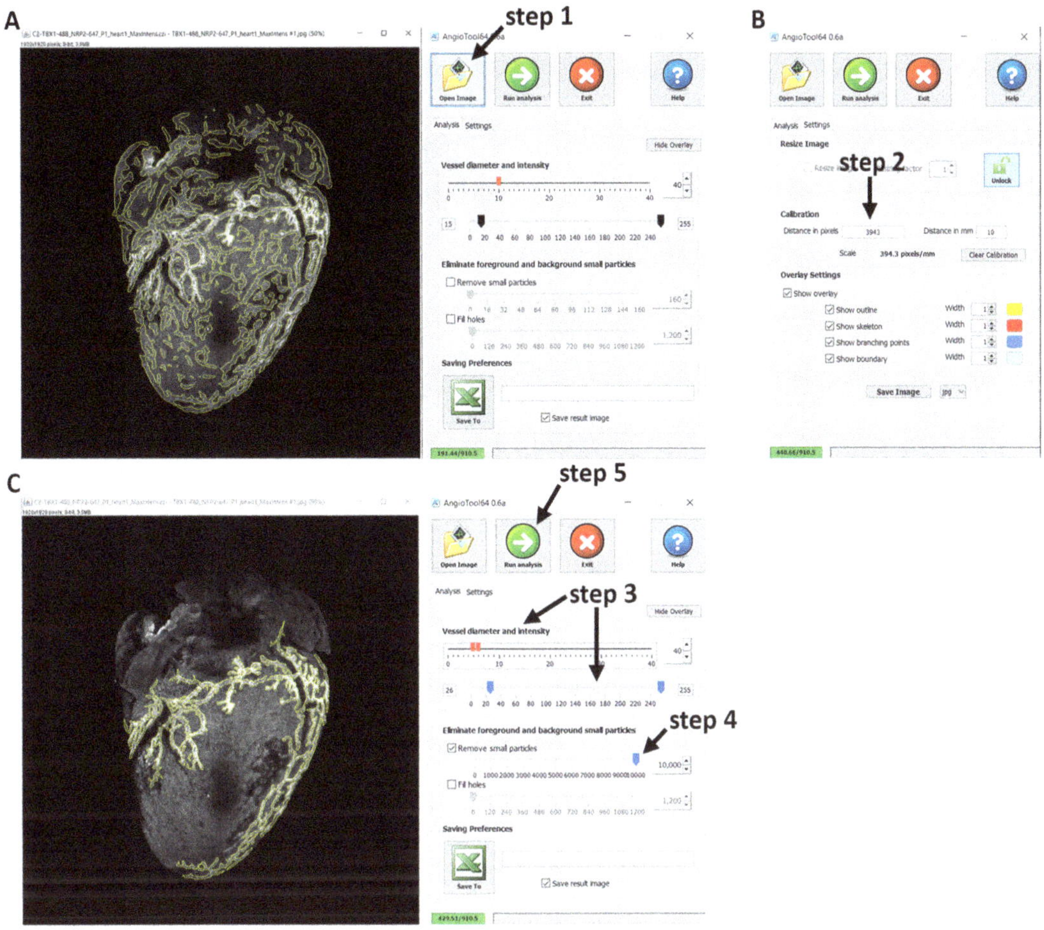

Fig. 3 Using AngioTool to characterize the cardiac lymphatic vasculature in a few steps. (**a**) The first step requires opening the 8-bit image. (**b**) In the "Settings" tab, insert the correct scale value in pixels per mm. (**c**) For steps 3 and 4, use the preview window as a guide to define the vessel diameter and intensity, and remove small particles. In step 5, hit the "Run analysis" button, which generates an Excel file and a JPEG image of the analysis

7. There is the option to define and remove small particles, which is particularly useful if there is high background noise or if an anti-LYVE1 antibody is used to remove LYVE1-expressing macrophages (Fig. 3a, c).
8. When the outline overlay in the preview window matches the vessels click "Run analysis" (Fig. 3a, c).
9. An Excel file and a processed image is then generated and stored at the folder where the 8-bit image was initially saved. Use the parameters in the Excel file to proceed with statistics (e.g., GraphPad Prism, Fig. 4 and Table 1).

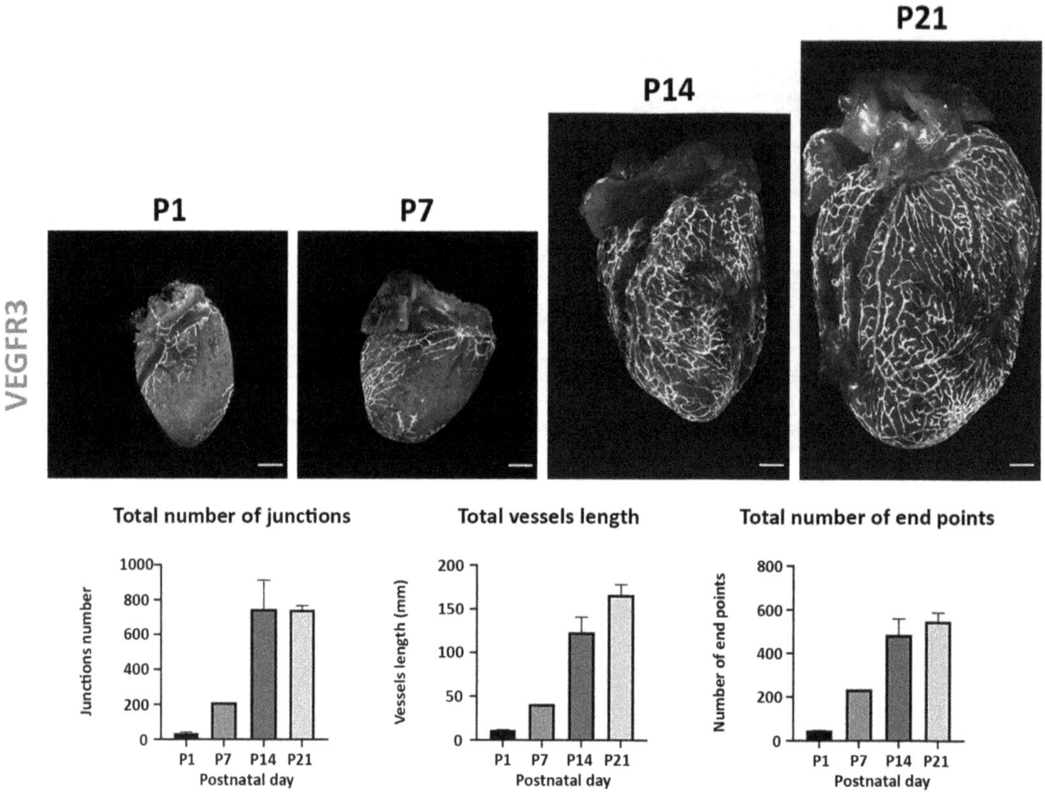

Fig. 4 Quantitative assessment of cardiac lymphatics at different postnatal stages using AngioTool. After running the AngioTool analysis, the parameters from the output Excel file can be used for further analysis. Based on VEGFR3 immunoreactivity, a progressive increase in junction number, vessel length, and number of vessel end points was detected from postnatal day (P)1 to P21. $n = 1–3$ hearts per developmental stage. Scale bars 0.5 mm

Table 1
AngioTool Excel parameters and their description

Parameters	Description
Explant area	Area occupied by the convex envelope containing the vessels
Vessels area	Area of segmented vessels
Vessels percentage area	Percentage of area occupied by vessels in the explant area
Total number of junctions	Total number of vessel junctions
Junctions density	Number of vessel junctions normalized per unit area
Total vessels length	Sum of Euclidean distances between the pixels of all the vessels
Average vessels length	Mean length of all the vessels
Total number of end points	Number of open-ended segments
Mean E lacunarity	Mean lacunarity over all size boxes

4 Notes

1. Primary antibodies raised in the same host species should not be combined in a mix to co-stain samples, as the secondary-specific antibodies will cross-react with each of the primary antibodies.
2. Always use a glass flask that can fit at least twice the volume of PBS to be used.
3. This step must be performed by trained and licensed individuals.
4. It is recommended that an optimization step should be performed to determine the best working dilution for each primary antibody.
5. As fluorophores are photosensitive, it is important that from this point on samples are protected from light either by covering the tubes with aluminum foil or by using LightSafe tubes (e.g., Sigma-Aldrich).
6. Always wear appropriate protection as the agarose solution can overheat in the microwave and erupt when swirled.
7. Here we used a 5× NA 0.16 air objective, 488 nm and 638 nm lasers, and the mounting chamber was filled with PBS.
8. It is important to convert the scale of the image obtained via the analysis software (e.g., ImageJ/FIJI) to pixel per mm, as this is the required input format for AngioTool.

5 Summary

The identification of specific markers for cardiac lymphatics has become increasingly important to aid in the characterization of this vascular network in the context of the developing and/or postnatal heart, intact versus injury/disease models. Here, we showed that VEGFR3 is the most specific marker to study cardiac lymphatics in the postnatal heart. This is agreement with previous reports suggesting that VEGFR3 while initially expressed in blood endothelium, becomes progressively restricted to lymphatic endothelial cells during embryogenesis [4, 21]. Having said that, the selection of lymphatic markers is ultimately dependent on the working hypothesis and aims of the study. For instance, LYVE1 provides an excellent choice for studying cardiac lymphatics in combination with tissue-resident macrophages [5] while NRP2 allows the combined characterization of sympathetic axons [22] and lymphatics [23] (Fig. 2). In contrast, as PDPN is expressed in the epicardium [24], this makes it challenging to distinguish lymphatic vessels expanding in the sub-epicardial compartment

(Fig. 2). Likewise, an analysis based on the use of antibodies against PROX1 is not endorsed here as it results in poorly defined images of lymphatic vessels with significant immunoreactivity background arising from PROX1 expression in the myocardium [25] (not shown). Moreover, as PROX1 antibodies generate a nuclear signal, downstream analysis with AngioTool is difficult for the reason that nuclei are considered small individual particles rather than vessel segments. Following whole-mount immunostaining and imaging using a multi-view light-sheet microscope, we recommend using the semi-automated software AngioTool [20] to quantify several parameters of the lymphatic vasculature (e.g., vessel length, number of junctions, number of vessel end points).

Acknowledgments

K.K. is funded by the Wellcome Trust Foundation (4-year PhD studentship 215103/Z/18/Z). J.M.V. is supported by a British Heart Foundation Intermediate Basic Science Research Fellowship (FS/19/31/34158). P.R.R. is supported by a British Heart Foundation chair award (CH/11/1/28798) and programme grant (RG/08/003/25264). We thank the Micron Oxford Advanced Bioimaging Unit for access to and training in the use of light-sheet microscopy: https://micronoxford.com.

Competing Interests *P.R.R. is co-founder and equity holder in OxStem Cardio, an Oxford University spin-out that seeks to exploit therapeutic strategies stimulating endogenous repair in cardiovascular regenerative medicine. The other authors declare no competing interests.*

References

1. Dongaonkar RM, Stewart RH, Geissler HJ, Laine GA (2010) Myocardial microvascular permeability, interstitial oedema, and compromised cardiac function. Cardiovasc Res 87: 331–339
2. Laine GA, Allen SJ (1991) Left ventricular myocardial edema. Lymph flow, interstitial fibrosis, and cardiac function. Circ Res 68: 1713–1721
3. Klaourakis K, Vieira JM, Riley PR (2021) The evolving cardiac lymphatic vasculature in development, repair and regeneration. Nat Rev Cardiol 18(5):368–379. https://doi.org/10.1038/s41569-020-00489-x
4. Klotz L et al (2015) Cardiac lymphatics are heterogeneous in origin and respond to injury. Nature 522:62–67
5. Cahill TJ et al (2021) Tissue-resident macrophages regulate lymphatic vessel growth and patterning in the developing heart. Development 148(3):dev194563. https://doi.org/10.1242/dev.194563
6. Juszyński M, Ciszek B, Stachurska E, Jabłońska A, Ratajska A (2008) Development of lymphatic vessels in mouse embryonic and early postnatal hearts. Dev Dyn 237: 2973–2986
7. Flaht-Zabost A et al (2014) Cardiac mouse lymphatics: developmental and anatomical update. Anat Rec 297:1115–1130
8. Stone OA, Stainier DYR (2019) Paraxial mesoderm is the major source of lymphatic endothelium. Dev Cell 50:247–255.e3
9. Maruyama K, Miyagawa-Tomita S, Mizukami K, Matsuzaki F, Kurihara H (2019) Isl1-expressing non-venous cell lineage contributes to cardiac lymphatic vessel development. Dev Biol 452:134–143

10. Lioux G et al (2020) A second heart field-derived vasculogenic niche contributes to cardiac lymphatics. Dev Cell 52:350–363.e6
11. Henri O et al (2016) Selective stimulation of cardiac lymphangiogenesis reduces myocardial edema and fibrosis leading to improved cardiac function following myocardial infarction. Circulation 133:1484–1497
12. Houssari M et al (2020) Lymphatic and immune cell cross-talk regulates cardiac recovery after experimental myocardial infarction. Arterioscler Thromb Vasc Biol 40:1722–1737
13. Vieira JM et al (2018) The cardiac lymphatic system stimulates resolution of inflammation following myocardial infarction. J Clin Invest 128:3402–3412
14. Trincot CE et al (2019) Adrenomedullin induces cardiac lymphangiogenesis after myocardial infarction and regulates cardiac edema via Connexin 43. Circ Res 124:101–113
15. Hartikainen J et al (2017) Adenoviral intramyocardial VEGF-DDNDC gene transfer increasesmyocardial perfusion reserve in refractory angina patients: a phase I/IIa study with 1-year follow-up. Eur Heart J 38:2547–2555
16. Kataoka Y et al (2010) The first clinical pilot study of intravenous adrenomedullin administration in patients with acute myocardial infarction. J Cardiovasc Pharmacol 56:413–419
17. Tatin F et al (2017) Apelin modulates pathological remodeling of lymphatic endothelium after myocardial infarction. JCI Insight 2:1–16
18. Rueden CT et al (2017) ImageJ2: ImageJ for the next generation of scientific image data. BMC Bioinformatics 18:529
19. Schindelin J et al (2012) Fiji: an open-source platform for biological-image analysis. Nat Methods 9:676–682
20. Zudaire E, Gambardella L, Kurcz C, Vermeren S (2011) A computational tool for quantitative analysis of vascular networks. PLoS One 6: e27385
21. Kaipatnen A et al (1995) Expression of the fms-like tyrosine kinase 4 gene becomes restricted to lymphatic endothelium during development. Med Sci 92(8):3566–3570
22. Maden CH et al (2012) NRP1 and NRP2 cooperate to regulate gangliogenesis, axon guidance and target innervation in the sympathetic nervous system. Dev Biol 369:277–285
23. Yuan L et al (2002) Abnormal lymphatic vessel development in neuropilin 2 mutant mice. Development 129(20):4797–4806
24. Mahtab EAF et al (2008) Cardiac malformations and myocardial abnormalities in Podoplanin knockout mouse embryos: correlation with abnormal epicardial development. Dev Dyn 237:847–857
25. Risebro CA et al (2009) Prox1 maintains muscle structure and growth in the developing heart. Development 136:699

Chapter 14

Quantifying Lymphatic Vessel Density in Human Tissue Samples

Stewart G. Martin, Emad Rakha, and Sarah J. Storr

Abstract

The development of new blood and lymphatic vessels, through the process of angiogenesis and lymphangiogenesis, respectively, is critical to the development and growth of tumors, and integral to the process of metastasis. Lymphatic vessel density can be assessed as a surrogate measure of lymphangiogenesis in human tissue samples. Lymphatic vessel density has been shown to be associated with lymph node metastasis and patient survival in various solid tumor types. Here we describe a method for quantifying the number of lymphatic vessels within tumor tissue that can also be used to assess lymphatic vessel invasion, and compare with blood vessel density and invasion.

Key words Lymphangiogenesis, Lymphatic vessel density, D2-40

1 Introduction

The development of new blood and lymphatic vessels, through the process of angiogenesis and lymphangiogenesis respectively, is critical to the development and growth of tumors, and integral to the process of metastasis. Metastasis is the principal cause of mortality in cancer patients, and occurs through blood and lymphatic vessels. For many years, lymphatic vessels were thought to play a passive role in metastasis; however, this has since been shown to under-represent the dynamism displayed by the vasculature, including the process of lymphangiogenesis (reviewed in [1]). In fact, lymph nodes are often the first destination of metastatic cells in most carcinomas and typically occur via lymphatic vessel dissemination. Lymphatic vessel invasion has been demonstrated as the predominant mechanism of early metastatic spread in many tumor types including breast cancer [2–4].

The process of lymphangiogenesis within tumors is complex, and largely driven by the tumor itself; vascular endothelial growth

Andrew V. Benest (ed.), *Angiogenesis: Methods and Protocols*, Methods in Molecular Biology, vol. 2441,
https://doi.org/10.1007/978-1-0716-2059-5_14,

factor (VEGF)-C and VEGF-D are both closely implicated in the process (reviewed in [1]). In patient tumors, a surrogate measure of lymphangiogenesis can be assessed by determining lymphatic vessel density using immunohistochemistry. Over the last 20 years, the use of lymphatic specific markers, such as D2-40 and LYVE-1, rather than traditional Haematoxylin and Eosin (H&E) staining, has allowed the exploration of lymphatic specific vessel density and invasion and provides an accurate method of determining lympho-vascular invasion, being able to distinguishing between blood and lymphatic vasculature [5, 6]. Typically CD31 and CD34 are used to stain endothelial cells, discrimination of lymphatic vessels from blood vessels is achieved using D2-40 or LYVE-1.

While the importance of lymphatic vessel invasion in some solid tumors is well established, the importance of lymphatic vessel density within tumors remains less clear. Several studies have linked lymphatic vessel density with lymph node metastasis, and ultimately clinical outcome [2–4, 7]. Meta-analyses of lymphatic vessel density indicate a relationship with lymph node metastasis in breast cancer, and survival of gastric cancer patients [8–10].

Here we describe a protocol to determine lymphatic vessel density using an antibody against podoplanin, D2-40, in formalin-fixed paraffin-embedded tissue. Studies have assessed lymphatic vessel density by counting lymphatic vessels in three to five hotspot areas, in a similar method to determining blood vessel density, or by counting all lymphatic vessels across the tumor to accumulate additional information on their distribution, we describe both these methods.

2 Materials

1. 4 μm sections of formalin-fixed paraffin embedded (FFPE) tissue on glass slides.
2. Pencil or pen (xylene and ethanol resistant).
3. Hot plate.
4. Coplin jars (or autostainer).
5. Xylene.
6. Industrial methylated spirit (IMS)/ethanol.
7. TBS: Tris-buffered saline, pH 7.6.
8. Sequenza trays and Shandon (Sequenza) cover plates.
9. Leica Biosystems Novolink Polymer Detection kit.
10. Leica Biosystems Primary Antibody Diluent.
11. Anti-podoplanin (D2-40) antibody (Covance).
12. Glass coverslips.

13. DPX mounting agent.
14. Microscope.

3 Methods

Here we described the use of an antibody against podoplanin (D2-40), a 38 kDa glycoprotein, to quantify lymphatic vessel density in breast cancer tissue. Further consecutive sections can be cut to determine angiogenesis if direct correlations are required, or double immunohistochemistry can be utilized if additional characteristics of interest. Other tissue types may require further optimization of antibody dilution, antigen retrieval and/or general immunohistochemical procedure than that described here. We have used the Leica Biosystems Novolink Polymer Detection kit; however, many others are commercially available.

3.1 Dewaxing

1. Label slides with a pencil or pen (xylene and ethanol resistant), noting specimen details, and primary antibody to be used (*see* **Note 1**).
2. Place 4 μm formalin-fixed paraffin-embedded sections on a 60 °C hotplate for 10 min (*see* **Note 2**).
3. Allow to cool and place in slide rack.
4. Sections should be dewaxed and dehydrated by moving through a series of coplin jars (or autostainer should it be available), containing the following solutions: xylene (5 min), xylene (5 min), ethanol/IMS (2 min), ethanol/IMS (2 min), ethanol/IMS (2 min).
5. Slides should be placed in water for 5 min (*see* **Note 3**).

3.2 Blocking

1. Working in large bath of tap water, load slides onto the Shandon coverplates, then place in Sequenza trays (*see* **Note 4**).
2. Add 100 μl of Peroxidase Block (Novolink Polymer Detection kit) solution to each slide and incubate for 5 min.
3. Rinse the slides three times with TBS (each rinse should be 5 min).
4. Add 100 μl of Protein Block (Novolink Polymer Detection kit) solution to each slide and incubate for 5 min.

3.3 Primary and Secondary Antibodies

1. Apply 100 μl primary antibody (anti-podoplanin (D2-40) 1: 100 (Covance) in Leica Primary Antibody Dilluent) per slide, and incubate at room temperature for 1 h (*see* **Note 5**).
2. Rinse the slides three times with TBS (each rinse should be 5 min).

3. Apply 100 μl Post Primary Block (Novolink Polymer Detection kit) solution to each slide and incubate for 30 min.
4. Rinse the slides three times with TBS (each rinse should be 5 min).
5. Apply 100 μl Novolink Polymer (Novolink Polymer Detection kit) solution to each slide and incubate for 30 min.
6. Rinse the slides three times with TBS (each rinse should be 5 min).

3.4 Staining

1. Prepare DAB solution by diluting DAB substrate (1:20 DAB chromogen (Novolink Polymer Detection kit) in DAB substrate (Novolink Polymer Detection kit) (*see* **Note 6**).
2. Apply 100 μl diluted DAB solution to each slide and incubate for 5 min.
3. Rinse the slides three times with TBS (each rinse should be 5 min).
4. Apply 100 μl hematoxylin solution to each slide and incubate for 6 min.
5. Rinse the slides immediately with water, remove from Shandon coverplates and place slides into a slide rack.
6. Slides should be moved through a series of coplin jars containing the following solutions: water (5 min), ethanol/IMS (2 min), ethanol/IMS (2 min), ethanol/IMS (2 min), xylene (5 min) xylene (5 min).
7. Apply one or two drops of DPX to glass coverslips.
8. Place the slides tissue face down onto the coverslips and allow the weight of the slide to spread the DPX mounting agent.
9. Allow slides to dry overnight.

3.5 Quantifying Lymphatic Vessel Density

Studies have assessed lymphatic vessel density by counting lymphatic vessels in three or five areas with the highest density of vessels (hotspots), in a similar methodology to that of determining microvessel density of vascular endothelium [11]. The authors typically determine total lymphatic vessel density across the full section; although this method of assessment is difficult and time consuming, it can enable detailed distribution information to be obtained if required [2]. Here we provide both methodologies to determine total lymphatic vessel density.

3.5.1 Quantifying Total Lymphatic Vessel Density

1. Examine the section at low magnification (40×) and divide into three areas, the central 2/3 core, the intratumoral (IT) area, the outer 1/3 of the tumor tissue, the peripheral tumoral area (PP) and the normal area surrounding the tumor, the peritumoral area (PT) (Fig. 1).

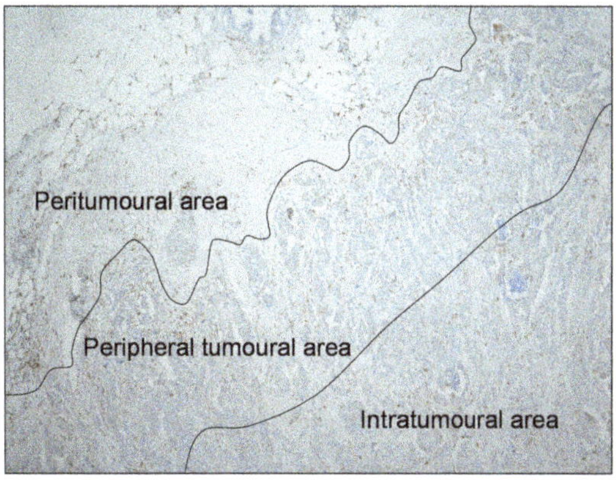

Fig. 1 Representative photomicrograph of breast cancer tissue with a representation of how tumor sections are divided into intratumoral, peripheral, and peritumoral areas (40× magnification). Figure adapted from [12]

2. Count vessels in each area using 100× magnification (with a surface area of 3.46 mm^2).
3. The sum of lymphatic vessel density in all fields in each zone is divided by the total surface areas of all fields to give the lymphatic vessel density for that zone, presented as vessels/mm^2. This method gives the lymphatic vessel density in the peritumoral area, in the peripheral tumor area, and in the intratumoral area with the sum of three giving the total lymphatic vessel density for the specimen.
4. Vessels in the whole tumor section are counted (*see* **Note** 7).

3.5.2 Quantifying Lymphatic Vessel Density Hotspots

A 25 dot Chalkley graticule microscope eyepiece is required to quantify lymphatic vessel density within tissue using the hotspot method. The Chalkley overlap morphometric technique has been successfully implemented to measure blood microvessel density in an objective manner and is also suitable for quantifying lymphatic vessel density.

1. Examine the section at low magnification (40×) and divide into three areas, the central two-third core, the intratumoral (IT) area, the outer one-third of the tumor tissue, the peripheral tumoral area (PP) and the normal area surrounding the tumor, the peritumoral area (PT) (Fig. 1).
2. The tissue is scanned to identify possible "hot spot" areas at 100× magnification.
3. The eyepiece graticule is orientated on the tissue so that the maximum number of dots overlap stained cells at 200× magnification (*see* **Note** 7).
4. This should be done three to five times in different regions within the tumor (*see* **Note 8**).

4 Notes

1. In each run a positive and negative control must be included to confirm that the experiment has been successful. For positive control, lymph node sections can be used. For negative control, sections are stained following the same procedure but omitting primary antibody.
2. Sectioned tissues should be stored at 4 °C until staining, and ideally used as soon after sectioning. If sections cannot be stained soon after sectioning, they should be overlaid with wax to reduce tissue oxidization.
3. Once the tissue has been deparaffinized and rehydrated it must not be allowed to dry as this may interfere with staining.
4. Fill the Sequenza reservoir half full with TBS to rinse the slides, and watch it flow through to ensure there are no air bubbles (as this will impair staining). If the TBS runs through a Sequenza plate either very quickly or very slowly, this indicates an air bubble, in which case the slide must be re-loaded into the plate. Humidity trays can be used instead of the Sequenza system; however the use of humidity trays will require more antibody and reagents.
5. The concentration of primary antibody should be optimized to ensure ideal staining, this may differ by supplier and antibody batch. Here we use D2-40, however other antibodies can also be used, including those against LYVE-1; each has different limitations and should be carefully considered.
6. Once prepared DAB solution should be kept in the dark and used within 6 h.
7. To confirm the scoring of the first assessor, a percentage of slides should be evaluated by an independent assessor, blinded to the first assessor's data and clinicopathological criteria of the patients; typically greater than 30%.
8. In some situations, depending on the frequency of lymphatic vessels, more hotspots, or total fields of view may need to be utilized.

Acknowledgments

S.J.S. is funded by the University of Nottingham Research Vision.

References

1. Christiansen A, Detmar M (2011) Lymphangiogenesis and cancer. Genes Cancer 2: 1146–1158
2. Mohammed RA, Ellis IO, Elsheikh S, Paish EC, Martin SG (2009) Lymphatic and angiogenic characteristics in breast cancer: morphometric analysis and prognostic implications. Breast Cancer Res Treat 113:261–273
3. Mohammed RA, Ellis IO, Mahmmod AM, Hawkes EC, Green AR, Rakha EA, Martin SG (2011) Lymphatic and blood vessels in basal and triple-negative breast cancers: characteristics and prognostic significance. Mod Pathol 24:774–785
4. Mohammed RA, Martin SG, Mahmmod AM, Macmillan RD, Green AR, Paish EC, Ellis IO (2011) Objective assessment of lymphatic and blood vascular invasion in lymph node-negative breast carcinoma: findings from a large case series with long-term follow-up. J Pathol 223:358–365
5. Schacht V, Dadras SS, Johnson LA, Jackson DG, Hong YK, Detmar M (2005) Up-regulation of the lymphatic marker podoplanin, a mucin-type transmembrane glycoprotein, in human squamous cell carcinomas and germ cell tumors. Am J Pathol 166:913–921
6. Evangelou E, Kyzas PA, Trikalinos TA (2005) Comparison of the diagnostic accuracy of lymphatic endothelium markers: Bayesian approach. Mod Pathol 18:1490–1497
7. Storr SJ, Safuan S, Mitra A, Elliott F, Walker C, Vasko MJ, Ho B, Cook M, Mohammed RA, Patel PM, Ellis IO, Newton-Bishop JA, Martin SG (2012) Objective assessment of blood and lymphatic vessel invasion and association with macrophage infiltration in cutaneous melanoma. Mod Pathol 25:493–504
8. Zhang S, Zhang D, Yi S, Gong M, Lu C, Cai Y, Tang X, Zou L (2017) The relationship of lymphatic vessel density, lymphovascular invasion, and lymph node metastasis in breast cancer: a systematic review and meta-analysis. Oncotarget 8:2863–2873
9. Zhang S, Yi S, Zhang D, Gong M, Cai Y, Zou L (2017) Intratumoral and peritumoral lymphatic vessel density both correlate with lymph node metastasis in breast cancer. Sci Rep 7:40364
10. Liang L, Huang WT, He RQ, Liang HW, Huang CQ, Zhou H, Wei FL, Zhou SS, Peng ZG, Chen G, Chen JQ, Qin XG (2018) A meta-analysis of the lymphatic microvessel density and survival in gastric cancer with 1809 cases. Oncotarget 9:5406–5415
11. Vermeulen PB, Gasparini G, Fox SB, Colpaert C, Marson LP, Gion M, Belien JA, de Waal RM, Van Marck E, Magnani E, Weidner N, Harris AL, Dirix LY (2002) Second international consensus on the methodology and criteria of evaluation of angiogenesis quantification in solid human tumours. Eur J Cancer 38:1564–1579
12. Ahmad NS, Martin SG, Storr SJ (2016) Immunohistochemical assessment of leukocyte involvement in angiogenesis. Methods Mol Biol 1430:49–57

Chapter 15

Quantifying Vascular Remodeling in the Mouse Spinal Cord

Marlene E. Da Vitoria Lobo, Lydia Hardowar, and Richard P. Hulse

Abstract

The spinal cord, a compartment of the central nervous system, is made up of a number of architecturally distinct neural centers that influence an array of neurophysiological systems. The primary role of the spinal cord is the modulation of sensory and motor function by acting as a relay station between the periphery and the brain. Inherently these are considered as neural networks, however the functional dynamics of these tissues consist of a heterogenic population of cell types, all working in harmony to maintain physiological function. Part of this cellular diversity comprises of the vascular network that delivers essential nutrients and oxygen to the spinal cord tissue, whilst also protecting it from potentially tissue damaging substances such as foreign entities including toxic pharmacological agents or pathogens. The viability of the spinal cord is dependent upon the harmonious balance between opposing angiogenic processes; vascular remodeling and vascular regression, tipping the balance to either side contributes to neurodegeneration. Exploring vascular remodeling in the central nervous system requires consideration of the anatomical landscape of the spinal cord and the dynamic nature of the microvasculature. Utilizing immunofluorescent staining and 3D image rendering analysis of the endothelium and mural cell population allows for investigation of cellular as well as molecular mediation of vascular remodeling in the spinal cord. This method can be utilized in a range of rodent models (utilizing pharmacological, disease models, transgenic and/or viral approaches) offering extensive appreciation of the blood–spinal cord barrier.

Key words Pain, Angiogenesis, Endothelial, Spinal cord, Astrocyte, Pericyte

1 Introduction

The dialogue between the vascular and neuronal networks is of fundamental importance to the maintenance of neurophysiological outcomes. The microvessels deliver essential provision (such as nutrients) to neural centers to support healthy neurological function such as those associated with cognition, movement, and nociception [1–3], whilst providing a protective barrier from potentially harmful agents [4]. Disturbances in the viability of this system are implicated in disease, whereby an impairment of the microvessel network demonstrated by alterations in vessel architecture, vessel perfusion or vessel leakage within neural tissues, are associated with

Andrew V. Benest (ed.), *Angiogenesis: Methods and Protocols*, Methods in Molecular Biology, vol. 2441,
https://doi.org/10.1007/978-1-0716-2059-5_15,

neurodegenerative disease such as Alzheimer's and Stroke [1, 3, 5]. The vascular network consists of a heterogenous population of cell types including endothelial cells, pericytes, smooth muscle cells and astrocytes, which are essential to supporting a healthy organism and contribute to such disturbances in microvessel function. These cellular interactions have been investigated in human tissues, as well in rodent models (disease models and transgenic models) allowing for interrogation of the differing cellular compartments that make up the neurovascular unit. For example, alterations in endothelial cell tight junction maintenance and/or loss of pericytes have been proven to be associated with vascular complications such as in the retina [6] and brain [7].

The role of angiogenesis in tissue development and pathology has been extensively discussed. However, the role of microvessel development and function are increasingly important in relation to those physiological processes associated with the spinal cord and consequent impact of disease [8]. The spinal cord is essential to the processing of incoming sensory information and coordination of motor function in relation to the organisms surrounding environment. Typically disturbances in microvessel function defined as prominent pro-angiogenic actions are associated with exacerbated vascular leakage in the spinal cord [9–12], with a breakdown in the endothelial junctional barrier a causative factor in the development of neurodegenerative disease (e.g., multiple sclerosis [13]) and motor impairment (e.g., spinal cord traumatic injury [14]). This is in addition to disturbances in nociceptive processing and resulting chronic pain manifestation (such as in arthritis, traumatic injury [9, 15]). However, pro-angiogenic activity depicted as heightened function and/or expansion of the vascular network in neural tissues such as highlighted in the spinal cord [9–12], necessitates opposing actions to curtail aberrant vascular remodeling processes such as vessel regression and occlusion [16]. For example, impaired blood flow through vessel constriction and/or occlusion is now regarded as a prominent feature of neurodegenerative processes in relation to cognitive function [1, 5, 17] as well as sensory [18] and motor [14] disturbances in age and disease related neuropathologies [19]. In such instances the endothelium degenerates leading to vessel pruning, mediated in part via vascular endothelial growth factor receptor 2 signaling [19].

To understand the dynamics of this vascular remodeling of the blood spinal cord barrier, the resulting impact upon neurophysiological processes and development of pathological states, an appreciation of vessel density or leakage is needed. Typically image acquisition of microvessels in the spinal cord provide a 2D representation of the capillary network in the dorsal-ventral plane of the spinal cord providing an understanding of the angiogenic processes in play. Microvessel adaptation to changes in their environments, as in this instance the spinal cord, do not occur in a single plane as represented by 2D image analysis but in the 3D dimension, with vessels expanding and retracting in a multitude of directions. In the central nervous system the microvessel network also extends medial-laterally as well as in the anterior to posterior axis. Furthermore, the extension and regression of endothelium sprouting is tortuous in nature, potentially moving in out of the chosen visual plane. Utilizing fluorescent-based immunohistochemistry in combination with image rendering analysis allows for a greater unbiased appreciation of the microvessel architecture and function within nervous tissues. Image segmentation of the blood vessel networks enables extraction of experimental data including volume of the vascular network in the region of interest, blood length, branch number, and vessel diameter.

2 Materials

2.1 Anesthesia

- Sodium pentobarbital (Euthatal; 200 mg/ml).
- 25 Gauge needle and 1 ml syringe (all syringes and needles manufactured by Terumo).

2.2 Transcardial Perfusion and Tissue Preparation

- Electric clippers.
- All surgical tools from World precision instruments. Surgical tools include Dumont Tweezers, Swiss Curved Tweezers, Hemostats, Vannas scissors, Iris straight scissors, Scalpel handle and blade (#10).
- 21G needle.
- 50 ml Syringes.
- 4% Paraformaldehyde.
- Phosphate buffered saline.
- (Optional) Dissection microscope and light source.
- 30% Sucrose (in PBS).

2.3 Slide Preparation

- Superfrost slides (VWR international or Equivalent).
- Cryostat (Leica CM1860 UV Cryostat or Equivalent).
- Optimal Cutting Tek (O.C.T or equivalent).
- Triton x-100.
- Hydrophobic Pen (Imm pen, Vectorlabs).
- Humidified Chamber (Sigma-Aldrich).

2.4 Immunofluorescence and Slide Imaging

- Primary antibody—list of suitable antibodies, concentrations/dilutions and suppliers.
  - Endothelial Cell marker; biotin conjugated isolectin B_4 (IB_4; 1:100, Sigma-Aldrich);
  - Endothelial Cell marker; goat anti-CD31 (1:200, R&D systems);
  - Pericyte Marker; rabbit anti NG2 proteoglycan, (1:200, dilution);
  - Smooth Muscle marker; mouse anti-α smooth muscle Cy3 (1:200, Sigma-Aldrich);
  - Basement Membrane marker; Rabbit anti collagen IV (1:200; Abcam);
  - Astrocytic marker; rabbit anti-GFAP (1:500, Abcam);
  - Additional neuronal markers can be used in combination rabbit anti-fos (1:100, Santa Cruz); mouse anti-NeuN (1: 200; Millipore, Billerica, MA, USA).
- Secondary fluorescent antibodies—utilize Alexafluor fluorescent dyes (conjugated where appropriate with appropriate species IgG—typically mouse and rabbit or streptavidin) sourced from Thermo Fisher. Secondary antibodies typically raised in donkey (or chicken) utilizing Alexfluor 488, 555 and 647. Consideration made with reference to the imaging capabilities of the microscope in relation to fluorescent dye emission wavelength. Used at 1:1000 dilution in PBS/0.2% Triton x-100.
- Nuclear stains (such as DAPI, Hoescht).
- Confocal microscope (Leica SP8 or equivalent).
- Glass coverslips (VWR International).
- Antifade Mounting medium (Vectorshield, Vectorlabs; Fluorosave, Sigma-Aldrich) to preserve fluorescence.

2.5 Software Analysis

- Imaris.
- Statistical Package (such as GraphPad Prism v8).

3 Methods

3.1 Surgical Preparation

All surgical equipment (e.g., surgical tools) and consumables (e.g., PBS) are to be sterilized and suitable provision for appropriate aseptic surgical practice is needed (e.g., drapes, surgical gloves).

3.2 Animal Anesthesia and Transcardial Perfusion

1. Terminally anesthetize mice (adult 25–30 g male C57bl6) via intraperitoneal injection of Euthatal (Sodium Pentobarbital) (10 mg/kg). Anesthetic maintenance should be regularly checked and anesthetic depth monitored e.g., foot and corneal reflex.

2. Additional anesthetic maybe provided via intraperitoneal injection of Euthatal where appropriate if a sufficient depth of anesthesia hasn't been reached.
3. Once the mouse has reached sufficient depth of anesthesia (no corneal reflex, no righting reflex, no pedal response to pinch) it should be placed on its back in a stainless steel tray.
4. Using dissection scissors and forceps, an incision is made across the base of the rib cage. Starting midline at the base of the sternum through the abdominal wall terminating at each lateral aspect of the mouse. This incision should be approximately 4–6 cm in length.
5. Using forceps secure the sternum and the dissection scissors should be used to cut along the rib cage penetrating the diaphragm exposing the heart and lungs.
6. Dissection scissors should be used to cut through the rib cage from the base up toward the mouse shoulder. This should be repeated on both sides. Using hemostats secure the base of the sternum and move the rib cage to position toward the head exposing the heart and lungs.
7. A small incision is made in the right atrium. Fill a 50 ml syringe with ringers or PBS and secure a 21G needle. Prepare an additional 50 ml syringe and needle. Blunt forceps should be used to secure the left ventricle and the 50 ml syringe is inserted into the left ventricle. The forceps (or alternatively hemostats) should be used to secure the needle in the ventricle. Ensure the needle tip does not penetrate other chambers of the heart as it will compromise blood circulatory system preventing efficient fixation.
8. Introduce up to 50 ml of ringers or PBS (or until the blood has been removed) and at this stage remove the syringe containing ringers and insert the syringe containing 4% paraformaldehyde and repeat the process?

3.3 Tissue Extraction and Preparation

1. A laminectomy is performed to expose the spinal cord. Make an insertion along the vertebral column on the dorsal surface using a scalpel blade. This would be approximately ~5 cm in length from the base of the tail to the neck.
2. Separate skin from underlying tissue using blunt dissection scissors exposing the muscle and connective tissue.
3. Use hemostats to secure the skin to allow access to the dorsal surface of the vertebral column.
4. Use sharp dissection scissors and the scalpel blade to remove the muscle and connective tissue and expose the vertebral processes. Remove any attached muscle and connective tissue from lateral aspects of the vertebrae T12-L1.

5. Perform a laminectomy using vannas scissors between T12-L1 to allow exposure of lumbar segment L3-L5 of the spinal cord. Insert a single arm of the vannas scissors between vertebra T12-T13 into the epidural space. With careful movements cut the bone away on both distal sides. Be careful to minimize injury to the spinal cord i.e., pressure applied or scissors penetrating the spinal cord. Peel the bone off gently. Once the vertebra are removed the L3-L4 segment can be visualized.
6. Using a scalpel blade and forceps the lumbar spinal cord can be isolated by cutting either side of the lumbar enlargement. Carefully remove the lumbar region of the spinal cord and submerge in 4% paraformaldehyde for 24 h at 4 °C.
7. Replace the paraformaldehyde with 30% sucrose solution for cryoprotection and leave for 24 h at 4 °C.
8. Remove the spinal cords using forceps, handle the tissue with care to minimize damage and dry the tissue by placing on tissue paper.
9. Submerge spinal cord tissue in optimal cutting temperature compound and freeze. Tissue is ready for cryosectioning or storage at −80 °C.

3.4 Immuno-fluorescence

1. Cryosection the spinal cords at a minimum of 40 μm thickness and mount onto superfrost slides. These can be used immediately or stored in a slide box at −80 °C until needed.
2. Using a hydrophobic pen draw a hydrophobic barrier around the tissue sections and place in a humidified chamber.
3. Wash the slides for 5 min three times with PBS at room temperature. Subsequently, wash the slides with PBS/0.2% Triton x-100 at room temperature.
4. Incubate for 60 min in blocking solution (5% bovine serum albumin, 10% fetal calf serum PBS/0.2% Triton) at room temperature.
5. Incubate slides in primary antibody (or combination) in blocking solution (5% bovine serum albumin, 10% fetal calf serum) overnight at 4 °C for 72 h. Endothelial Cell marker goat anti-CD31 to be used at 1:200 dilution.
6. Wash the slides for 5 min three times with PBS at room temperature.
7. Secondary antibodies (1:1000 dilution) to be made up in PBS + 0.2% Triton X-100. Add secondary antibody (or combination) to the slides and incubate for 2 h at room temperature.
8. Wash the slides for 5 min three times with PBS at room temperature. Add a nuclear stain in the final wash step (i.e., DAPI 1 μg/ml) to identify cellular nuclei.
9. Remove excess fluid from the slides and mount glass coverslips with antifade medium. Store at 4 °C until needed.

3.5 Image Analysis

1. Slides are imaged on a confocal microscope (Leica SP8 or equivalent) (Fig. 1a–e) to generate Z stacks of region of interest. Vessel structures and cellular makeup should be visible through fluorescent staining with a random selection acquired focusing upon neural center of interest i.e., dorsal horn for nociception. Consideration regarding controls needs to be made with reference to omission of primary antibodies, blocking peptides or accessing knockout animal tissue of your protein target.
2. Z stacks are processed using 3D image rendering software such as Bitplane Imaris. The software allows you to visualize and perform analyses. Images are opened in surpass view. The filament tracer tool is used to trace microvessels (Fig. 2a, b). Paths are traced manually using autopath with diameter correction. This allows segmentation of the individual microvessels

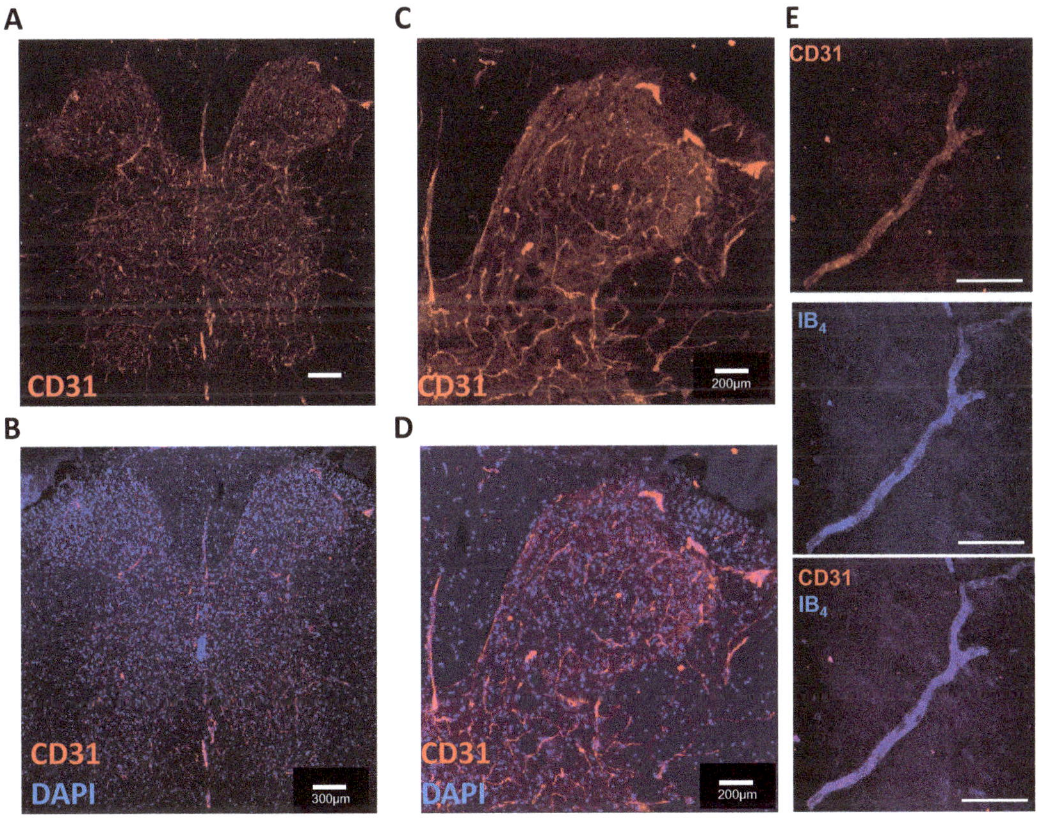

Fig. 1 Spinal cord microvessel identification. Blood vessel endothelium is identified following fluorescent labeling of the spinal cord. Lumbar spinal cord sections from C57.bl6 adult mice are cryosectioned (40 μm thickness) following cardiac perfusion with paraformaldehyde. Cryosections are incubated with endothelium marker [**a–e**] CD31 as well as with [**b** and **d**] DAPI (nuclear label) allow identification of the spinal cord vasculature [**a**; scale bar = 300 μm & **c**; scale bar = 200 μm]. [**e**] High magnification images of spinal cord microvessels identified with CD31 and IB_4 (scale bar = 30 μm)

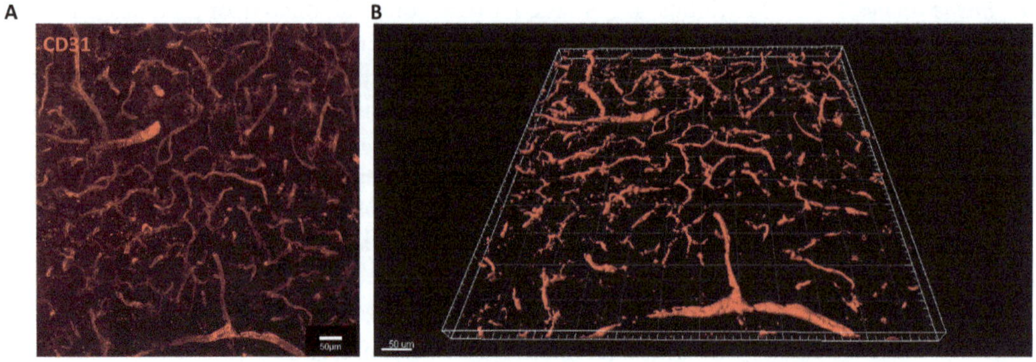

Fig. 2 Image rendering. Cryosectioned lumbar spinal cord sections (50 μm thickness) prepared from adult C57.bl6 mice were incubated with endothelial marker [**a**] CD31 allowing visualization of the microvasculature (scale bar = 50 μm). Using Imaris a surface rendering of the microvessel can be produced to produce a 3D digital representation of the section (scale bar = 50 μm)

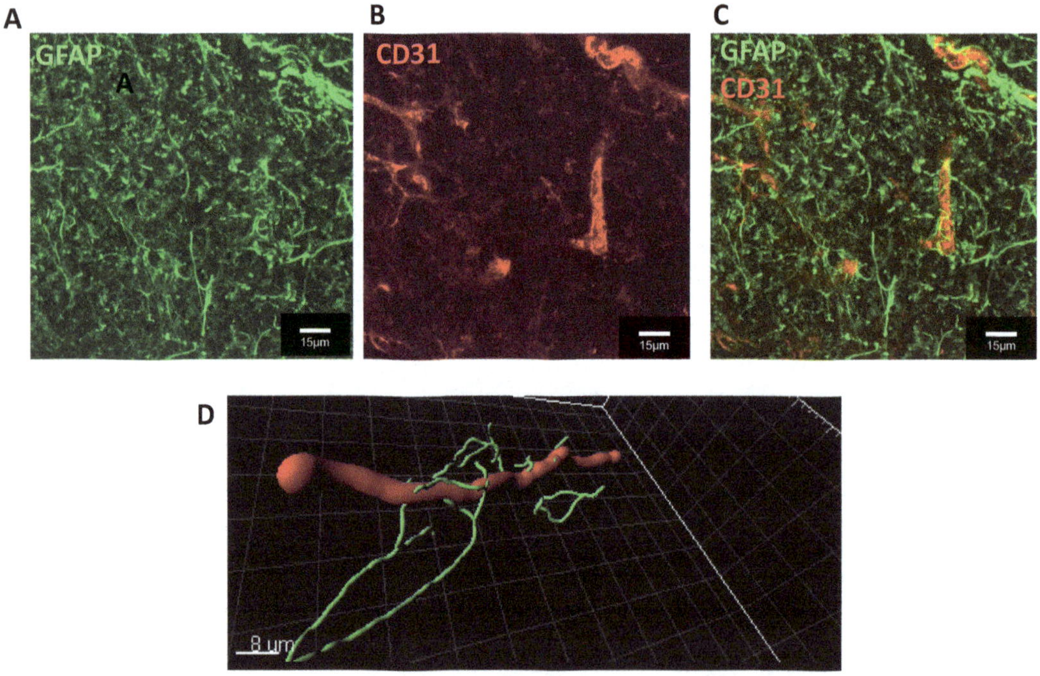

Fig. 3 Mural cell-endothelial cell interaction. Lumbar spinal cord cryosections prepared from adult C57.bl6 mice incubated with [**a**] astrocyte marker (GFAP) and [**b**] endothelial marker CD31 utilized to identify the proximity of astrocytes and the endothelium ([**c**] merge) scale bar = 15 μm). [**d**] Imaris is used to 3D render the endothelial-astrocyte interaction (scale bar = 8 μm)

and extraction of a multitude of datapoints including length and diameter of vessel, number of branch points and volume of microvessels as well as interaction with mural cells (Fig. 3a, b).

4 Notes

1. Depending on species of animal consideration of differing blocking agents is required such as species specific serum.
2. Concentration of primary antibody should be determined from manufacturers guidance and/or reputable literature in line with inhouse validation of optimal antibody concentration. In certain instances amplification steps maybe required through employing biotin-avidin binding to enhance image acquisition.
3. Measurement of vascular leakage can also be determined. Quantification of vascular permeability following infusion of a fluorescent tracer (e.g., TRITC dextran, Evan's blue [9]) or fluorescently labeled cell types (e.g., monocyte [20]) administered either via intravenous or systemic approaches can be evaluated [9, 21]. Introduction of fluorescent dye during the perfusion step allows vessel leakage to acquired inline with 3D architectural datapoints.
4. If rats are utilized, a larger cardiac perfusate volume is required. Peristaltic pump can be used to perfuse the animal.
5. Spinal cord tissue should not be submerged in 4% paraformaldehyde for greater than 24 h as it can hinder antibody binding.

Acknowledgments

R.P.H., M.D., and L.W. performed the experimental work. R.P.H., M.D., and L.W. contributed to the conception or design of the work in addition to acquisition, analysis, or interpretation of data for the work. R.P.H., M.D., and L.W. drafted the work or revised it critically for important intellectual content. R.P.H. drafted the manuscript with contributions from all authors. All authors approved the final version of the manuscript, agree to be accountable for all aspects of the work in ensuring that questions related to the accuracy or integrity of any part of the work are appropriately investigated and resolved all persons designated as authors qualify for authorship, and all those who qualify for authorship are listed.

Funding *This work was supported by the European Foundation for the Study of Diabetes Microvascular Programme supported by Novartis to R.P.H. (Nov 2015_2 to R.P.H.), the EFSD/Boehringer Ingelheim European Research Programme in Microvascular Complications of Diabetes (BI18_5 to R.P.H.), the Rosetree Trust (A1360 to R.P.H.)*

References

1. Nortley R, Korte N, Izquierdo P et al (2019) Amyloid beta oligomers constrict human capillaries in Alzheimer's disease via signaling to pericytes. Science 365(6450):eaav9518. https://doi.org/10.1126/science.aav9518
2. Weiss N, Miller F, Cazaubon S, Couraud PO (2009) The blood-brain barrier in brain homeostasis and neurological diseases. Biochim Biophys Acta 1788:842–857. https://doi.org/10.1016/j.bbamem.2008.10.022
3. Wardlaw JM, Smith C, Dichgans M (2019) Small vessel disease: mechanisms and clinical implications. Lancet Neurol 18:684–696. https://doi.org/10.1016/S1474-4422(19)30079-1
4. Harilal S, Jose J, Parambi DGT et al (2020) Revisiting the blood-brain barrier: a hard nut to crack in the transportation of drug molecules. Brain Res Bull 160:121–140
5. Korte N, Nortley R, Attwell D (2020) Cerebral blood flow decrease as an early pathological mechanism in Alzheimer's disease. Acta Neuropathol 140(6):793–810
6. Cai J, Boulton M (2002) The pathogenesis of diabetic retinopathy: old concepts and new questions. Eye 16:242–260. https://doi.org/10.1038/sj/eye/6700133
7. Sa-Pereira I, Brites D, Brito MA (2012) Neurovascular unit: a focus on pericytes. Mol Neurobiol 45:327–347. https://doi.org/10.1007/s12035-012-8244-2
8. Montague-Cardoso K, Malcangio M (2021) Changes in blood–spinal cord barrier permeability and neuroimmune interactions in the underlying mechanisms of chronic pain. Pain Rep 6:e879. https://doi.org/10.1097/PR9.0000000000000879
9. Beggs S, Liu XJ, Kwan C, Salter MW (2010) Peripheral nerve injury and TRPV1-expressing primary afferent C-fibers cause opening of the blood-brain barrier. Mol Pain 6:74. https://doi.org/10.1186/1744-8069-6-74
10. Slomnicki LP, Myers SA, Saraswat Ohri S et al (2020) Improved locomotor recovery after contusive spinal cord injury in Bmal1 −/− mice is associated with protection of the blood spinal cord barrier. Sci Rep 10:14212. https://doi.org/10.1038/s41598-020-71131-6
11. Echeverry S, Shi XQ, Rivest S, Zhang J (2011) Peripheral nerve injury alters blood-spinal cord barrier functional and molecular integrity through a selective inflammatory pathway. J Neurosci 31:10819–10828. https://doi.org/10.1523/JNEUROSCI.1642-11.2011
12. Lee JY, Choi HY, Ahn HJ et al (2014) Matrix metalloproteinase-3 promotes early blood-spinal cord barrier disruption and hemorrhage and impairs long-term neurological recovery after spinal cord injury. Am J Pathol 184: 2985–3000. https://doi.org/10.1016/j.ajpath.2014.07.016
13. Fung A, Vizcaychipi M, Lloyd D et al (2012) Central nervous system inflammation in disease related conditions: mechanistic prospects. Brain Res 1446:144–155. https://doi.org/10.1016/j.brainres.2012.01.061
14. Li Y, Lucas-Osma AM, Black S et al (2017) Pericytes impair capillary blood flow and motor function after chronic spinal cord injury. Nat Med 23:733–741. https://doi.org/10.1038/nm.4331
15. Costigan M, Moss A, Latremoliere A et al (2009) T-cell infiltration and signaling in the adult dorsal spinal cord is a major contributor to neuropathic pain-like hypersensitivity. J Neurosci 28:14415–14422
16. Korn C, Augustin HG (2015) Mechanisms of vessel pruning and regression. Dev Cell 34: 5–17. https://doi.org/10.1016/j.devcel.2015.06.004
17. Cruz Hernandez JC, Bracko O, Kersbergen CJ et al (2019) Neutrophil adhesion in brain capillaries reduces cortical blood flow and impairs memory function in Alzheimer's disease mouse models. Nat Neurosci 22:413–420. https://doi.org/10.1038/s41593-018-0329-4
18. Ved N, Lobo MEDV, Bestall SM et al (2018) Diabetes-induced microvascular complications at the level of the spinal cord: a contributing factor in diabetic neuropathic pain. J Physiol 596(16):3675–3693. https://doi.org/10.1113/JP275067
19. Reeson P, Choi K, Brown CE (2018) VEGF signaling regulates the fate of obstructed capillaries in mouse cortex. eLife 7:e33670. https://doi.org/10.7554/eLife.33670
20. Montague K, Simeoli R, Valente J, Malcangio M (2018) A novel interaction between CX 3 CR 1 and CCR 2 signalling in monocytes constitutes an underlying mechanism for persistent vincristine-induced pain. J Neuroinflammation 15(1):101. https://doi.org/10.1186/s12974-018-1116-6
21. Hulse RP, Beazley-Long N, Ved N et al (2015) Vascular endothelial growth factor-A_{165} b prevents diabetic neuropathic pain and sensory neuronal degeneration. Clin Sci 129(8):741–756. https://doi.org/10.1042/CS20150124

Chapter 16

Quantitative Methods to Assess Adipose Vasculature

Martina Rudnicki, Alexandra Pislaru, and Tara L. Haas

Abstract

Adipose tissue depots are invested with an extensive capillary network that is closely associated with maintenance of adipose functions and enables healthy tissue expansion. The capillary network displays a high level of plasticity, demonstrating either growth (angiogenesis) or regression (rarefaction) under various physiological/pathological conditions, which has significant consequences for cardiometabolic health. Thus, the visualization and quantification of adipose vascular networks is an important aspect of studying factors that regulate adipose tissue health. This chapter provides an overview of several methods to quantify adipose vascularization. In-depth protocols are provided for the visualization of vascular structures by staining and imaging of whole-mount adipose tissues or paraffin-embedded adipose tissue sections, together with the quantitative analysis of vascularization from these images.

Key words Adipose imaging, Angiogenesis, Capillary density, Capillary morphology, Endothelial markers, Whole-mount adipose

1 Introduction

The homeostasis of adipose tissue, both white and brown depots, relies on local vasculature. In this regard, it has been shown that the density of the capillary networks is a major determinant of the growth and function of these tissues [1, 2]. Thus, both formation and proper expansion of the adipose are supported by concurrent formation of new vessels from pre-existing ones, a process called angiogenesis. Adipogenesis (the formation of new adipocytes) and angiogenesis appear to be tightly coordinated processes [3, 4]. Moreover, it has been demonstrated that capillary density can affect tissue phenotype in distinct adipose depots. In fact, a greater vascular density has been associated with "beiging" of white adipose depots [5, 6]. Conversely, brown adipose tissue, responsible for thermogenesis, is characterized by greater metabolic activity and higher vascularization than white adipose depots and decreased

Andrew V. Benest (ed.), *Angiogenesis: Methods and Protocols*, Methods in Molecular Biology, vol. 2441,
https://doi.org/10.1007/978-1-0716-2059-5_16,

vascularity in this depot has been associated with a "whitening" phenotype [7].

Low capillary density due to impairments in the angiogenic process has been associated with features of adipose tissue dysfunction, such as unresolved inflammation and inappropriate extracellular matrix (ECM) deposition [8]. The importance of the vascular density to adipose health is also supported by the fact that forced stimulation of vascular growth in white adipose tissue leads to improved adipose tissue functions [6, 9]. Substantial evidence indicates that angiogenesis is impaired in white adipose tissue during obesity [1]. Thus, impaired angiogenesis has been indicated as a critical contributor to the development of obesity-related disorders and much attention has been placed on strategies to improve adipose vascularization. Establishing the effectiveness of such strategies is contingent on accurate assessments of the vascular content of the adipose tissue.

Several experimental approaches can be employed to assess the vascular content of white and brown adipose depots, such as quantitative/real-time PCR, Western Blot, flow cytometry, and imaging analysis. The detection of endothelial cells, which are the predominant cell type within the microvascular compartment, is the basis of all techniques. Among them, imaging analysis is the most effective approach mainly because it provides morphometric information about the adipose microcirculation. We will provide detailed protocols for different imaging analysis procedures below.

Other techniques (i.e., quantitative/real-time PCR, Western Blot, flow cytometry) present some technical advantages over imaging analysis procedures, such as being less time-consuming or requiring less optimization (e.g., of immunostaining parameters). However, they also have important limitations, which are mainly related to the information obtained with each technique. The main drawback of these methods is the inability to relate the obtained results to morphometric changes in vascular structure. For example, an increase in endothelial cell number (as assessed by flow cytometry) may represent a greater vessel number but also could indicate enlargement of existing vessels.

Western blot and PCR-based technologies perform an indirect assessment of endothelial cell content by the analysis of endothelial-specific genes and proteins. Of note, this type of analysis relies on the assumptions that (a) the marker measured is not expressed by other cell types and (b) that it is not up- or downregulated in the endothelial cells in response to the experimental conditions. Platelet endothelial cell adhesion molecule 1 (Pecam-1—also known as CD31) is the most frequently used endothelial cell marker in PCR-based, Western Blot, and flow cytometry analyses. However, Pecam-1 is also expressed by leukocytes (macrophages, T cells, B cells, and NK cells) [10], cell types that reside in the adipose tissue and can be increased during nutritional stress. Furthermore, endothelial cell expression of Pecam-1 may fluctuate during inflammation [11]. Thus, accurate assessment of the vascular content of adipose tissue employing techniques other than imaging analysis requires the examination of multiple endothelial cell markers to provide reliable evidence about endothelial cell number. The most common endothelial markers employed in PCR-based methods when assessing adipose depots are listed in Table 1.

Imaging analysis usually is performed on pieces of whole-mount samples using confocal microscopy and/or on paraffin-embedded samples. Although cryo-sectioned samples also can be used, the high lipid content of white adipose tissue makes the collection of high-quality frozen sections from fresh tissue very difficult.

The use of whole-mount samples has several technical advantages. The main one is that the three-dimensional structure of the tissue remains intact and can be visualized by capturing a z-series of images through confocal microscopy. These images provide additional details on microvascular network topology and capillary structure that are not captured in tissue sections, thus enabling vascular assessments beyond vessel density. Importantly, vascular remodeling in the adipose tissue can involve blood vessel enlargement either concomitant with increased density [14] or independently of changes in vascularity [21]. Moreover, the tissue preparation time is minimal, because no processing, embedding, or sectioning is required following tissue fixation (~45 min). However, the disadvantage of examining the adipose vasculature by this technique is that the vascular assessment is not normalized to adipocyte number, which can lead to either overestimation or underestimation of the vascular content when the adipocyte size differs in the groups assessed. Thus, histological analysis to assess adipocyte size should be performed in parallel. When changes in adipocyte sizes are detected, it is advisable to complement the whole-mount analysis with either assessment of mRNA levels of markers for endothelial cells or additional imaging analysis using paraffin-embedded samples.

While vascular network morphology cannot be visualized in paraffin-sectioned samples, this technique has the advantage of allowing the use of the same preparation to assess adipocyte cross-sectional areas, which enables the researcher to normalize the adipose vasculature relative to adipocyte number or area. However, imaging analysis from paraffin-sectioned sample is laborious and time-consuming due to the longer processing time associated with paraffin tissue embedding and sectioning. Moreover, antigen detection may be compromised by the paraffin embedding process.

Visualization of blood vessels in either whole-mount or paraffin-sectioned samples involves the specific labeling and detection of the vasculature, which can be achieved by antibody-mediated detection of endothelial cells or by the use of lectins

Table 1
Genes frequently assessed as markers of endothelial cell content in murine adipose tissue

Target	Gene	Synonyms	References
Vascular endothelial cells	Pecam1	Pecam-1, CD31	All[a]
	Cdh5	VE-cadherin, CD144	[12, 13]
	Vwf	von Willebrand factor	[2, 13, 14]
	Nos3	eNOS, endothelial nitric oxide synthase	[2, 14]
	Tek	Tie-2, tyrosine-protein kinase receptor	[15, 16]
	Kdr	Vegfr2, vascular endothelial growth factor receptor 2	[2, 17, 18]
	Flt1	Vegfr1, vascular endothelial growth factor receptor 1	[15, 17]
	Tie1	Tyrosine kinase with immunoglobulin like and EGF like domains 1	[18]
	Eng	Endoglin, CD105	[12]
Lymphatic endothelial cells	Lyve1	Lyve-1, lymphatic vessel endothelial hyaluronic acid receptor-1	[19, 20]
	Flt4	Vegfr3, vascular endothelial growth factor receptor 3	[19, 20]
	Prox1	Prox-1, prospero-related homeobox-1	[19, 20]
	Pdpn	Podoplanin	[19, 20]

[a]All references in this table assess Pecam1

labeled with fluorochromes or with biotin (Fig. 1). Lectins are proteins, usually derived from plants, that recognize carbohydrate moieties found in glycoproteins and glycolipids present on endothelial cells. The choice of which lectin to use is dependent on the species of the tissue to be imaged. While Griffonia (Bandeiraea) simplicifolia lectin I, isolectin B4 (GSL-I) is used to stain rodent endothelial cells [22], labeling of human endothelial cells is achieved with Ulex europaeus I lectin (UEA-I) [23]. Moreover, counterstaining of adipocytes and/or nuclei is employed during the imaging of capillary networks of adipose depots to provide visualization of additional tissue details. A list of common antibodies and lectins used for vascular assessment, together with lipid and nuclear stains that can be used for the complementary visualization of adipocytes and nuclei is provided in Table 2.

2 Materials

2.1 General

All solutions are prepared using deionized water and stored at room temperature unless otherwise indicated.

- 1× Phosphate-buffered saline, pH 7.4 (v/v).
- 37% Formaldehyde solution (Sigma).
- Fixing buffer: 3.7% formaldehyde in PBS, pH 7.4 (v/v).
- Triton X-100 (Sigma).

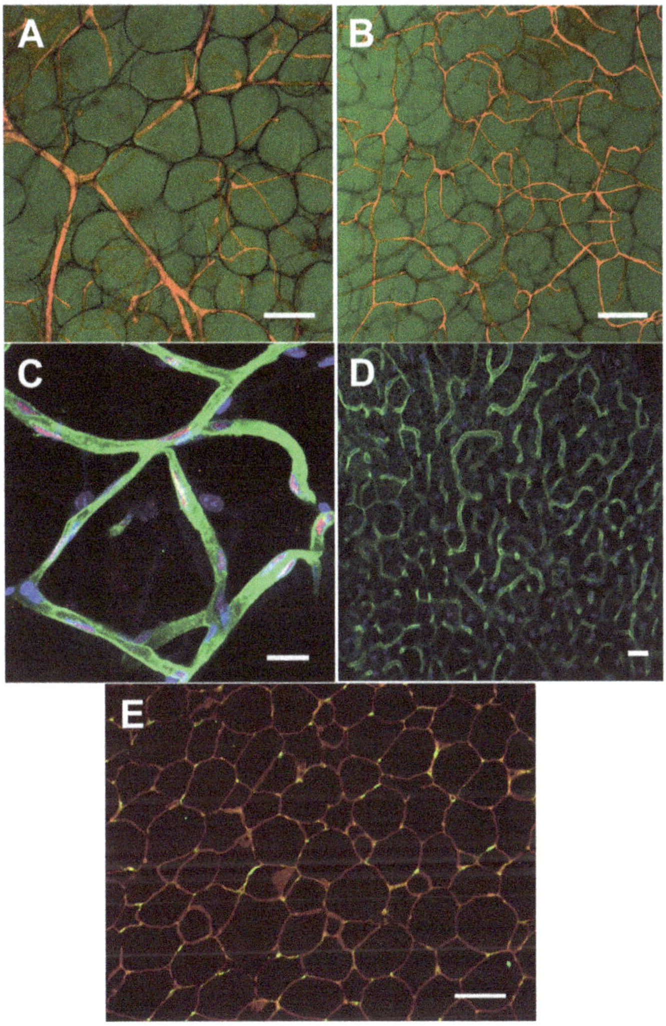

Fig. 1 Detection of adipose tissue vasculature employing whole-mounted and paraffin-sectioned samples. (**a** and **b**) Confocal images of whole-mount samples from human (**a**) and murine (**b**) subcutaneous depots stained with BODIPY 493/503 (green) and UEA-I lectin (red, **a**) or GSL-I lectin (red, **b**). Scale bar = 100 μm, ×10 magnification. (**c**) 3D image from whole-mounted murine visceral adipose tissue stained with ERG antibody (red) to label endothelial nuclei and GSL-I lectin (green) to label vascular structures and counterstained with DAPI (blue). Scale bar = 20 μm, ×40 magnification. (**d**) Single optical slice of whole-mount murine brown adipose tissue stained with GSL-I lectin (green) and DAPI (blue). Scale bar = 20 μm, ×20 magnification. (**e**) GSL-I lectin (green) and Wheat germ agglutinin (red) staining of paraffin-sectioned from murine visceral adipose tissue. Scale bar = 100 μm, ×10 magnification

- Goat serum (Sigma G9023) *Depending on the species of secondary antibody.
- Blocking solution: 5% goat serum/PBS.

Table 2
Lectins, antibodies, and counter-stains commonly used for capillary detection in whole-mount and paraffin-sectioned samples of adipose tissue

Target	Antibody lectin/stain	Dilution ratio	Company	Species	References
Endothelial cells	CD31	1: 500 (WM) 1:100 (IHC) 1:100 or 1:200 (IHC)	BD Biosciences (Clone MEC 13.3, 553370) Agilent Dako (Clone JC70A)	Mouse Human	[24, 25] [26, 27]
	Cd34 (in combination with Cd31)	1:500 (IHC)	Thermo Scientific Neomarkers (Clone QBeEnd/10)	Human	[24]
	Endomucin	NR (WM)	Santa Cruz (sc-65495)	Mouse	[6, 19]
	ERG	1:200 (WM)[a]	Cell signaling (97249) or Abcam (ab133264)	Mouse	This protocol
	Griffonia Simplicifolia Lectin I (GSL-I) isolectin B4	1:100 (WM) 1:50 (IHC)	Vector Laboratories (FL-1101, DL-1207 or DL-1208)[b]	Mouse	[2, 14]
	Ulex Europaeus Agglutinin I (UEA-I) lectin	1:100 (WM) 1:50 (IHC)	Vector Laboratories (FL-1061, DL-1067, DL-1068)[b]	Human	This protocol
	Von Willebrand factor	1:200 (IHC)	Agilent Dako (A0082)	Human	[26]
Lymphatic vessels	Lyve1	N.R. (WM)[a] 1:100 (IHC) 1:100 (IHC)	R& D Research Systems (AF2125) Rabbit IgG Novus (N.R.) Rabbit IgG R&D Research Systems (N.R.)	Mouse Human	[19] [27] [27]
	Podoplanin	1:50 (IHC)	Agilent Dako (Mouse IgG clone D2–40)	Human	[27]
Pericytes	CD13	1:200 (WM)	AbD Serotec (Clone ER-BMDM1)	Mouse	[28]
	NG-2	1:200 to 1:500 (WM)	Millipore Sigma (AB5320CH)	Mouse	[25, 28]
Lipids	BODIPY	0.25 μg/mL	Molecular Probes (D3922)	All	[14]
Basement membranes	Wheat germ agglutinin	1:100	Vector Laboratories (RL-1022)	Mouse, Human	This protocol
Nuclei	DAPI		Sigma-Aldrich (D9542)	All	

N.R. not reported
BODIPY (4,4-difluoro-1,3,5,7,8-pentamethyl-4-bora-3a,4a-diaza-s-indacene)
DAPI (40,6-diamidino-2-phenylindole dihydrochloride)
[a]requires permeabilization
[b]depending on desired fluorochrome

- Microscope slides (Fisher Scientific).
- Coverslips (Fisher Scientific; 22 mm × 50 mm, standard thickness (#1.5)).
- Hydrophobic pen (ImmEdge, Vector Labs Cat).
- Antifade Mounting Medium (VECTASHIELD® Vibrance™ Antifade Mounting Medium; Vector Labs, H-1700) or similar product.
- Image J (https://imagej.nih.gov/ij/).

2.2 Whole-Mount Samples

1. Blocking/permeabilization solution: 0.5% Triton X-100/5% goat serum/PBS (when intracellular staining is desired).
2. Antibodies/Stains: Refer to Table 2 for details.
3. 96-well plates or 1.5 mL microcentrifuge tubes (for tissue collection and staining incubation steps).
4. Rotating/rocking platform.
5. Rapid dry nail polish.

2.3 Paraffin Sections

1. 1.5 mL microcentrifuge tubes or 48-well plate for tissue collection.
2. 100% Xylene (or Histoclear II; National Diagnostics).
3. 100% Ethanol.
4. Sodium citrate buffer [10 mM sodium citrate pH 6.0, 0.05% Tween 20] (2.94 g tri-sodium citrate (dihydrate) in 1000 mL water. Dissolve salt, pH to 6.0 with HCl, then add 0.5 mL Tween 20).
5. Blocking solution: 0.05% Triton X-100/5% goat serum/PBS.
6. Antibodies/Stains: Refer to Table 2 for details.
7. Heat-proof glass slide rack and staining dish for dewaxing and antigen retrieval (i.e., Wheaton TM glass staining dish; Fisher Scientific 02-912-617 and 22-309-245).
8. Humid chamber for slide staining. This can be assembled using a tissue culture tray with foil-covered lid. Line the bottom with a dampened paper towel to provide the humidity. A raised platform, on which to set the slides, is created by placing parallel—1 mL (or similar caliber) pipettes (Fig. 2).

3 Methods

1. Harvest adipose tissue depots (e.g., brown, perigonadal, subcutaneous/inguinal) and cut them into small pieces (~7 × 7 mm).

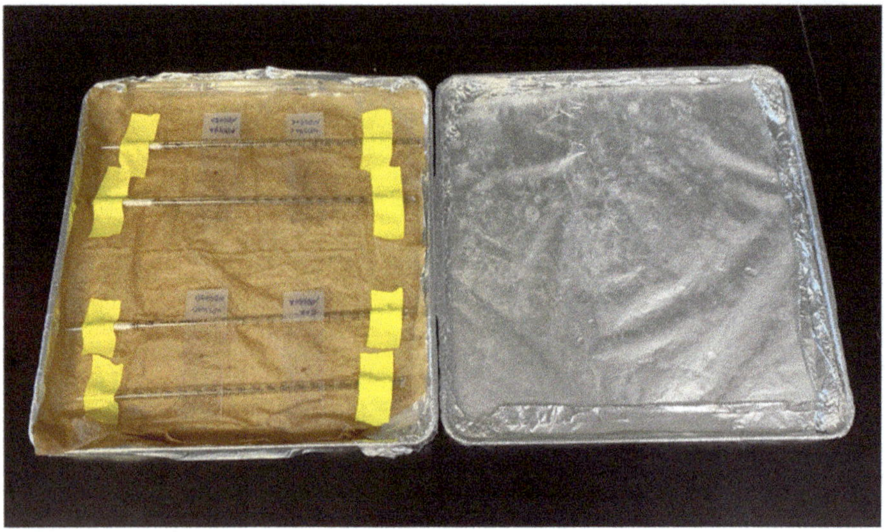

Fig. 2 Humid chamber for slide staining. A humid chamber can be prepared from a tissue culture tray with lid (lined with foil to protect from the light). A dampened paper towel on the bottom provides the humidity. Serological pipettes (1 mL or similar caliber) can be used to create a platform on which the slides sit

3.1 Assessment of Whole-Mount Samples

3.1.1 Sample Preparation

2. Incubate adipose pieces in fixing buffer for 30–45 min at room temperature. Ensure that the tissue is completely submerged in the fixing buffer.
3. Rinse away the fixing buffer with three, 5 min washes with PBS, pH 7.4, under gentle rocking at room temperature. Tissues may be stored at 4 °C in PBS, pH 7.4 until staining (*see* **Note 1**).
4. Cut fixed samples into smaller pieces (2 × 2 mm, Fig. 3a) and submerge them in PBS, pH 7.4 until having all samples prepared. Make sure the tissue does not dry. Either 96-well plates or microcentrifuge tubes can be used to incubate tissue (*see* **Notes 2** and **3**).

For staining performed only using lectin, proceed to staining step; for staining using antibodies, proceed to blocking step.

3.1.2 Blocking

Incubate the samples in 150 μL of blocking solution for 30 min at room temperature with gentle rocking. If permeabilization is required, perform this step using the blocking/permeabilization solution.

3.1.3 Staining

1. (a) When staining of the vasculature is performed with lectin ONLY, incubate samples with 150 μL lectin diluted in PBS, pH 7.4, for 1 h protected from light, with gentle rocking. In case counterstaining with BODIPY is performed, this stain can be added to the same solution.

(b) When staining samples with antibodies, incubate samples with 150 μL primary antibodies diluted in blocking or blocking/permeabilization solution overnight at 4 °C, protected from light, with gentle rocking. Refer to Table 2 for suggested stains and dilutions.

2. (a) When conducting staining with lectin, wash samples with 150 μL of PBS, pH 7.4, three times (3× 5 min per wash) at room temperature with gentle rocking. In case counterstaining with DAPI is performed, stain should be diluted in PBS, pH 7.4, used for the first wash. Proceed to the mounting step or store samples in PBS, pH 7.4, at 4 °C, protected from light.

   (b) When staining samples with antibodies, wash samples in 150 μL of PBS, pH 7.4, three times (10–15 min per wash) at room temperature with gentle rocking and proceed to **step 3**.

3. Incubate samples with 150 μL of appropriate secondary antibodies diluted in blocking solution for 1 h at room temperature, protected from light, with gentle rocking. In case counterstaining with lectin and/or BODIPY is performed, these stains can be added to the secondary antibody solution.

4. Wash samples in 150 μL of PBS, pH 7.4, three times (5 min for the first wash followed by two washes of 10–15 min each) at room temperature with gentle rocking. In case counterstaining with DAPI is performed, stain should be diluted in PBS, pH 7.4, used for the first wash. Proceed to the mounting step or store samples in PBS, pH 7.4, at 4 °C, protected from light.

3.1.4 Mounting

1. Use a hydrophobic pen to draw a limited area on a microscope slide. Then place the small pieces of stained tissue within this area and add one to two drops of mounting medium onto the tissue samples (Fig. 3b). When mounting the slide with more than one sample, ensure that space is left between them.

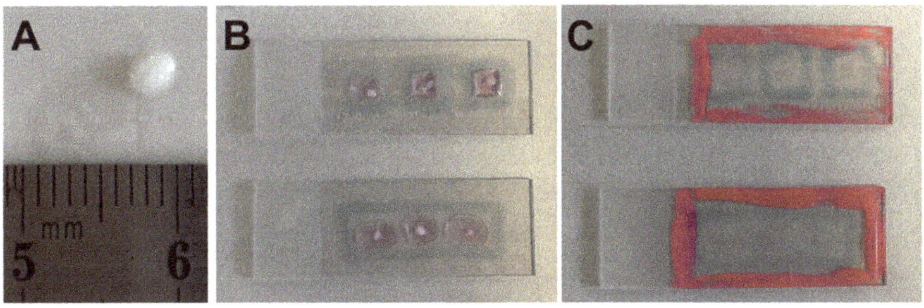

Fig. 3 Preparation of whole-mount samples. (**a**) Sample of a small piece of adipose tissue that can be used for whole-mount staining. (**b**) After staining, draw limited areas on a microscope slide using a hydrophobic pen, place the stained samples within the area, and add one to two drops of mounting medium onto the tissue samples. (**c**) Seal coverslips to the slides by applying nail polish to all edges of the coverslip

2. Cover the samples by gently placing a coverslip onto them and lightly pushing the coverslip down. For best results, push the coverslip down evenly on all sides, pressing the sides of the coverslip, and avoid adding too much pressure on top of the tissue samples. Mounting media should disperse evenly around all tissue samples.
3. To seal coverslips to the slides, apply nail polish to all edges of the coverslip and allow nail polish to dry (Fig. 3c).
4. Store the slides in the dark until examination under confocal microscope, which must be done within a few days for best results. Depending on the mounting medium used, slides can be stored at room temperature or will require refrigeration.

3.1.5 Whole-Mount Confocal Image Acquisition

Specific parameters for the acquisition of images of whole-mount-stained adipose tissues will vary dependent on the experimental conditions and the available confocal microscope and image acquisition software.

Adjustments for optimal gain and black level settings should be defined per staining condition and fluorochromes used. Similar exposure settings should be maintained across all experimental conditions during acquisitions. Moreover, be aware that the scan speed during acquisition will affect the quality of the images.

For vascular density assessment, capture confocal images at ×10 or ×20 magnification, avoiding regions with larger blood vessels or visible tissue defects. While a single scan can be captured at the best focal plane, the ideal is to collect a series of 5–10 z-stack images (6–20 μm depth, depending on the size of tissue piece) that can be reconstructed as a 3D image. If single scans are to be captured, set the focal plane carefully as vascular density can vary greatly depending on the focal plane selected, as exemplified in the series of z-stack images presented in Fig. 4.

Acquire five to six random and non-overlapping fields of view (FOV) per sample, which can be achieved by capturing two to three images per piece of tissue prepared. For the high-resolution analysis of vascular structures, acquire confocal images at higher magnification (×40 or ×60).

3.1.6 Whole-Mount Image Analysis

Confocal images can be easily analyzed using the open source image analysis software Image J (https://imagej.nih.gov/ij/) [29]. Below, we will describe step-by-step analysis for the assessments of vascular density and vascular diameter.

A. Assessing vascular density:

1. Open the image (TIF file format) captured at ×10 or ×20 magnification in Image J.
2. If you have captured a multi-channel image (of multiple stains), you first need to split the channels (Image → Color → Split

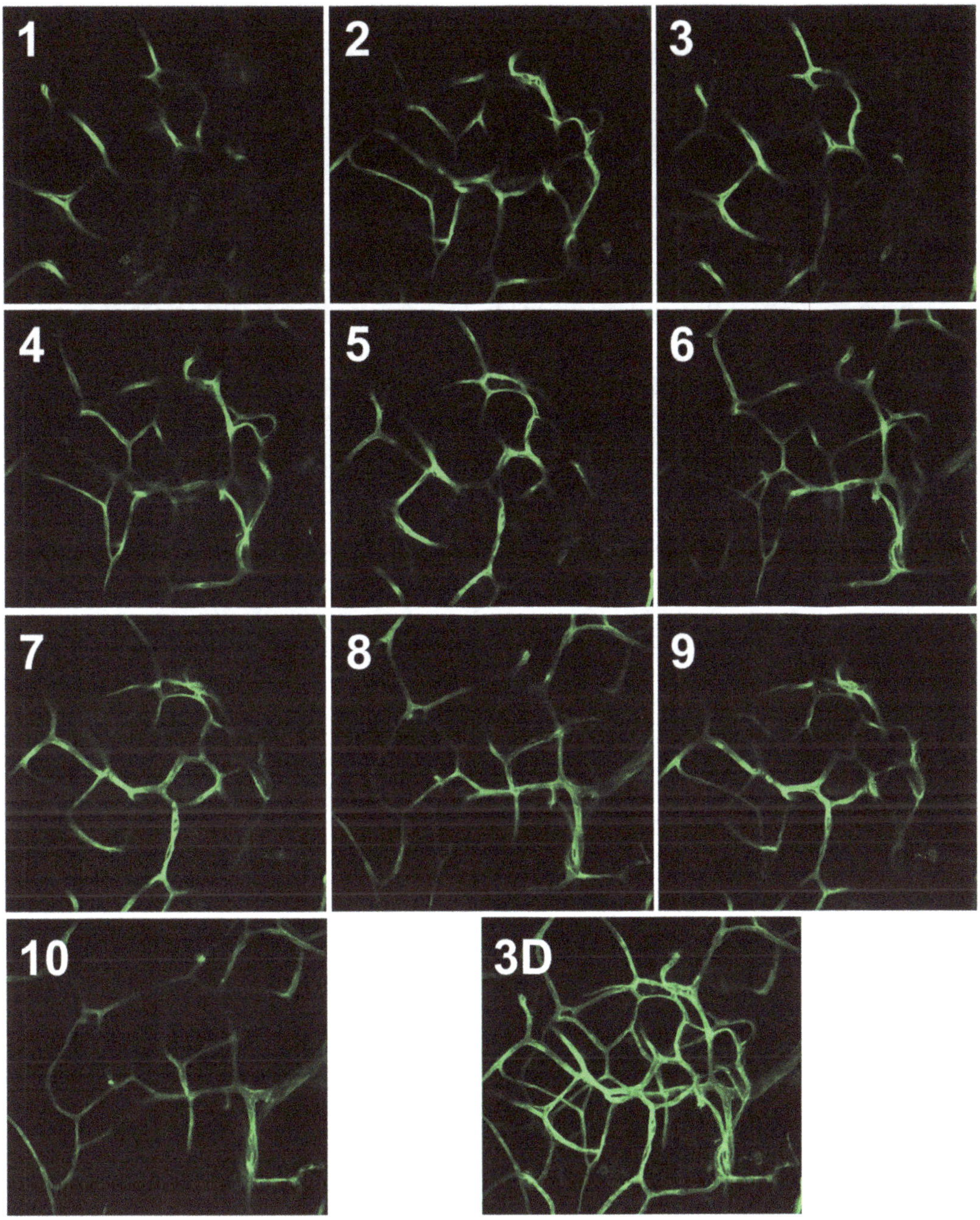

Fig. 4 Example of the acquisition of a series of optical sections from whole-mounted stained sample reconstructed as a 3D image. Z-stack images from adult murine visceral adipose tissue stained with GSL-I lectin were captured at ×40 magnification (0.5 zoom) in 18.45 μm depth. Individual optical slices (1–10) show partial density compared to the 3D reconstructed image

Channels) (Fig. 5a–c). If you captured a single-channel image (of a single stain), you should already be working with an 8-bit single-channel image. If it is a 16-bit image, you must first convert to 8-bit prior to **step 3** (Image → Type → 8 bit).

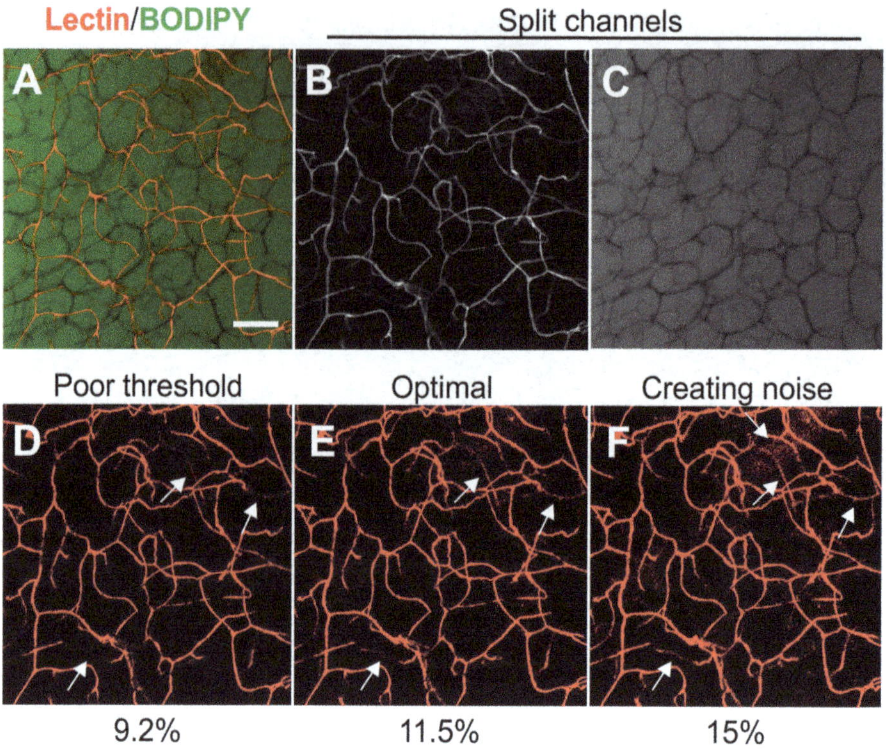

Fig. 5 Analysis of vascular density from whole-mounted stained adipose samples. (**a**) Single confocal ×10 scan from adult murine adipose tissue (subcutaneous) stained with GSL-I lectin (red) and BODIPY (green). (**b** and **c**). Split the color channels of multi-stained samples. (**d–f**) Carefully set appropriate threshold, paying attention to whether the threshold captures (**e**) or fails to capture (**d**) structures of interest without introducing additional noise (**f**) indicated by arrows. The threshold setting directly affects the results as exemplified by the different % area values listed using the distinct thresholds. Refer to Subheading 3.1.6.A for further details

3. Make sure you have *Area*, *Area fraction*, and *Limit to threshold* selected before proceeding with your measurements (Analyze → Set Measurements) (*see* **Note 4**).
4. Carefully define a suitable threshold (Image → Adjust → Threshold) or (Command/Ctrl+Shift+T) by selecting "dark background" and adjusting the sliders on the threshold window until the features of interest (*see* **Note 5** and Fig. 5d–f).
5. Select "Measure" from the analyze menu (or Command/Ctrl +M). A popup "Results" box will appear reporting values for total area, threshold settings, and % area. Use the % area values for your assessments. Copy measurements into an Excel spreadsheet.
6. Repeat **steps 1–5** for all five to six random and non-overlapping FOV acquired for each sample.

7. When finished vascular density assessments. Calculate the average % area for each sample and express the results as vascular area (% of FOV).

B. Assessing microvessel diameter:

1. Open the image (TIF file format) for which you want to analyze the vessel diameters (*see* **Note 6**).
2. Set scale (Analyze → Set Scale). Consult the microscopy software, microscope handbook, or use an image of a stage micrometer to determine the appropriate conversion factor of pixel to micrometer (μm) for the images that you will analyze. Enter the distance in pixels with the corresponding μm distance. Select "micrometer" for the Unit of length box. Select "Global" to apply this calibration scale to all additional images opened in the Image J session.
3. Select the straight segment tool.
4. Proceed to draw a straight line across the diameter of the vessel.
5. Click Image → Overlay → Add selection (or Command/Ctrl +B) to save the measurement.
6. Click Analyze → Measure or (Command/Ctrl+M) to obtain the line length (i.e., vessel diameter).
7. Measure the diameter of every vessel segment, each time repeating **steps 4–6**. It is advisable to measure long vessel segments in more than one site (Fig. 6).
8. When finished measuring vessels in that image, copy measurements into an Excel spreadsheet.
9. Repeat **steps 1–7** for all fields of view for each experimental sample.
10. When finished assessment, proceed to analysis.
11. Calculate the average vessel diameter for each experimental sample or use the individual diameter measurements to generate a frequency histogram (*see* **Note 7**).

3.2 Assessment of Paraffin-Embedded Sections

3.2.1 Sample Preparation

1. Harvest adipose tissue depots (e.g., brown, perigonadal, subcutaneous/inguinal) and cut them into small pieces (~7 × 7 mm).
2. Place tissue in 1.5 mL tubes or in wells of a 48-well plate containing fixing buffer. Ensure that the tissue pieces are completely submerged in fixation buffer. Incubate them in fixing buffer overnight at 4 °C.
3. Rinse away the fixing buffer with three × 5 min washes with PBS, pH 7.4, at room temperature with gentle rocking. Samples should be stored in 70% ethanol at 4 °C until tissue processing.

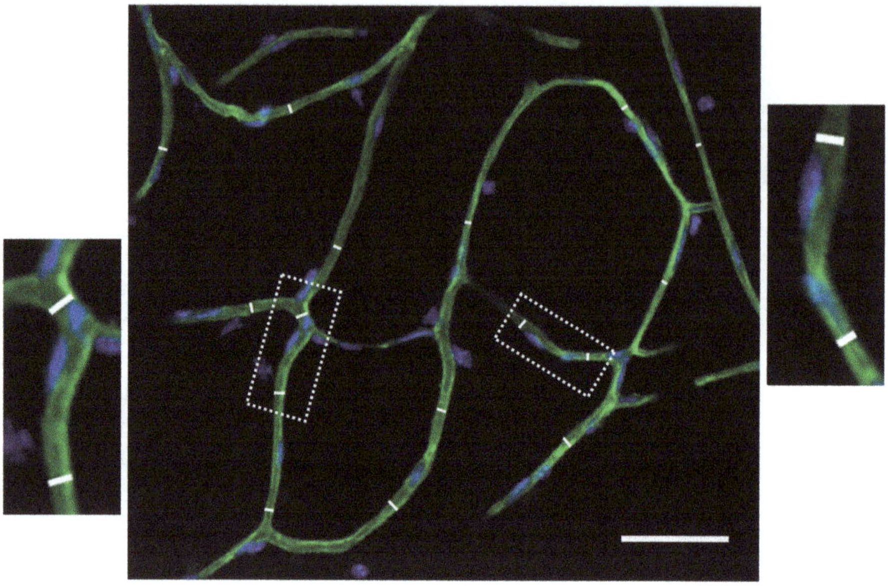

Fig. 6 Digital analysis of vessel diameter. Using the freeform tool in Image J, draw straight lines across the diameter of each vessel segment. It is advisable to measure long vessel segments in more than one site to capture diameter variability. Representative diameter measurements are illustrated by white lines across vessels, with zoomed images shown for increased resolution. 3D image from adult murine adipose tissue (visceral) stained with GSL-I lectin (green) and counterstained with DAPI (blue) captured at ×40 magnification (0.5 zoom) in 6.15 μm depth. Scale bar = 50 μm. Refer to Subheading 3.1.6.B for further details

4. Tissue processing for paraffin embedding using standard protocols [30], followed by microtome sectioning to generate 5 mm thick tissue sections.

3.2.2 Dewaxing and Epitope Retrieval

1. Dewaxing of paraffin and re-hydration of sections, by incubating slides for 5 min in each of the following (at room temperature):
   (a) Xylene/Histoclear × 2
   (b) 100% Ethanol × 2
   (c) 95% Ethanol × 1
   (d) Water × 2
2. Place ~600–800 mL of epitope retrieval buffer within a large heat-proof dish and bring to a boil on a hot plate (*see* **Note 8**).
3. Place slides in slide box and add to the hot epitope retrieval buffer. Boil for 10 min (cover loosely).
4. Remove slide box from heat and let sit 30 min. Proceed to blocking step.

3.2.3 Blocking

1. Use a hydrophobic pen to draw a complete barrier around the tissue sections to limit the spread of staining solutions. Limit the size of the area as much as possible but do not touch the tissue section(s). If multiple sections are on the same slide, they can be circled independently, allowing for multiple distinct stains to be conducted on the same slide.
2. Incubate sections with PBS, pH 7.4, at room temperature for 5 min.
3. Remove excess liquid gently. Tissue sections must remain wet from this point forward.
4. Cover sections with blocking/permeabilization buffer. Incubate the slides in a humid chamber for 30 min at room temperature.

   Volumes used will depend on size of area circled on the slide. Ensure solution covers the tissue section thoroughly, without overflowing the hydrophobic barrier.
5. Gently vacuum or wick away blocking/permeabilization buffer and proceed to staining step.

3.2.4 Staining

1. Add primary antibody solution or lectin solution. Refer to Table 2 for suggested stains and dilutions. Incubate slides in a humid chamber, protected from light, for 2 h at room temperature.
2. Gently vacuum or wick away the antibody/lectin solution. Wash slides three times (5 min per wash) with PBS, pH 7.4. Washes can be achieved by pipetting PBS onto each tissue section or by immersing slides in a glass staining dish filled with PBS.

   If staining with lectin only, proceed to mounting step.
3. Incubate samples with secondary antibody solution for 1 h at room temperature, protected from light by covering the humid chamber with foil.
4. Wash samples in PBS, pH 7.4, three times, 5 min per wash, at room temperature. If counterstaining nuclei with DAPI, stain should be diluted in PBS used for the first wash.

3.2.5 Mounting

1. Following removal of the final PBS wash, add one or two drops of mounting medium to the surface of tissue sections and carefully place coverslip over the top, avoiding formation of air bubbles. Mounting media should disperse evenly, covering the entire area within the hydrophobic barrier.
2. Mounting media may require time to set, according to the manufacturer's directions. Slides may be imaged immediately or stored in a light-protected container at 4 °C until the following day. The fluorescence intensity may decline if stored longer than 1 day.

3.2.6 Image Acquisition of Adipose Tissue Sections

1. Images can be acquired using a conventional inverted or upright fluorescence microscope that is equipped with excitation/emission filter sets for DAPI (if using as counter-stain), FITC and rhodamine and a cooled digital CCD camera. Typically, ×10 or ×20 objectives are sufficient for vascular quantification.
2. Optimize exposure settings for each wavelength to avoid pixel saturation or signals that are too weak. Maintain similar exposure settings (rather than using automatic exposure settings) between each specimen examined to ensure the highest accuracy of quantification.
3. Optimize slide positioning so that the tissue section fills each field of view, when possible. Collect matched images of each label (i.e., capillaries and adipocyte basement membrane) for every field of view. Collect images from multiple non-overlapping fields of view, ensuring representation of various regions of the tissue section (*see* **Note 9**).

3.2.7 Image Analysis of Adipose Tissue Sections

A. *Analysis of vascular area*:

1. The GSL-I lectin (or UEA-I or another endothelial cell marker)-stained image is used to assess vascular area. Open image (TIF format) in Image J. If it is a 16-bit image, you must first convert to 8-bit prior to **step 2** (Image → Type → 8 bit).
2. Set scale (Analyze → Set Scale). Consult the microscopy software, microscope handbook, or use an image of a stage micrometer to determine the appropriate conversion factor of pixel to micrometer (μm) for the images that you will analyze. Enter the distance in pixels with the corresponding μm distance. Select "micrometer" for the unit of length box. Selecting "Global" will ensure that this calibration scale is applied to all additional images opened in the Image J session.
3. Set threshold (Image → Adjust → Threshold) or (Command/Ctrl+Shift+T).

To define a suitable threshold, select "dark background" and adjust the sliders on the threshold window until the features of interest (capillaries) are highlighted but not exaggerated (*see* **Note 10**).

4. Next, you will calculate the area represented by the thresholded pixels. To do this, make sure you have *Area*, and *Limit to threshold* selected before proceeding with your measurements (Analyze → Set Measurements). Select "Measure" from the analyze menu (or Command/Ctrl+M). A popup box (Results) will appear (Fig. 7a). Copy the vascular area measurements into an Excel spreadsheet (*see* **Note 11**).

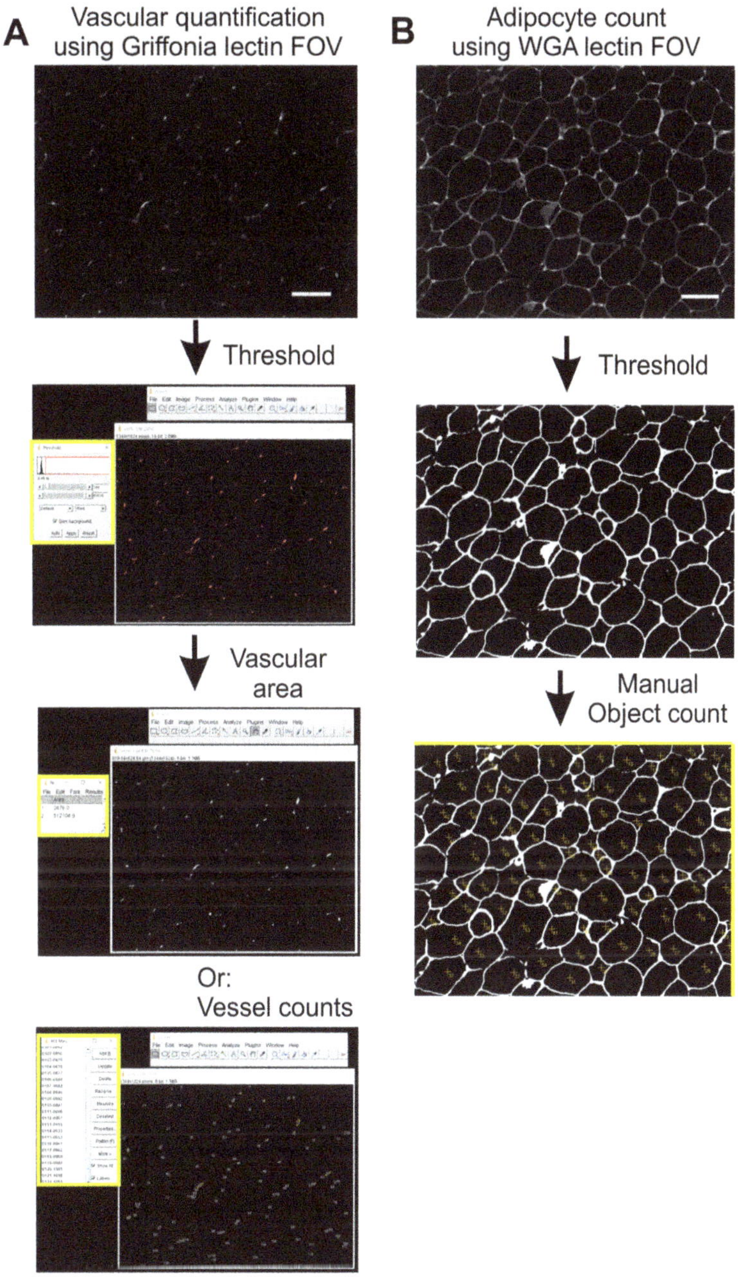

Fig. 7 Analysis of adipose vascular content from paraffin-embedded samples. (**a**) Example of a ×10 image of murine visceral adipose tissue stained with GSL-I lectin (**a**) and WGA (**b**) that can be used for the analysis of adipose vasculature and adipocyte counting. Analysis of vascular area or vessel counts can be performed from images of paraffin-embedded adipose samples stained with lectin (or another endothelial cell stain) after definition of a suitable threshold. Yellow boxes highlight the relevant Image J analysis windows. (**b**) Adipocyte counting within a field of view can be performed by either manual object count or by using the object counter tool in Image J from WGA-stained images. Yellow lines indicate the two sides of the image on which partial adipocytes are counted. Scale bar = 100 μm. Refer to Subheading 3.2.7 for further details

5. Determine the area of the FOV (in μm^2) by de-selecting Limit to Threshold (Analyze → Set Measurements), then measure the area (Analyze → Measure; or Command/Ctrl+M). Copy the adipose area (μm^2) measurements into an Excel spreadsheet. Convert to adipose area (mm^2) by dividing the μm^2 values by 10^6 (*see* **Note 4**).
6. Repeat **steps 1–3** for all non-overlapping FOV acquired for each sample.
7. When finished assessments from all FOV, calculate the average vascular area/mm^2 adipose area for each sample.

B. Analysis of capillary to adipocyte ratio:

1. The GSL-I lectin (or UEA-I or another endothelial cell marker)-stained images are used to assess capillary number (Fig. 7a). Open image (TIF file format). If working with the thresholded image that was generated to calculate vascular area, proceed directly to **step 3**.
2. Set threshold (Image → Adjust → Threshold) or (Command/Ctrl+Shift+T).
3. Capillaries can be quantified manually (using the "point tool") or can be quantified using the "Analyze Particles" function (Analyze → Analyze Particles) (*see* **Note 12**).
   (a) Select *Display results*, *Exclude on edges*, *Summarize*, *Add to Manager*.
   (b) "Display results" provides the area of each individual object; "exclude on edges" disregards objects that extend past the field of view; "Summary" provides the total object count and additional details such as total area of counted objects, average object size, and % area (of FOV); "Add to manager" provides additional measurement options and the opportunity to "Flatten" the count overlay onto the image to save.
4. WGA images are used to count the adipocytes, using the object counter tool in Image J. Count all adipocytes that are fully visible within the field of view. Adipocytes that overlap the edge of the image should be counted on two sides only (i.e., top and right side). This will provide you with the number of adipocytes per FOV (Fig. 7b).
5. The capillary to adipocyte ratio (C:A) is calculated by using the "total object count" divided by the number of adipocytes. This calculation should be performed on each independent FOV per experimental sample.
6. The C:A for one sample is calculated by averaging the C:A for each individual FOV for that experimental sample.

4 Notes

1. Storage longer than 2 weeks may compromise staining quality.
2. 96-well plate is recommended for the preparation of multiple samples.
3. Volumes mentioned in the following steps are recommended for preparation using 96-well plate.
4. If the field of view is only partially covered by your sample, select the area of interest using any of the drawing/selection tools (i.e., rectangle, circle, polygon, or freeform). No selection is necessary in situations where the sample fills the full field of view.
5. Thresholding is used to identify pixels above or below a particular (threshold) value. There is no always-applicable strategy to determine a threshold and it is unlikely that a single threshold setting can be applied to all images, as variability in staining intensity will be observed between experiments and between images within a single experiment. Thus, threshold will be set for each image. To define a suitable threshold, segment the image into features of interest and background by carefully setting the lower and upper threshold values using the sliders on the threshold window. Carefully observe what the threshold detects. Poor threshold settings will either over- or underestimate vascular area (Fig. 5e, f).
6. ×20 or ×40 images are needed for better visualization of vessel structure.
7. If you did not set the scale conversion from pixels to micrometers prior to making the diameter assessments, you may apply the conversion factor in the excel spreadsheet by multiplying the diameter in pixels by the number of pixels per μm.
8. Numerous procedures exist for epitope retrieval. Alternative buffer solutions (Tris-EDTA, or commercially purchased) and heating methods (microwave, pressure cooker) may be utilized.
9. The number of fields of view will depend on the magnification and the size of the tissue. Generally, analysis of three independent FOV is appropriate for quantification if using ×10, whereas five to six FOV should be captured if using ×20.
10. The same thresholded image can be used for quantifying vascular area and the capillary number (as described in Subheading 3.2.7.B).
11. The thresholded area also can be represented as a % of the field of view. If this is your intention, then select *Area Fraction* (under Analyze → Set Measurements). The area fraction values represent the % vascular area within the FOV.

12. When using the "analyze particles" function, size and circularity of objects can be specified. The default size is 0 to Infinity, meaning that all items within the thresholded image will be counted. Specifying a minimum size can be used to avoid punctate background signals and maximum size can be used to exclude large vessels. The default circularity is 0.00–1.00, meaning that all items within the thresholded image will be counted regardless of their shape. Adjusting the minimum allowed "circularity" may be useful to disregard irregular objects or transverse vessels. Several iterations of adjusting the "size of objects" and "circularity" parameters until you are satisfied that the appropriate objects are captured in the analysis.

References

1. Pasarica M, Sereda OR, Redman LM et al (2009) Reduced adipose tissue oxygenation in human obesity: evidence for rarefaction, macrophage chemotaxis, and inflammation without an angiogenic response. Diabetes 58: 718–725. https://doi.org/10.2337/db08-1098
2. Rudnicki M, Abdifarkosh G, Rezvan O et al (2018) Female mice have higher angiogenesis in perigonadal adipose tissue than males in response to high-fat diet. Front Physiol 9: 1452. https://doi.org/10.3389/fphys.2018.01452
3. Fukumura D, Ushiyama A, Duda DG et al (2003) Paracrine regulation of angiogenesis and adipocyte differentiation during in vivo adipogenesis. Circ Res 93:e88–e97. https://doi.org/10.1161/01.RES.0000099243.20096.FA
4. Han J, Lee J-E, Jin J et al (2011) The spatiotemporal development of adipose tissue. Development 138:5027–5037. https://doi.org/10.1242/dev.067686
5. Seki T, Hosaka K, Lim S et al (2016) Endothelial PDGF-CC regulates angiogenesis-dependent thermogenesis in beige fat. Nat Commun 7:12152. https://doi.org/10.1038/ncomms12152
6. Zhao Y, Li X, Yang L et al (2018) Transient overexpression of vascular endothelial growth factor A in adipose tissue promotes energy expenditure via activation of the sympathetic nervous system. Mol Cell Biol 38:e00242-18. https://doi.org/10.1128/MCB.00242-18
7. Shimizu I, Aprahamian T, Kikuchi R et al (2014) Vascular rarefaction mediates whitening of brown fat in obesity. J Clin Invest 124:2099–2112. https://doi.org/10.1172/JCI71643
8. Crewe C, An YA, Scherer PE (2017) The ominous triad of adipose tissue dysfunction: inflammation, fibrosis, and impaired angiogenesis. J Clin Invest 127:74–82. https://doi.org/10.1172/JCI88883
9. Sung H-K, Doh K-O, Son JE et al (2013) Adipose vascular endothelial growth factor regulates metabolic homeostasis through angiogenesis. Cell Metab 17:61–72. https://doi.org/10.1016/j.cmet.2012.12.010
10. Muller WA (2003) Leukocyte-endothelial-cell interactions in leukocyte transmigration and the inflammatory response. Trends Immunol 24:327–334. https://doi.org/10.1016/s1471-4906(03)00117-0
11. Woodfin A, Voisin M-B, Nourshargh S (2007) PECAM-1: a multi-functional molecule in inflammation and vascular biology. Arterioscler Thromb Vasc Biol 27:2514–2523. https://doi.org/10.1161/ATVBAHA.107.151456
12. Geys L, Scroyen I, Roose E et al (2015) ADAMTS13 deficiency in mice does not affect adipose tissue development. Biochim Biophys Acta 1850:1368–1374. https://doi.org/10.1016/j.bbagen.2015.03.008
13. Senol-Cosar O, Flach RJR, DiStefano M et al (2016) Tenomodulin promotes human adipocyte differentiation and beneficial visceral adipose tissue expansion. Nat Commun 7:10686. https://doi.org/10.1038/ncomms10686
14. Rudnicki M, Abdifarkosh G, Nwadozi E et al (2018) Endothelial-specific FoxO1 depletion prevents obesity-related disorders by increasing vascular metabolism and growth. eLife 7: e39780. https://doi.org/10.7554/eLife.39780

15. Halberg N, Khan T, Trujillo ME et al (2009) Hypoxia-inducible factor 1alpha induces fibrosis and insulin resistance in white adipose tissue. Mol Cell Biol 29:4467–4483. https://doi.org/10.1128/MCB.00192-09
16. An YA, Sun K, Joffin N et al (2017) Angiopoietin-2 in white adipose tissue improves metabolic homeostasis through enhanced angiogenesis. eLife 6:e24071. https://doi.org/10.7554/eLife.24071
17. Kusminski CM, Park J, Scherer PE (2014) MitoNEET-mediated effects on browning of white adipose tissue. Nat Commun 5:3962. https://doi.org/10.1038/ncomms4962
18. Bonacina F, Moregola A, Porte R et al (2019) Pentraxin 3 deficiency protects from the metabolic inflammation associated to diet-induced obesity. Cardiovasc Res 115:1861–1872. https://doi.org/10.1093/cvr/cvz068
19. Chakraborty A, Barajas S, Lammoglia GM et al (2019) Vascular endothelial growth factor-D (VEGF-D) overexpression and lymphatic expansion in murine adipose tissue improves metabolism in obesity. Am J Pathol 189: 924–939. https://doi.org/10.1016/j.ajpath.2018.12.008
20. Bulut GB, Alencar GF, Owsiany KM et al (2021) KLF4 (Kruppel-Like Factor 4)-dependent perivascular plasticity contributes to adipose tissue inflammation. Arterioscler Thromb Vasc Biol 41:284–301. https://doi.org/10.1161/ATVBAHA.120.314703
21. Sawane M, Kajiya K, Kidoya H et al (2013) Apelin inhibits diet-induced obesity by enhancing lymphatic and blood vessel integrity. Diabetes 62:1970–1980. https://doi.org/10.2337/db12-0604
22. Laitinen L (1987) Griffonia simplicifolia lectins bind specifically to endothelial cells and some epithelial cells in mouse tissues. Histochem J 19:225–234. https://doi.org/10.1007/BF01680633
23. Holthöfer H, Virtanen I, Kariniemi AL et al (1982) Ulex europaeus I lectin as a marker for vascular endothelium in human tissues. Lab Investig 47:60–66
24. Goossens GH, Alessandro B, Nicolas V et al (2011) Increased adipose tissue oxygen tension in obese compared with lean men is accompanied by insulin resistance, impaired adipose tissue capillarization, and inflammation. Circulation 124:67–76. https://doi.org/10.1161/CIRCULATIONAHA.111.027813
25. Robciuc MR, Kivelä R, Williams IM et al (2016) VEGFB/VEGFR1-induced expansion of adipose vasculature counteracts obesity and related metabolic complications. Cell Metab 23:712–724. https://doi.org/10.1016/j.cmet.2016.03.004
26. Gealekman O, Guseva N, Hartigan C et al (2011) Depot specific differences and insufficient subcutaneous adipose tissue angiogenesis in human obesity. Circulation 123:186–194. https://doi.org/10.1161/CIRCULATIONAHA.110.970145
27. de Redondo PAG, Gubert F, Zaverucha-do-Valle C et al (2020) Lymphatic vessels in human adipose tissue. Cell Tissue Res 379: 511–520. https://doi.org/10.1007/s00441-019-03108-5
28. Onogi Y, Wada T, Kamiya C et al (2017) PDGFRβ regulates adipose tissue expansion and glucose metabolism via vascular remodeling in diet-induced obesity. Diabetes 66: 1008–1021. https://doi.org/10.2337/db16-0881
29. Schneider CA, Rasband WS, Eliceiri KW (2012) NIH image to ImageJ: 25 years of image analysis. Nat Methods 9:671–675. https://doi.org/10.1038/nmeth.2089
30. Im K, Mareninov S, Diaz MFP, Yong WH (2019) An introduction to performing immunofluorescence staining. Methods Mol Biol 1897:299–311. https://doi.org/10.1007/978-1-4939-8935-5_26

Chapter 17

Quantification of Angiogenesis in Laser Choroidal Neovascularization

Kathryn R. Green, Nicholas Beazley-Long, Amy P. Lynch, Claire L. Allen, David O. Bates, and Andrew V. Benest

Abstract

Murine laser-induced laser choroidal neovascularization is a widely used and robust model of wet (exudative) age-related macular degeneration (wAMD). wAMD is one of the leading causes of blindness in the Western world. In brief, a focused laser beam is used to penetrate Bruch's membrane, which separates the choriocapillaris (well-vascularized choroid layer) from the pigmented layers of the retina. Damage to the integrity of this membrane during diabetes leads to fluid accumulation and vascular invasion into the subretinal layers resulting in a progressive worsening of vision. Here we describe a 14-day model using untreated C57/Bl6 mice, but it is equally applicable to incorporation into transgenic studies and therapeutic agent development (such as eye drops), injection of therapeutic agents (including antibodies), and for longer time course studies. In vivo functional analysis or lesioned choroids can be studied with further immunohistochemical staining for further analyses.

Key words Angiogenesis, Choroid, Neovascularization, wAMD, Endothelial cell

1 Introduction

Studying pathological angiogenesis in the eye has revealed much about the key signaling molecules (such as vascular endothelial growth factor, VEGF-A) which in turn has led ocular disease to have been the first to benefit from licensed antiangiogenic therapy [1]. Over the subsequent decade and a half, ocular angiogenesis, which is often linked to underlying diabetes is a key focus of drug developers worldwide. One of the leading causes of blindness today is age-related macular degeneration (wet, wAMD). The choroid is a pigmented and highly vascularized membrane in the posterior region of the eye, sitting between the sclera and the retina [2]. Of note for oculo-vascular diseases, the capillary rich "choriocapillaris" is a fenestrated and sinusoidal microvascular network which is separated from the retina by the Bruch's membrane. Blood flow

Andrew V. Benest (ed.), *Angiogenesis: Methods and Protocols*, Methods in Molecular Biology, vol. 2441, https://doi.org/10.1007/978-1-0716-2059-5_17,

here is reported to account for nearly 90% of bloody supply to the eye. The highly metabolic retina and retinal pigmented epithelium demand a significant blood supply [2]. Lesions to Bruch's membrane can result in inflammation, fluid leakage, and vascular invasion through the membrane and into the subretinal space [3]. Left untreated, the dysfunction results in scarring, and progression edema and inflammation resulting in a progression vision loss.

Use of a laser to generate lesions in the Bruch's membrane is a well-tolerated and robust way to induce inflammation and vascular invasion into the subretinal space. Additionally, fluorescent fundus angiography (FFA) is a supplementary technique to monitor in vivo consequences of vessel leakage. The use of a fluorescent contrast medium allows the extent of the neovascularization and plasma leakage to be quantified in a noninvasive manner to allow time course experiments to take place. At the end of the procedures, the mouse can be sacrificed and the eye dissected to perform immunohistochemical staining to contribute to phenotypic characterization of the lesions themselves. Here we describe a 14-day model using untreated C57/Bl6 mice, but it is equally applicable to incorporation into transgenic studies and therapeutic agent development (such as eye drops), injection of therapeutic agents (including antibodies), and for longer time course studies [4].

2 Materials

2.1 Laser Photocoagulation and Sodium Fluorescein Angiography (FFA)

- C57BL/6 J mice (Charles River) (*see* **Note 1**).
- Meridian Merilas green laser photocoagulator (532 nm, class 3b).
- Micron IV™ ophthalmoscope (Phoenix Technology Group Inc.) and mouse specific objective lens piece.
- Imaging cradle with fitted heat mat (Phoenix Technology Group Inc.).
- Anesthetic induction: 0.125 mL ketamine hydrochloride (Ketasetâ, 100 mg/mL stock); 0.147 mL medetomidine hydrochloride (Sedastartâ, 1 mg/mL stock); 0.728 mL sterile water.
- 1 mL syringes and 24-gauge sterile needles.
- (5%, w/v) Phenylephrine hydrochloride eye drops.
- (0.8%. w/v) Tropicamide eye drops.
- Cotton tips.
- Viscotearsâ.
- Ten percent sodium fluorescein (Na–Fl) solution (w/v): Weigh 5 mg of Na–Fl salt and dissolve in (0.9%, w/v) 50 mL sodium chloride solution. Filter through a 0.45 μm Corning filter, protect from light, and store at 4 °C to avoid degradation.

- Anesthetic recovery: Add 0.6 mL of (5 mg/mL) atipamezole hydrochloride stock solution (Sedastopâ) to 2.4 mL sterile water and mix thoroughly.
- Heat mat (37 °C), a recovery cage with absorbent bedding, mash, and a freshwater bottle.
- Discover imaging software (Phoenix Technology Group Inc.).

2.2 Eye Enucleation and Choroid Flat-Mount Dissection

- Fine curved forceps.
- Sterile phosphate buffered saline (PBS).
- Paraformaldehyde (4%, w/v PFA): In a class II chemical safety cabinet dissolve a 10 mL vial of 16% PFA in 30 mL sterile PBS. Store at 4 °C.
- Fine curved Vannas iridectomy scissors.
- A microsurgical incision knife.
- Fine straight Vannas scissors.
- 48-well tissue culture dish.
- Sterile PBS.
- PBS-X (PBS 0.5%Triton-X).
- Permeabilizing blocking buffer (PBB) (4% BSA, 0.5% Triton-X in PBS, syringe filtered).
- Fluoroshield mounting medium.
- Dissection microscope (Leica).
- 30 mm Petri dishes.
- 96-well plate.
- SuperFrost™ slides.
- Coverslips.

3 Methods

3.1 Micron IV Ophthalmoscope and Meridian Laser Setup (Day 1)

1. Affix the mouse objective lens to the Micron unit along with the laser adaptor and fiber optic laser cable.
2. Turn on the computer, monitor, light source, and heating unit.
3. Initiate the software.
4. Check the white balance by holding a piece of white paper in front of the lens and click AWB (adjust white balance).
5. Set time to continuous and annotate the file with the mouse ID and left or right eye.
6. Ensure the file path is set correctly for saving files.

3.2 Laser Photocoagulation (Day 0)

1. Weigh the animal and anesthetize with a mix of ketamine and medetomidine hydrochloride as prepared in Subheading 2. Administer 100 μL per 25 g mouse.
2. Topically apply one drop of 5% phenylephrine hydrochloride and 0.8% tropicamide to both eyes, to dilate the pupils.
3. Once anesthetized transfer the mouse to the imaging cradle fitted with a heat mat.
4. Continually check the consciousness of the animal using the toe and/or eyelid reflex.
5. Add a drop of Viscotears to each eye to prevent the eyes from drying out. Re-administer throughout the experimental protocol.
6. Orientate the animal so that the eye is at right angles to the objective lens.
7. Remove any whiskers that may be in the way using a cotton tip.
8. Use the focus dial to adjust the focus until the optic disk can be visualized through the Viscotears, on the computer monitor (*see* **Note 2**).
9. Center the optic disc and take an image of the retina, checking for any ocular abnormalities.
10. Turn on laser and set the desired wavelength, time, and spot size; 450 mW, 130 ms, 50 μm (*see* **Note 3**).
11. Switch on the guide laser and adjust the focus and brightness.
12. Put on safety glasses (*see* **Note 4**).
13. Generate four laser burns per eye at equal distance from the optic nerve (approximately twice the diameter of the optic nerve), at 3, 6, 9, and 12 o'clock positions or in the center of the 4 individual retinal quadrants (*see* **Note 5**). Aim the red laser dot between major vessels and create a lesion by firing the laser with the foot pedal. Repeat this a further 3 times (*see* **Note 6**).
14. Recover the animal by administering an i.p. injection of 0.2 mL of atipamezole hydrochloride as prepared in Subheading 2.
15. Place the animal in a pre-warmed recovery cage containing fresh, soft absorbent bedding, mash, and a water bottle. Dim the lighting during recovery.

3.3 Sodium Fluorescein Fundus Angiography (FFA) (Days 3, 7, and 14)

1. Weigh the animal and anesthetize with a mix of ketamine and medetomidine hydrochloride as prepared in Subheading 2. Administer 100 μL per 25 g mouse.
2. Topically apply one drop of 5% phenylephrine hydrochloride and 0.8% tropicamide to each eye to dilate the pupils.

3. Once anesthetized transfer the mouse to the imaging cradle fitted with a heat mat.
4. Continually check the consciousness of the animal using the toe and/or eyelid reflex.
5. Add a drop of Viscotears to each eye to prevent the eyes from drying out. Re-administer throughout the experimental protocol.
6. Orientate the animal so that the eye is at right angles to the Micron IV lens.
7. Remove any whiskers that may be in the way using a cotton tip.
8. Advance the Micron IV ophthalmoscope towards the cornea until a seal is formed between the lens and Viscotears and align the optic disc in the center of the image.
9. Adjust the focus and brightness until vessels and lesions are clearly visible.
10. Once the eye is correctly aligned capture brightfield fundus images of the eye and check for any ocular abnormalities.
11. Administer 100 mL of 10% sodium fluorescein i.p., by gently lifting the left leg and injecting in the left quadrant (*see* **Note 7**).
12. Change the barrier and excitation filters to the FITC channel (2) and time for approximately 3 min before capturing a Tiff format image of the sodium fluorescein perfused lesions.
13. Recover the animal by administering an i.p. injection of 0.2 mL of atipamezole hydrochloride as prepared in Subheading 2.
14. Place the animal in a pre-warmed recovery cage containing fresh, soft absorbent bedding, mash and a water bottle. Dim the lighting during recovery.
15. Repeat FFA on days 3, 7, and 14.
16. Termination (day 14).

    On day 14 following imaging the mice will be immediately killed by cervical dislocation and tissues will be harvested.

3.4 Choroid Dissection

1. Enucleate the whole eye from the culled mouse using curved forceps, taking extra care not to cut too close to where the optic nerve meets the eye (*see* **Note 8**).
2. Pierce the cornea with a needle to ensure proper fixation of all components of eye.
3. Fix the eye whole in 4% PFA for 1–2 h.
4. Wash the eye twice with PBS to remove excess PFA.
5. Place eye in a Petri dish with PBS.
6. Carefully remove any fatty or connective tissue at the back of the eye, ensuring you do not pierce the sclera.

7. Using forceps to keep the eye steady make an incision in the cornea.
8. Using small scissors remove the cornea tissue from the eye (*see* **Note 9**).
9. Using small scissors again remove the Iris.
10. Petal the choroid by making 4–6 incisions directly towards the optic nerve (*see* **Note 9**).
11. If still attached, gently remove the retina by pealing it away from the choroid.

3.5 Choroid Flat-Mount Immunofluorescence

1. Retain choroids in separate wells of a 96-well plate for the duration of the staining procedure.
2. Wash the choroid flat mounts with PBS every 15 min for 2 h, on a rocking platform.
3. Block nonspecific binding sites with PBB for 2 h at room temperature, on a rocking platform.
4. Prepare antibodies in PBB at the following concentrations:
   (a) Rat anti-mouse CD31 1:100 (BD Bioscience).
   (b) Goat anti-mouse CD45 1:100 (R&D Systems).
5. Add 50–70 μL of antibody per choroid and incubate at 4 °C for 48 h, on a rocking platform.
6. Wash choroids every 30 min on a rocking platform with PBS-X, for 4 h.
7. Prepare secondary antibody conjugates in PBB at the following concentrations:
   (a) Anti-Rat Alexa Fluor 488, 1:500.
   (b) Anti-Rabbit Alexa Fluor 555, 1:500.
   (c) DAPI, 1:1000.
8. Add 50–70 μL of secondary antibody and incubate in the dark, at 4 °C for 48 h, on a rocking platform.
9. Wash choroids every 30 min on a rocking platform with PBS-X for 4 h and in the dark to prevent photobleaching.
10. Flat-mount choroids on SuperFrost glass slides with approximately 20 L Fluoroshield and a glass coverslip.
11. Seal the edges of the coverslip using clear nail varnish and leave to dry at room temperature for 30 min and away from direct light.
12. Store slides in the dark at 4 °C.
13. Image slides using a confocal microscope (Leica TCS SPE) and generate z stacks of each lesion.

3.6 Lesion Analysis from Immunofluorescent Staining

1. Import the LIF files with z stacks into ImageJ software and create a maximum intensity projection of each lesion. *See* Fig. 1.
2. Using the freehand selection tool, delineate each lesion.
3. Measure the area and mean gray value per lesion.
4. Repeat the above for each channel (CD31 and CD45).
5. Repeat for each lesion. Data points can be presented per lesion or as an average per mouse.

3.7 FFA Lesion Image Analysis

1. Import the FFA TIF file into ImageJ software.
2. Using the freehand selection tool, delineate each lesion.
3. Measure the area and mean gray value per lesion.
4. Repeat for each lesion. Data points can be presented per lesion or as an average per mouse.

3.8 Conclusions

Lesions begin to resolve at 7 days, as the inflammation and angiogenic response subside (Fig. 2). During this time window, the non-invasive of assay permits the user to study the use of potential anti-angiogenic agents such as (1) transgenes or other GM methods, (2) application of therapeutic eye drops, (3) intraocular injection of therapeutic agents such as antibodies, small molecule inhibitor, or other novel compounds. Following the end of the experiments, the eye can be dissected to permit immunohistochemical analyses to be made. Angiogenesis and inflammation can be studied by using endothelial and leukocyte staining, but naturally any characterization could be made at this point.

4 Notes

1. Both gender and age of animals can influence the outcome of laser-induced CNV. Previous data and studies suggest that mice of both genders, at 6–8 weeks of age are optimal for reproducibility.
2. A key aspect of generating a reliable and consistent lesion with the image-guided laser system is fine focus adjustment. Initially the lens should be positioned approximately 5 mm away from the cornea and the lens focus adjusted to visualize the major retinal vessels. The lens should be slowly advanced until it makes contact with the surface of the cornea and the optic nerve should be repositioned in the center of the field of view. It is important to align the axis of both the mouse and the lens for subsequent laser beam alignment. This latter is achieved when the reflections of the retinal nerve fibers are evenly bright and clear in the field of view.

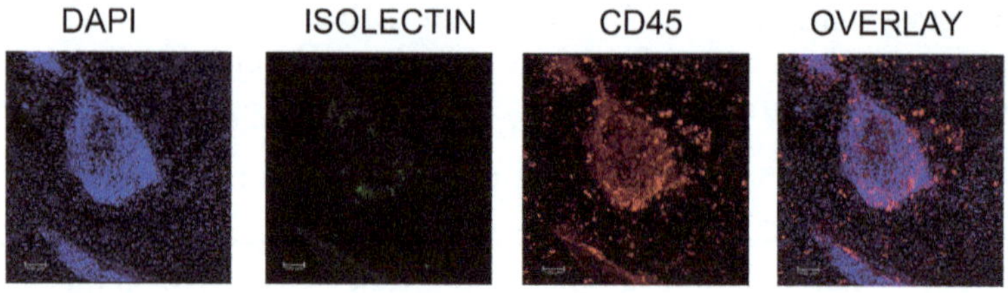

Fig. 1 Confocal imaging of dissected choroids. Immunohistochemical labeling of dissected choroid allows greater lesion characterization to be performed. Quantification of the vascular component (e.g. using the panEC marker Isolectin IB4) allows the vascularity to be studied alongside other aspects, such as inflammation (e.g. CD45). Scale bar =100 μm

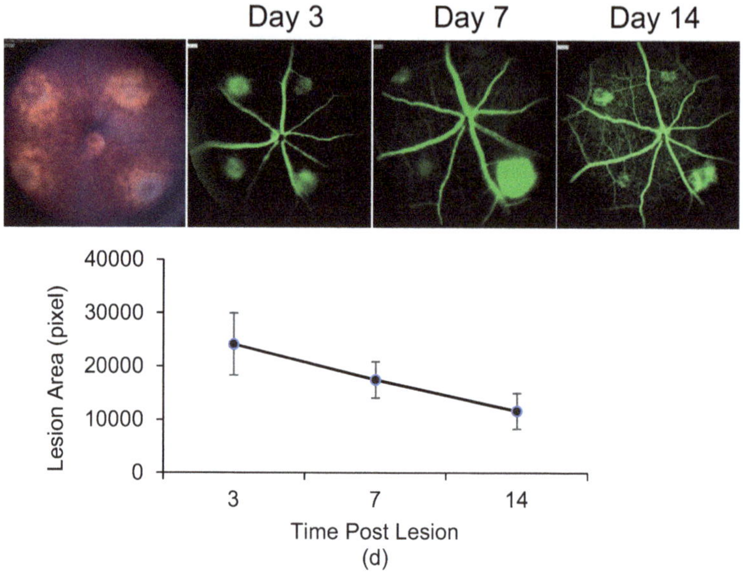

Fig. 2 Fluorescent fundus angiography in the murine eye. Sodium fluorescein injected mice when imaged with a fluorescent fundus camera can be used to image the retinal and choroidal vessels. The lesions appear as a green area in each quadrant of the eye' and represent a combination of leakage and vascular invasion. Repeat imaging allows a time course of lesion area to be monitored, showing a general regression in lesion size over time

3. Laser power is positively correlated to lesion area. Lower laser power may lead to unsuccessful rupture of the Bruch's membrane and higher laser power, bleeding, choroidal damage, fused lesions, and variation in lesion area.
4. Wear the correct goggles with high enough OD at operating wavelength. CE31-30103—Argon/KTP, UV-visible—filter 103 safety glasses.
5. The distance between laser burns must be at least double the diameter of the optic nerve to avoid fusion of lesions. Major

retinal and choroidal vessels should be avoided to prevent hemorrhage. The formation of a subretinal or vaporization bubble immediately following laser photocoagulation indicates successful rupture of the Bruch's membrane. Optical coherence tomography (OCT) can be used in conjunction with laser photocoagulation to confirm successful rupture.

6. Exclusion criteria.
   (a) The presence of cataracts, corneal abrasions, or epithelial edema prior to photocoagulation.
   (b) Exclude lesions with unsuccessful rupture of the Bruch's membrane.
   (c) Fused lesions.
   (d) Hemorrhage of the lesion.
   (e) The lesion is 5 times larger than the mean of the lesions under the same experimental conditions.
7. When administering Na–Fl inject in the lower quadrant at a 30–40° angle towards the head, to avoid damage to the urinary bladder, cecum, and other organs. Vary sides according to the anesthetic dose regimen.
8. Take extra care when removing the eye from the animal. Cutting to close to where the optic nerve meets the eye could make a hole in the sclera and cause damage to lesions. Leaving the optic nerve attached to the eye will also allow for better handling of the eye.
9. When removing the lens from the eye, the retina may sometimes come out attached to the lens. If the retina is still attached to the sclera, this can be removed in later steps after petaling by peeling away the retina from the sclera.

Acknowledgments

We would like to thank National Eye Research Charity, British Heart Foundation *PG/18/31/33759, and Royal Society RGS\R1 \191221* for their generous funding.

References

1. Gragoudas ES, Adamis AP, Cunningham ET Jr, Feinsod M, Guyer DR (2004) Pegaptanib for neovascular age-related macular degeneration. N Engl J Med 351:2805–2816
2. Liu C-H, Wang Z, Sun Y, Chen J (2017) Animal models of ocular angiogenesis: from development to pathologies. FASEB J 31:4665–4681
3. Lambert V et al (2013) Laser-induced choroidal neovascularization model to study age-related macular degeneration in mice. Nat Protoc 8: 2197–2211
4. Batson J et al (2017) Development of potent, selective SRPK1 inhibitors as potential topical therapeutics for neovascular eye disease. ACS Chem Biol 12:825–832

Part III

Reducing Complexity: Growing and Studying Endothelial Cells Outside the Body

Chapter 18

A Rapid Adaptable Method for Isolation of Endothelial Cells from Human Adipose Tissue

Jacqueline Herold and Joanna Kalucka

Abstract

Human adipose tissue is the largest endocrine organ and plays a role in whole-body metabolism. Dysfunction of this tissue is involved in multiple diseases, such as obesity, diabetes, and cardiovascular disease. An important factor in maintaining healthy adipose tissue is ensuring correct functioning of the blood vessels in this highly vascularized tissue. The endothelial cells (ECs) which line blood vessels show remarkable heterogeneity in structure and function in physiological and pathological conditions. While multiple studies have been performed to characterize ECs in different organs, the endothelium of adipose tissue remains poorly characterized. One of the significant challenges in working with adipose tissue is the separation and isolation of single viable cells, including ECs. This chapter describes a reliable and flexible approach for the isolation of adipose ECs that could be used for various analysis, including single-cell RNA sequencing, in vitro culture, and downstream applications.

Key words Endothelial cells, Adipose tissue, Isolation, Digestion, Purification, In vitro culture

1 Introduction

Adipose tissue is a highly plastic organ that plays a central role in regulating whole-body energy metabolism. It is mainly composed of adipocytes supported by stromal vascular tissue, which contains vascular endothelial cells (ECs), immune cells, and stem cells [1]. Recent studies highlighted the importance of angiogenesis (the process of new vessel formation) for the modulation of adipose tissue growth, mass, and metabolism. Importantly, adipose tissue is among the tissues that have the highest angiogenic capacities [1, 2]. During adipose tissue expansion, impaired vascular remodeling promotes hypoxia and inflammation, which leads to development of metabolic diseases, including type 2 diabetes, fatty liver disease, and cardiovascular diseases. Emerging evidence indicates that the vascular endothelium is a key player in the development of adipose tissue dysfunction [1, 2]. ECs are the main building blocks

Andrew V. Benest (ed.), *Angiogenesis: Methods and Protocols*, Methods in Molecular Biology, vol. 2441,
https://doi.org/10.1007/978-1-0716-2059-5_18,

of blood vessels and multiple studies have suggested that EC dysfunction contributes to increased inflammation and pathogenesis of conditions such as obesity. This could provide a hint as to why the risk of cardiovascular events is higher in obese patients.

In attempt to address the biology of adipose ECs (AdECs) and the mechanisms that lead to their dysfunction, several strategies have been reported for the isolation of AdECs [3–6]. However, none of these isolation techniques is tailored to cover the wide range of *state-of-the-art* technologies and multidisciplinary approaches (e.g., single-cell/nuclei RNA sequencing, cell culture molecular, and cell biology techniques) that researchers may wish to use. In this chapter, we describe protocols for the isolation of AdECs from fresh and nuclei from frozen samples. The protocols are either based on a combination of positive immunomagnetic selection of AdECs [7] or fluorescence-activated cell (FACS) sorting [8]. Immunomagnetic selection permits the consistent isolation of pure AdECs that could be further utilized for cell culture, OMICS, or molecular biology approaches (*see* Fig. 1), whereas FACS sorting of nuclei might be used for single-nuclei RNA sequencing (*see* Fig. 2). These easy-to-follow protocols outline a highly consistent and versatile method of isolating AdECs. Compared with previously published protocols, the proposed reproducible strategy allows researchers to use the isolated AdECs in various downstream applications. For instance, isolated AdECs can be grown in standard cell culture conditions or used for functional assays (e.g., vessel sprouting) while maintaining their EC-phenotype assessed by the expression of key EC genes, such as platelet endothelial cell adhesion molecule 1 (PECAM-1), cadherin 5 (CDH5), and vascular cell adhesion protein 1 (VCAM1) (*see* Fig. 3). Furthermore, although these protocols focus on the isolation of ECs from human subcutaneous adipose tissue, they could be further adapted to multiple adipose depots from human and mouse tissues.

2 Materials

The protocols were established using either fresh or snap-frozen human adipose tissue for live cell or nuclei isolation, respectively. If the isolated cells will be used for cell culture experiments, prepare all the reagents and solutions under the hood to ensure that they are sterile.

Before attempting this protocol, ensure all reagents and buffers are prepared in advance in a suitable quantity. To avoid unnecessary waiting during steps that are reliant on enzymatic activity, we recommend turning on your shaker-incubator and setting the temperature to 37 °C beforehand.

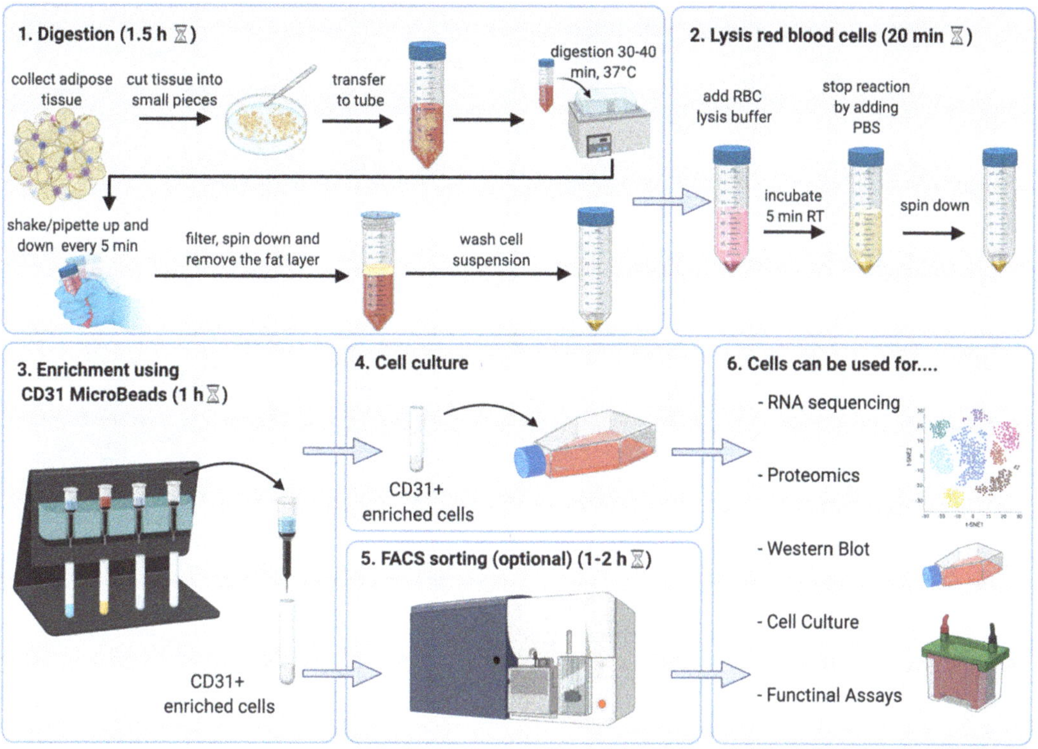

Fig. 1 Graphical abstract summarizing the main steps of the protocol for ECs isolation from fresh adipose tissue

2.1 General Reagents and Materials

Table 1 provides an overview of all reagents used for both protocols.

2.2 Media and Buffers

2.2.1 PBS-based Wash Buffer

PBS-based wash buffer contains: 0.5% (w/v) BSA (BSA Fraction V, Sigma-Aldrich) and 2 mM EDTA (Sigma-Aldrich) in PBS (Thermo Fisher Scientific) (*see* Table 2). Filter and store at 4 °C.

2.2.2 Supplemented KnockOut™ DMEM-Medium for Isolation of AdECs from Fresh Adipose Tissue

Supplemented KnockOut™ DMEM-medium is an FBS-free medium which is used to prepare adipose tissue digestion solution. It contains the following components: KnockOut™ DMEM (Thermo Fisher Scientific) supplemented with 1% (v/v) Penicillin/Streptomycin (e.g., Thermo Fisher Scientific), 2× Antibiotic–Antimycotics (200 units/mL of pen., 200 μg/mL of strep., 0.5 μg/mL of Gibco Amphotericin B, e.g., Thermo Fisher Scientific), 1 mM Sodium Pyruvate (Thermo Fisher Scientific), 1× MEM Non-Essential Amino Acids Solution (MEM-NEAA) (Thermo Fisher Scientific), 2 mM L-Glutamine (e.g., Thermo Fisher Scientific), and 1× Endothelial Cell Growth Factor supplements (ECGS/ Heparin) (PromoCell) (*see* **Notes 1** and **2** and Table 3).

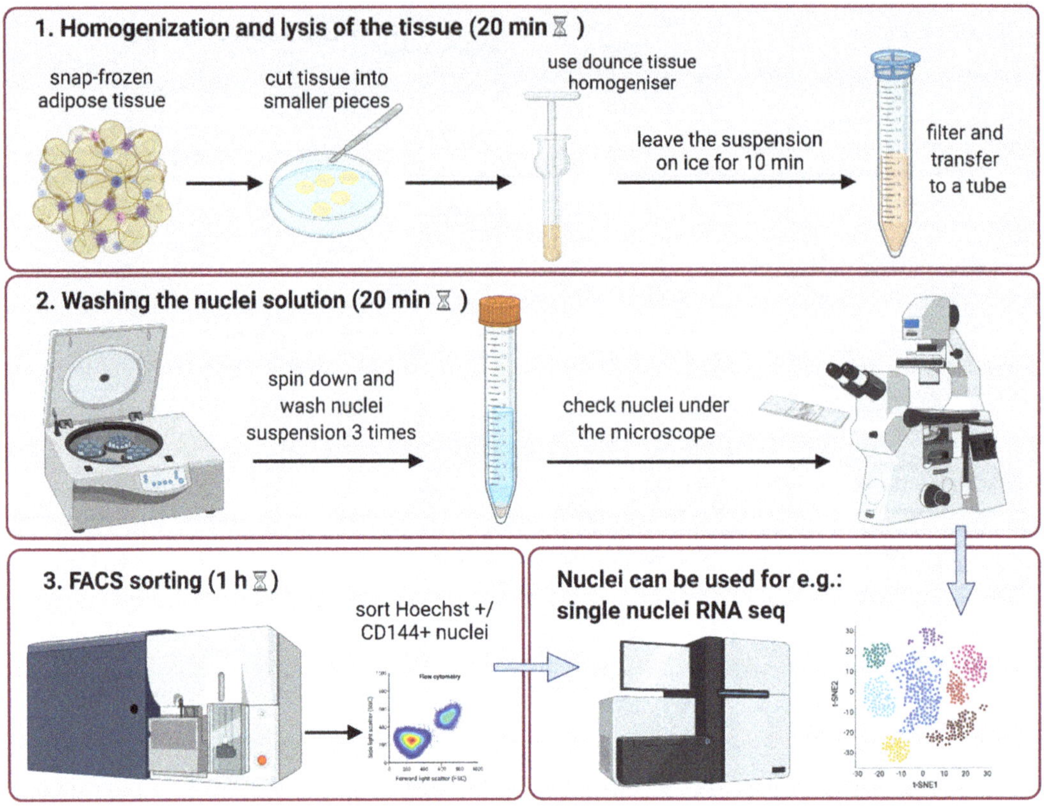

Fig. 2 Graphical abstract summarizing the main steps of the protocol for nuclei isolation from frozen adipose tissue

2.2.3 Adipose Tissue Digestion Buffer for AdECs Isolation from Fresh Adipose Tissue

Shortly before the start of the isolation procedure prepare the adipose tissue digestion buffer as outlined in Table 4. The proposed quantities are sufficient for 10 mL of digestion medium, which suffices for 5 g of fresh adipose tissue. Scale the volumes up or down according to your needs (*see* **Note 3** and Table 4).

2.2.4 Salt-Tris Solution (ST) Buffer for Nuclei Isolation from Frozen Adipose Tissue

Prepare ultrapure water-based 2× ST buffer as outlined in Table 5. The buffer can be prepared in advance. Autoclave if required. It is recommended to supplement ST buffer with 1 U/μL RNase inhibitor.

2.2.5 Tween ST (TST) Buffer for Nuclei Isolation from Frozen Adipose Tissue

TST buffer contains: 1× ST buffer, 0.03% (w/v) Tween-20 (Sigma-Aldrich), *and* 0.01% (w/v) BSA (BSA Fraction V, Sigma-Aldrich) in nuclease free water (*see* Table 6). It is recommended to supplement TST Buffer with 1 U/μL RNase inhibitor.

2.2.6 FACS Sorting (FANS) Buffer for Both Fresh and Frozen Adipose Tissue Isolation

FANS buffer contains: 1% (w/v) BSA (BSA Fraction V, e.g., Sigma-Aldrich) and 1 mM EDTA in nuclease free water (*see* Table 7). If the downstream application is single nuclei sequencing, supplementing of TST Buffer with 1 U/μL RNase inhibitor is recommended.

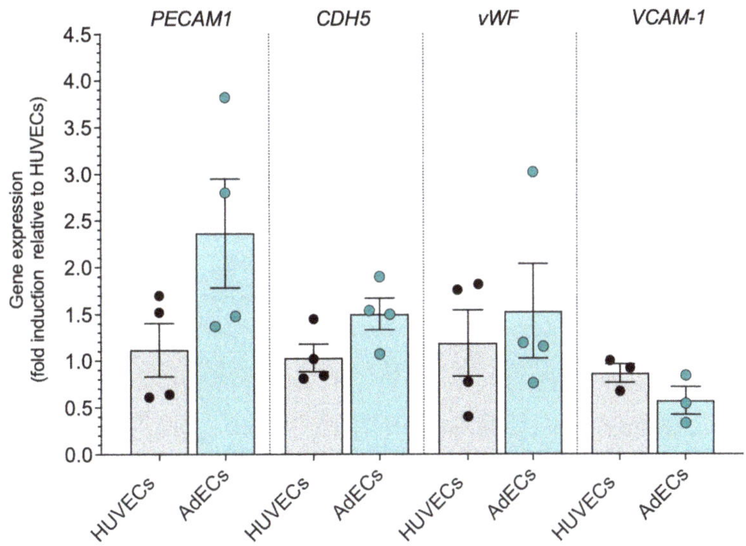

Fig. 3 Effect of cell culture conditions on AdECs vs. human umbilical vein endothelial cells (HUVECs) phenotype marker gene expression. Freshly isolated AdECs and HUVECs were cultured simultaneously and RNA was extracted at P5. Expression of PECAM1, CDH5, vWF, and VCAM-1 was assessed by qPCR. Data are mean ± SEM; p is non-significant. $N = 3$ or 4 donors per tested gene

3 Methods

3.1 Isolation of Viable ECs from Fresh Adipose Tissue

3.1.1 Preparation of Fresh Adipose Tissue for AdECs Isolation

Figure 1 shows a scheme for AdECs isolation from fresh tissue.

1. Prepare adipose tissue digestion buffer for digestion of fresh tissue according to the indication in Subheadings 2.2.2 and 2.2.3 (*see* **Note 1**).
2. Prepare cold wash buffer as indicated in Subheading 2.2.1.
3. Weight approximately 5 g of adipose tissue and pre-cut the tissue using scalpels. Keep the samples on ice during cutting (*see* **Note 4**).
4. Transfer approximately 5 g of fat tissue to a tube containing 10 mL of adipose tissue digestion buffer supplemented with enzymes.
5. Incubate the mixture of tissue and digestion buffer for 30–45 min in 37 °C (e.g., water bath) and shake the tube every other minute (*see* **Note 5**).
6. During the digestion pipette the solution using 25 mL, 10 mL, and 5 mL pipettes sequentially. Only if the homogeneous solution is easy to pipette, continue with the next step (*see* **Note 6**).

Table 1
Complete list of all reagents and materials

Reagent or resource	Source
Antibodies	
CD144 (VE-Cadherin) Antibody, anti-human, PE, REAfinity	Miltenyi Biotec
CD31 Antibody, anti-human, APC, REAfinity	Miltenyi Biotec
CD45 Antibody, anti-human, PE-Vio® 770, REAfinity	Miltenyi Biotec
Chemicals	
Antibiotic-antimycotic	Thermo Fisher Scientific
Bovine serum albumin (BSA Fraction V)	Sigma-Aldrich
Calcium chloride	Sigma-Aldrich
Collagenase type I	Thermo Fisher Scientific
Dispase	Thermo Fisher Scientific
DNase I	Sigma-Aldrich
EDTA	Sigma-Aldrich
Endothelial Cell Media (C-22022)	PromoCell
Endothelial cell growth factor supplements (ECGS/Heparin)	PromoCell
Fetal bovine serum (FBS)	Thermo Fisher Scientific
Fixable Viability Dye eFluor™ 450	Thermo Fisher Scientific
Gelatin	Sigma-Aldrich
Hoechst 33258 solution	Sigma-Aldrich
KnockOut™ DMEM	Thermo Fisher Scientific
Magnesium chloride	Sigma-Aldrich
MEM–NEAA	Thermo Fisher Scientific
Penicillin/streptomycin	Thermo Fisher Scientific
Phosphate buffered saline (DPBS)	Thermo Fisher Scientific
Red Blood Cell Lysis Buffer Hybrid Max	Sigma-Aldrich
Sodium chloride	Sigma-Aldrich
RNase inhibitor (New England Biolabs)	
Sodium pyruvate	Thermo Fisher Scientific
Tris hydrochloride	Sigma-Aldrich
Trypan blue	Thermo Fisher Scientific
Tween-20	Sigma-Aldrich
Ultrapure water	Thermo Fisher Scientific
Critical commercial assays	
CD31 MicroBeads, human	Miltenyi Biotec
Software	

(continued)

Table 1
(continued)

Reagent or resource	Source
FlowJo (version 8.8.6)	https://www.flowjo.com
GraphPad Prism	GraphPad Software
Other	
100 μm cell strainer	Sigma-Aldrich
40 μm cell strainer	Sigma-Aldrich
BD FACSAria™ III sorter	BD Biosciences
Brand™ Bürker Counting Chambers	Thermo Fisher Scientific
Centrifuge tube, conical, HDPE CentriStar™, PP, 15 mL	VWR
Centrifuge tube, conical, HDPE CentriStar™, PP, 50 mL	VWR
LS columns	Miltenyi Biotec
MACS MultiStand	Miltenyi Biotec
Multipurpose- and micro-centrifuge	N/A
QuadroMACS™ Separator	Miltenyi Biotec
Surgical Scalpel Blade No 10	Swann Morton

Table 2
Components and final concentrations of PBS-based wash buffer

Reagent	Final concentration	Amount
BSA (Sigma-Aldrich)	0.5%	2.5 g
EDTA (Sigma-Aldrich)	2 mM	0.1 mL (e.g., from 10 M preprepared filtered/autoclaved stock solution)
PBS	–	499.9 mL

7. Pipette the solution to 50 mL by filtering it through a 100 μm nylon strainer to remove the undigested material. Then, allow the adipocytes to float to the surface and remove the fat layer.
8. Stop the reaction with the same volume of wash buffer and filter through a 40 μm strainer into a 50 mL conical tube.
9. Spin down the solution for 7 min at 300 × *g* and remove the layer of fat.
10. Remove the supernatant and resuspend the cells in 1 mL wash buffer before transferring it to a 15 mL conical tube.

Table 3
Components and final concentrations of supplemented KnockOut™ DMEM-medium

Reagent	Final concentration	Amount
KnockOutTM DMEM-medium (Thermo Fisher Scientific)	–	477 mL
Penicillin/Streptomycin (Thermo Fisher Scientific)	1% (v/v)	1 mL
Antibiotic-Antimycotic (Thermo Fisher Scientific)	2×	10 mL
Sodium Pyruvate (Thermo Fisher Scientific)	1 mM	5 mL
MEM Non-Essential Amino Acids Solution (Thermo Fisher Scientific)	1×	5 mL
Optional: Endothelial Cell Growth Factor supplements (ECGS/ Heparin) (PromoCell)	1×	One vial to 2 mL

Table 4
Components and final concentration of Adipose tissue digestion buffer

Reagent	Final concentration	Amount
Supplemented KnockOut™ DMEM-medium (for details *see* Table 3)	–	8.925 mL
Collagenase I (Thermo Fisher Scientific)	0.1% (w/v)	10 mg
Dispase (Thermo Fisher Scientific)	0.25 U/mL	1 mL (from 2.5 U/mL stock solution prepared according to manufacturer's instructions)
DNAse I (Sigma-Aldrich)	7.5 mg/mL	75 mg (from 1 mg/mL stock solution prepared according to manufacturer's instructions)

11. Add 5 mL wash buffer to wash the cells and spin down at 300 × *g* for 5 min, then repeat twice or until the supernatant is clear.

12. Discard the supernatant.

3.1.2 Purification Using Red Blood Cell Lysis Buffer

Depending on the tissue sample residual blood may be present. If there are a large number of red blood cells (RBCs) in the cell solution, removing them with an RBC lysis buffer will improve binding of the magnetic beads. This step is optional. Proceed directly to Subheading 3.1.3 if this step will not be performed.

Table 5
Components and final concentration of Salt-Tris solution (ST) buffer

Reagent	Final concentration	Amount
Sodium chloride (Sigma-Aldrich)	292 mM	2.92 mL (e.g., from 5 M preprepared filtered/autoclaved stock solution)
Tris hydrochloride pH 7.5 (Sigma-Aldrich)	20 mM	1 mL (e.g., from 1 M preprepared filtered/autoclaved stock solution)
Calcium chloride (Sigma-Aldrich)	2 mM	0.1 mL (e.g., from 1 M preprepared filtered/autoclaved stock solution)
Magnesium chloride (Sigma-Aldrich)	42 mM	2.1 mL (e.g., from 1 M preprepared filtered/autoclaved stock solution)
Ultrapure water (Thermo Fisher Scientific)		43. 88 L

Table 6
Components and final concentration of TweenST buffer

Reagent	Final concentration	Amount
2× ST buffer	1×	5 mL
Tween-20 (Sigma-Aldrich)	0.03% (w/v)	0.3 mL (e.g., from 1% preprepared stock solution)
BSA (Sigma-Aldrich)	0.01% (w/v)	0.1 mL (e.g., from 10% preprepared stock solution)
Nuclease free water		4.69 mL

Table 7
Components and final concentration of FANS buffer

Reagent	Final concentration	Amount
BSA (Sigma-Aldrich)	1% (w/v)	5 mL (e.g., from 10% preprepared stock solution)
EDTA (Sigma-Aldrich)	1 mM	100 μL (e.g., from 0.5 M preprepared stock solution)
PBS		45 mL

1. Resuspend the cell pellet in 1 mL RBC lysis buffer (e.g., Red Blood Cell Lysis Buffer, Sigma-Aldrich Hybrid Max) and incubate the cells for 5 min at room temperature.
2. Add 10 mL PBS to stop the reaction.
3. Centrifuge the cell solution at 300 × *g* for 7 min.
4. Discard the supernatant.

3.1.3 CD31 Selection Using Microbeads and a Magnet Separator

1. Resuspend samples in an appropriate amount of wash buffer and add the appropriate volume of CD31 human MicroBeads and FcR Blocking Reagent, according to the manufacture's protocol (CD31 MicroBeads, human, Miltenyi Biotec, Cat#130-091-935) (*see* **Notes 7** and **8**).
2. Mixing well and incubate for 30 min at 4 °C.
3. Add 5 mL of wash buffer and centrifuge at 300 × *g* for 5 min.
4. Remove the supernatant and add again 3 mL of wash buffer before centrifuging at 300 × *g* for 5 min.
5. Discard supernatant and resuspend the cells in 1 mL wash buffer.
6. Place the LS column (Miltenyi Biotec, Cat#130-042-401) in the magnetic field of MACS MultiStand (Miltenyi Biotec, Cat#130-042-303) equipped with QuadroMACS™ Separator (Miltenyi Biotec, Cat#130-091-051) and activate the LS column by rinsing with 3 mL wash buffer.
7. Place a 40 μm cell strainer directly above the column and load the samples on the column by filtering them through the strainer.
8. Wash the LS column 3 times with 3 mL of wash buffer and collect the eluent in the conical tube. This fraction contains unlabeled cells.
9. Remove the column from the magnetic separator and place it in a 15 mL conical tube.
10. Pipette 5 mL wash buffer onto the column. Immediately flush out the magnetically labeled cells by firmly pushing the plunger into the column. The labeled cell fraction contains CD31-positive cells (ECs). The cells can be immediately put into culture (*see* Subheading 3.1.4) or further purified using FACS sorting (*see* Subheading 3.1.5).

3.1.4 AdECs Cell Culture

1. Pre-coat the 6 well-plate with 0.1% gelatine.
2. Centrifuge AdECs cell suspension at 300 × *g* for 5 min.
3. Resuspend the cell pellet in Endothelial Cell Medium (e.g., PromoCell) supplemented with 2× Antibiotic–Antimycotic solution (e.g., Thermo Fisher Scientific).

4. If the cell population seems to be overgrown by other cell types, the re-purification of AdECs should be performed using CD31 human MicroBeads as described in Subheading 3.1.3.

This protocol achieves a purity of 95% (*see* Fig. 4). The morphology and size of AdECs resemble the one of human umbilical vein endothelial cells. AdECs can be grown to confluency and passaged several times without losing EC-markers (e.g., *PECAM1*, *CDH5*, *vWF*, *VCAM*-1) (*see* Fig. 3). Furthermore, isolated AdECs can be used for functional assays including the scratch wound or sprouting assay (*see* Fig. 5).

3.1.5 Staining with Antibodies for FACS Sorting

1. Centrifuge the cells at 300 × *g* for 5 min.
2. Prepare antibody mix in FANS buffer by combining: CD31 Antibody (e.g., CD31, anti-human, APC, REAfinity, Miltenyi Biotec; 1:50); CD45 Antibody (e.g., CD45, anti-human, PE-Vio® 770, REAfinity, Miltenyi Biotec, Cat#130-110-634; 1:700); and Viability Dye (e.g., Fixable Viability Dye eFluor™-450, Thermo Fisher Scientific; 1:1000) (*see* **Note 9**).
3. Resuspend the cells in antibody mix.
4. Incubate the cell suspension at 4 °C for 30 min.
5. Add 5 mL of wash buffer and centrifuge 300 × *g* for 5 min.
6. Remove the supernatant and resuspend the cells in a desired volume.
7. Filter the cells through a 40 μm cell strainer before sorting (*see* **Note 10**).
8. Sort viable, $CD45^-$, $CD31^+$ cells to collection tubes (*see* **Notes 11** and **12** and Fig. 6).

3.2 Isolation of Nuclei from Frozen Adipose Tissue

Figure 2 shows a detailed scheme for nuclei isolation from frozen tissue.

3.2.1 Homogenization and Lysis of the Tissue

1. Place a piece (600 mg) of frozen adipose tissue onto a glass plate.
2. Cut the frozen adipose tissue sample into five to six smaller pieces and then use the Dounce tissue grinder with 4 mL of TST buffer. When the tissue is homogenized, add an additional 2 mL TST buffer and leave it for 10 min on ice.
3. Filter the homogenized solution through a 40 μm cell strainer into a 15 mL conical tube.
4. Wash the Dounce tissue grinder and filter with 2 mL of TST buffer.
5. Add 5 mL of 1× ST buffer to the sample solution.

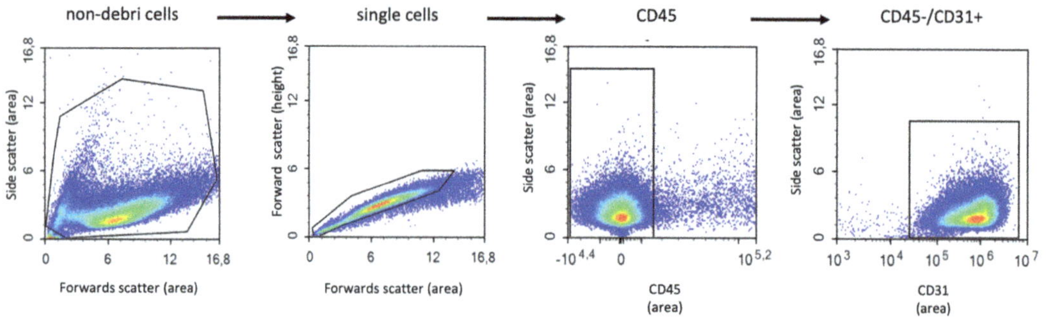

Fig. 4 The purity of cultured AdECs (approximately 95%) was determined by FACS on the basis of CD45 and CD31

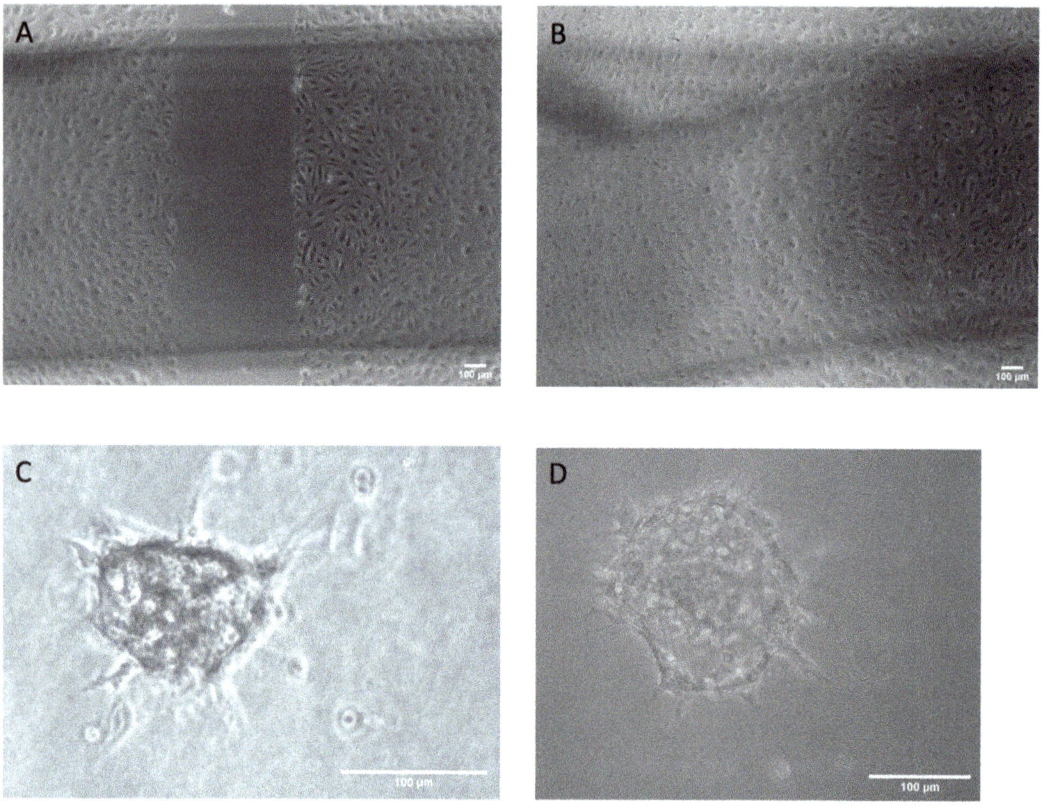

Fig. 5 Representative phase contrast images of AdECs used for functional assays: (**a**, **b**) A scratch wound was applied on confluent AdECs monolayers (pretreated overnight with 2 μg/mL Mitomycin C to inhibit cell proliferation (e.g., Sigma-Aldrich, #CatM0503). After wounding the AdECs pictures were taken at 0 h (**a**) and 18 h (**b**). (**c**, **d**) Representative phase contrast images of ECs sprouting assay using AdECs. Pictures were taken 24 h after embedding the spheroids in a fibrin gel

6. Spin down the sample at 4 °C for 5 min at 500 × *g* in a swinging bucket centrifuge.
7. Wash the nuclei solution 3 times with 5 mL 1× ST buffer to remove impurities and debris.

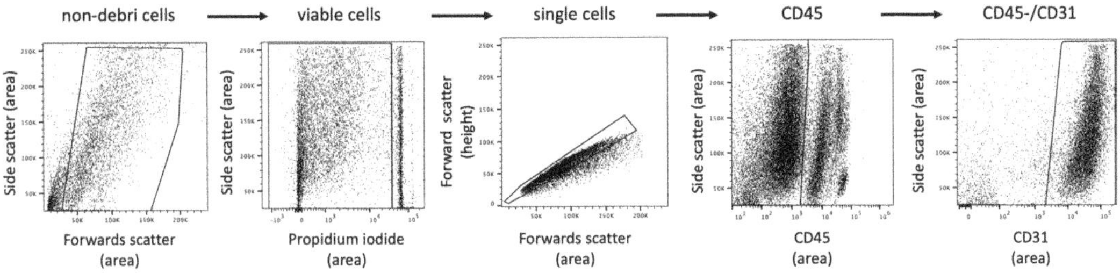

Fig. 6 Representative FACS plots for the gating strategy to sort ECs from adipose tissue based on sorting live/ $CD45^{-}/CD31^{+}$ cells

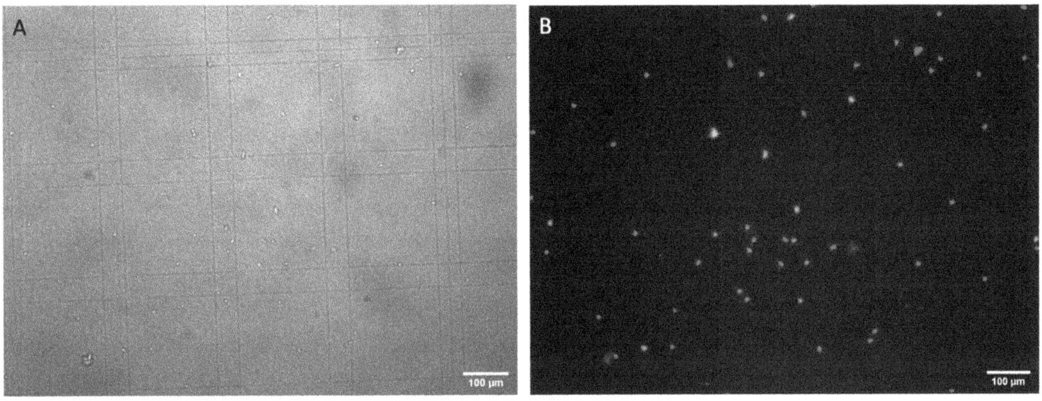

Fig. 7 Representative bright field (**a**) and fluorescent (**b**) images of nuclei isolated from frozen adipose tissue. Nuclei were stained with Hoechst stain solution

8. Check for the presence of nuclei under the microscope. For better visualization of nuclei, stain with Trypan Blue or Hoechst solution (*see* Fig. 7).

3.2.2 Sorting of CD144 Positive Nuclei

1. Spin down the sample for 5 min at 500 × *g* and remove the supernatant.
2. Resuspend the nuclei solution in the staining solution containing: CD144 (VE-Cadherin), Antibody (e.g., CD144, anti-human, PE, REAfinity, Miltenyi Biotec; 1:50), and Hoechst Staining Solution (e.g., Hoechst 33258 solution, Sigma-Aldrich; 1:2000) in FANS buffer (also *see* **Note 9**).
3. Incubate the nuclei for 30 min at 4 °C to avoid extensive exposure to the light.
4. Wash the nuclei with 2–3 mL of FANS buffer and centrifuge for 5 min at 500 × *g*.
5. Resuspend the nuclei in a desired volume of FANS buffer and filter shortly before performing FACS sorting.
6. To sort CD144 positive nuclei sort $Hoechst^{+}/CD144^{+}$ events. For gating strategy *see* Fig. 8. (also *see* **Note 9**).

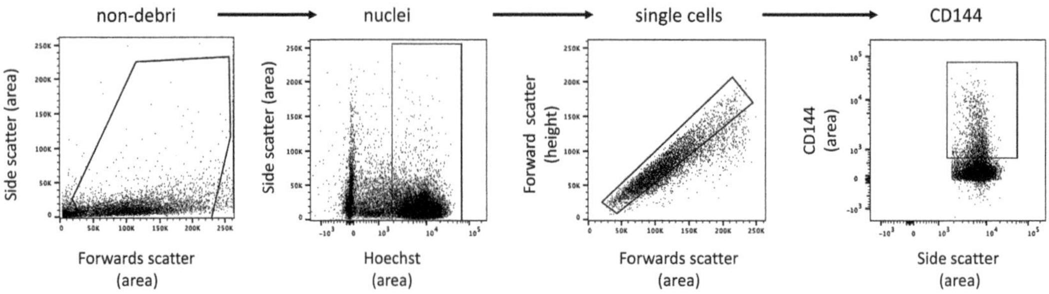

Fig. 8 Representative FACSspiepr com[spi]contText--FACSspiepr A3B2 show [spi3] plots for the gating strategy to sort ECs nuclei from frozen adipose tissue based on sorting Hoechst+/CD144+ signal

4 Notes

1. Supplemented KnockOut™ DMEM-medium can be stored up to several months in a sterile manner at 4 °C. The digestion enzymes (collagenase I, dispase, and DNase) can be added fresh to the digestion medium before initiation of the isolation procedures.
2. In order to maintain good EC viability, ECGS/Heparin is added to KnockOut™ DMEM-medium. This step is optional and if there is any indication that the growth factors included in ECGS might affect isolated ECs and alter the results, the endothelial cell growth factors can be left out.
3. Store the adipose tissue digestion buffer (freshly prepared) at 4 °C until it is needed.
4. Depending on the tissue sample, carefully remove/dissect out any non-adipose tissue (e.g., connective or necrotic tissue).
5. Digestion efficiency may depend on the vendor and/or the LOT number of the enzymes used. We recommend testing all newly purchased reagents before performing final experiments.
6. Try to avoid under- or over-digestion of the tissue, which may result in low numbers of isolated ECs. To avoid the extensive loss of AdECs, we recommend optimizing the mechanical steps of digestion (vigorous shaking by hand, pipetting, cutting with scalpel blade) to obtain the maximum possible number of ECs.
7. We suggest to resuspend the cells in 70 μL of wash buffer, determine the cell number, and follow the manufacturer's instructions, that is, for up to 1×10^7 total cell, resuspend the cells in 60 μL of wash buffer. Add 20 μL of FcR Blocking Reagent per 1×10^7 cells. Vortex briefly, then add 20 μL of CD31 MicroBeads to the mixture. When working with higher cell numbers, scale up all reagent volumes and total volumes accordingly (e.g., for 2×10^7 total cells use twice the volume

described). If fewer than 1×10^7 cells are available use the same volumes as indicated above.

8. Process samples fast in order to avoid nonspecific binding of beads.
9. Prepare all mandatory controls for flow cytometry analysis beforehand (e.g., unstained cell control, viability control, Fluorescence Minus One (FMO) control, isotype controls, as well as the compensation control (one for each fluorophore) for proper set up of the cytometer. If the cell number is limited, we would recommend to use for compensation controls the compensation beads (e.g., OneComp eBeadsTM Compensation Beads, ThermoFisher Scientific, Cat#01-1111-41) to ensure clear fluorescence signal and sufficient compensation. Based on the information provided in The Human Protein Atlas, we propose to enrich for EC-nuclei using CD144 antibody. However, we would recommend further validation of this strategy (e.g., by checking additional endothelial markers) with additional staining or qPCR.
10. The BD FACSAria™ III sorter was used to perform fluorescent activated sorting. However, any other sorter that allows for sorting of the cells at the required optical paths can be used.
11. To ensure high viability of the AdECs, sort the ECs to collection medium (10% (v/v) FBS (e.g., Thermo Fisher Scientific) in PBS).
12. To further improve the purity of isolated AdECs one could perform depletion of immune cells using CD45 MicroBeads (e.g., Miltenyi Biotec, according to manufacturer's instructions) before Subheading 3.1.3. However, the addition of this step might result in reduced viability of isolated ECs due to prolonged duration of the protocol. We recommend performing optimization of the depletion of immune cells.

Acknowledgements

We would like to acknowledge the help of FACS Core Facility at the Department of Biomedicine at Aarhus University for their technical assistance. J.K. is supported by the Lundbeckfonden by AIAS-COFUND II: GA: MSCA: 754513, Lundbeckfonden: R307-2018-3667, Carlsberg Fonden: CF19-0687, Kræftens Bekæmpelse: R302-A17296; A.P. Møller Fonden: 20-L-0317; Riisfort Fonden, and Steno Diabetes Center Aarhus (SDCA). The figures were created with BioRender.com.

References

1. Herold J, Kalucka J (2021) Angiogenesis in adipose tissue: the interplay between adipose and endothelial cells. Front Physiol 11
2. Graupera M, Claret M (2018) Endothelial cells: new players in obesity and related metabolic disorders. Trends Endocrinol Metabol 29:781–794
3. Kajimoto K, Hossen MN, Hida K et al (2010) Isolation and culture of microvascular endothelial cells from murine inguinal and epididymal adipose tissues. J Immunol Methods 357:43–50
4. Arts CHP, Heijnen-Snyder GJ, Hedeman Joosten PPA, Verhagen HJM, Eikelboom BC, Sixma JJ, de Groot PG (2001) A novel method for isolating pure microvascular endothelial cells from subcutaneous fat tissue ideal for direct cell seeding. Lab Investig 81:1461–1465
5. Frye CA, Patrick CW (2002) Isolation and culture of rat microvascular endothelial cells. In Vitro Cell Dev Biol Anim 38:208–212
6. Hearnden R, Sandhar B, Vyas V, Longhi MP (2021) Isolation of stromal vascular fraction cell suspensions from mouse and human adipose tissues for downstream applications. STAR Protoc 2:100422
7. Sokol L, Geldhof V, García-Caballero M et al (2021) Protocols for endothelial cell isolation from mouse tissues: small intestine, colon, heart, and liver. STAR Protoc 2:100489
8. Slyper M, Porter CBM, Ashenberg O et al (2020) A single-cell and single-nucleus RNA--Seq toolbox for fresh and frozen human tumors. Nat Med 26:792–802

Chapter 19

Simple Gene Knockdown in Endothelial Cells Using Short Interfering RNA Oligonucleotides

Zarah B. Tabrizi, Kathryn R. Green, Amy P. Lynch, Nada S. Ahmed, Nicholas Beazley-Long, and Andrew V. Benest

Abstract

Cell transfection using short interfering RNAs (siRNAs) is a widely used technique to perform loss of function studies by "knocking down" genes of interest. Oftentimes, primary cells can be difficult to transfect, but here we provide a simple and robust method using cultured endothelial cells and routine transfection reagents. Knockdown studies can be used to complement overexpression studies and validate biochemical pathway analysis, as well as functional assays. The enclosed protocol will compliment other in vitro assays detailed in this edition.

Key words Knockdown, RNA interference, siRNA, Transfection, HDLEC, HUVEC

1 Introduction

Short interfering RNAs (siRNAs) are 21–23 nucleotide sequence-specific RNA which can mediate posttranscriptional gene silencing [1, 2]. Knockdown of gene expression in mammalian cells enables the study of specific gene function. RNA interference is a multistep process initiated by the addition of double stranded RNA (dsRNA), which is processed by an RNase III family ribonuclease, Dicer, into 21–23 base pairs dsRNA with 3′ dinucleotide overhangs. The RNA-inducing silencing complex (RISC) incorporates the guide strand of siRNA, deleting the target mRNA and ultimately responsible for silencing endogenous gene expression [1]. SiRNAs oligonucleotides can be chemically synthesized and are commercially available for specific gene targeting in mammalian cells, these can be newly designed or purchased already validated to knock down specific genes. Lymphatic vessels are made up of lymphatic endothelial cells which differ in structure and function to blood endothelial cells [3]. And consequently there is much interest to identify

Andrew V. Benest (ed.), *Angiogenesis: Methods and Protocols*, Methods in Molecular Biology, vol. 2441,
https://doi.org/10.1007/978-1-0716-2059-5_19,

novel markers of either population. Both populations are primary cells, and therefore the expense of optimizing transfection techniques is greater than simpler cell models. This protocol gives a robust method to knock down genes of interest with minimal optimization required.

Human dermal lymphatic endothelial cells or human umbilical vein endothelial cells were initially isolated from lymphatic capillaries of adult skin or the umbilicus (purchased, isolated, and characterized from PromoCell), these can be cultured and used to study aspects of the human endothelium in vitro. Here we demonstrate how siRNAs oligonucleotides can be utilized to significantly knock down a target gene with confirmation at an RNA and protein level.

2 Materials

2.1 Culturing

1. Human dermal lymphatic endothelial cells (HDLECs) or human umbilical vein endothelial cells (HUVECs) both from PromoCell.
2. Endothelial cell growth medium (PromoCell).
3. Endothelial cell growth supplements (MV2 for HDLECs), (PromoCell).
4. Sterile PBS.
5. Trypsin/EDTA (0.04%/0.03%) (PromoCell).

2.2 Knockdown by siRNA

1. Lipofectamine RNAiMAX (Thermo Fisher).
2. Opti-MEM reduced serum medium.
3. Silencer select siRNAs oligos for your gene of interest (20 μM) (*see* **Note 8**).
4. Non-silencing oligos (20 μM).

2.3 Confirmation of Knockdown by PCR

Multiple techniques exist; however, validation of target of interest normalized to a housekeeping gene at the RNA level is recommended.

2.4 Confirmation of Knockdown by Western Blot

Multiple techniques exist; however, validation of target of interest normalized to a housekeeping gene at the protein level is recommended.

3 Methods

3.1 Culturing HDLECs and HUVECs

All the steps must be done in a category 2 microbiological safety cabinet to ensure sterility and routine mycoplasma testing is recommended.

3.1.1 Thawing

1. Create complete growth medium by adding 50 mL endothelial cell growth supplements to 500 mL bottle of endothelial cell growth medium.
2. Add 15 mL complete medium to a T75 flask and place in an incubator set at 37 °C and 5% CO_2 for 30 min to warm the medium.
3. Thaw 1 vial of HDLECs or HUVECs (500,000 cells per vial) by removing from liquid nitrogen and placing in a water bath set to 37 °C for 2 min (*see* **Note 1**).
4. Transfer all the contents of the vial into the pre-warmed T75 flask with complete medium.
5. Incubate at 37 °C and 5% CO_2 overnight.
6. The next day check attachment of cells under a microscope and change the medium (*see* **Notes 2** and **3**).
7. Change medium every 2–3 days, split cells once they reach 80–90% confluency.

3.1.2 Splitting

1. Remove medium from cells.
2. Wash twice with sterile PBS.
3. Add 1 mL of trypsin/EDTA per T75 flask, rock gently from side to side and incubate at 37 °C for 1 min.
4. Tap the side of the flask firmly to detach cells (*see* **Note 4**).
5. Add 4 mL medium to the flask to neutralize trypsin, transfer cell suspension to a falcon, and centrifuge at 1100 RPM (200 × *g*) for 5 min to pellet cells.
6. Resuspend cell pellet in a desired amount of cell medium, a cell count can be performed if required.
7. Re-plate in appropriate flask or well plate with appropriate volumes of complete medium.

3.2 Knockdown by siRNA

The following protocol is for transfecting cells within a six well plate with three gene knockdown wells and three non-silencing control wells (*see* **Note 5**). This can be scaled up and down as required. All steps must be done in a category 2 safety cabinet to ensure sterility.

1. Plate 250,000 cells per well and incubate with 3 mL complete medium per well overnight (*see* **Note 6**).
2. The following day check attachment of HDLECs/HUVECs and ensure confluency is above 80% (*see* **Note 7**).
3. For each condition (knockdown and non-silencing control) create the transfection solutions below:
   (a) Solution A: 30 μL oligos (at a concentration of 20 μM) and 270 μL Opti-MEM.
   (b) Solution B: 15 μL lipofectamine RNAiMAX and 285 μL Opti-MEM.

4. Gently mix and incubate mixes at room temperature for 10 min.
5. Mix solutions A and B for each condition together gently, creating a 600 μL solution, and incubate at room temperature for 30 min.
6. Aspirate the medium from the cells and gently wash cells twice with 1 mL PBS by pipetting on the side of the well.
7. Add 800 μL of Opti-MEM to each well, taking care again not to disturb the cells.
8. Add 200 μL dropwise of oligo/lipofectamine mix to each well for the appropriate condition.
9. Incubate at 37 °C for 4 h.
10. Remove transfection mixture and replace with 3 mL of complete medium per well.
    (a) If analyzing knockdown by RNA incubate for 48 h.
    (b) If analyzing knockdown by protein it is recommended to incubate for 72 h with a medium change on the second day.

3.3 Confirmation of Effective Gene Knockdown

1. RNA can be isolated at 48 h, and protein can be taken at 72 h post-transfection. Normalization to at least one housekeeping gene is essential.
2. One can expect RNA knockdown to be >95% and protein knockdown of >90% (*see* **Notes 9** and **10**).
3. RNAi cells can be employed into multiple other in vitro assays.

4 Notes

1. Take care that the cryovial does not explode due to the pressure. Releasing the pressure can be done by opening the cryovial under the hood and then resealing the lid before placing in the water bath.
2. It is normal for some cells not to recover and therefore will float within the cell culture medium.
3. Healthy HDLECs will grow in colony patches across the dish and have a typical endothelial cell cobblestone appearance, although when less confluent this cobblestone appearance is less profound. HUVECs do not grow in a similar pattern, but achieve a cobblestone appearance much sooner.
4. Check detachment under a microscope as times can vary, generally, the more confluent the cells, the longer they will take to detach. Cells can be placed back in the incubator for a further minute to aid detachment if required.

5. It is recommended to perform a "transfection reagent only" samples too, which serves as a non-transfected control.
6. Homogenous even plating of cells among the wells is accomplished by moving the plate backward and forward and leave the plate to rest for 2–3 min, then right to left and leave the plate to rest for 2–3 min before placing in the incubator.
7. If the confluency is too low, wait until the confluency is reached.
8. Routinely use three oligos to confirm on-target vs off-target effects. These can be pooled or purchased as a "smart pool" which reduces the off-target effects of any individual oligonucleotide. It is fundamental that off-target effects are excluded from subsequent interpretation of any results [4].
9. Confirmation of both RNA and protein knockdown is biologically fundamental and allows both transcriptomic and functional assays to be functionally relevant.
10. Normalizing your target gene expression to a housekeeping gene and then calculating the percentage knockdown compared with the non-targeting control allows a % knockdown to be established.

References

1. Elbashir SM et al (2001) Duplexes of 21-nucleotide RNAs mediate RNA interference in cultured mammalian cells. Nature 411: 494–498
2. Han H (2018) RNA interference to knock down gene expression. Methods Mol Biol 1706: 293–302
3. Wick N et al (2007) Transcriptomal comparison of human dermal lymphatic endothelial cells ex vivo and in vitro. Physiol Genomics 28: 179–192
4. Abid MR et al (2008) Whither RNAi? Arter Thromb Vasc Biol 455:489–490

Chapter 20

Transcription Factor Chromatin Immunoprecipitation in Endothelial Cells

Philip Kitchen, Kevin Gaston, and Padma-Sheela Jayaraman

Abstract

Interactions between DNA and proteins are crucial for the regulation of gene expression. Chromatin immunoprecipitation (ChIP) is a powerful technique that allows the study of specific protein–DNA interactions in cultured cells and fresh or fixed tissue. Chromatin is isolated and sheared, and antibodies against the protein(s) of interest are used to isolate specific protein–DNA complexes. Subsequent analysis by real-time polymerase chain reaction (qPCR) or next-generation sequencing (NGS) allows identification and quantification of the co-purified DNA fragments, and NGS also gives insight into the genomic binding sites of a protein. Here we describe a cross-linking ChIP (X-ChIP) protocol, based around the example of a myc-tagged Proline-Rich Homeodomain (PRH) protein expressed in human umbilical vein endothelial cells. We also describe how to analyse specific known or suspected binding sites using quantitative PCR as well as how to analyse genome-wide binding from ChIP sequencing data.

Key words ChIP sequencing, qPCR, Chromatin, Immunoprecipitation, Transcription factor, PRH, HHEX

1 Introduction

Chromatin is a DNA–protein complex intimately associated with structural and regulatory RNA molecules [1]. Generally, chromatin is thought of as histone protein octamers bound to around 147 bp of DNA to form nucleosomes. The histones have N-terminal tails that can be modified by histone modifying enzymes and the tails can also be bound by histone binding proteins [2]. Nucleosomes self-associate to form higher order complexes in the presence of specific modifications and histone binding proteins. Specific histone modifications and histone binding proteins are associated with regions of low gene density, decreased gene expression, and densely packed nucleosomes; these regions are known as heterochromatin. More open chromatin, known as euchromatin, occurs where there is less dense ordering of nucleosomes, increased gene density, and

Andrew V. Benest (ed.), *Angiogenesis: Methods and Protocols*, Methods in Molecular Biology, vol. 2441,
https://doi.org/10.1007/978-1-0716-2059-5_20,

gene activity and this is also associated with different but specific sets of histone modifications and histone modifying enzymes. Transcription factors (TFs) are proteins that are more loosely associated with chromatin and which regulate gene expression. TFs include DNA binding proteins and also co-regulatory proteins (co-activators and co-repressors) that bind DNA indirectly via DNA binding proteins, RNA polymerase complexes, or RNA molecules [3, 4].

Chromatin immunoprecipitation is a technique that allows the identification of DNA sequences associated with histones, histone binding proteins, or TFs (Fig. 1).

1.1 Chromatin Immunoprecipitation

Chromatin immunoprecipitation (ChIP), as the name suggests, involves antibodies that bind to specific features of chromatin or TFs (even after mechanical (sonication) or enzymatic fragmentation of chromatin) [5]. Following binding of the antibody to the specific protein or protein modification of interest the antibody–protein–DNA complex is purified using protein A/G, a fusion of two bacterial proteins that bind to antibodies from multiple species, or individual protein A or protein G, beads. Protein A/G has high affinity for antibody IgG subclasses and when covalently complexed to a magnetic bead functions as an affinity matrix for the precipitation of immunocomplexes. Subsequent digestion of the antibody and precipitated protein leaves only the DNA previously associated with the chromatin feature. The DNA can be further precipitated and used for PCR amplification with known primer sequences against regions of interest to allow identification of DNA regions associated with a specific chromatin feature. Alternatively the DNA can be used to make a DNA library of all regions in the genome that are associated with the specific feature and then sequenced by massive parallel sequencing (next-generation sequencing) to allow identification of the location of the chromatin feature across the genome (Fig. 1). These chromatin features can include specific histone modifications or specific histone variants, or specific histone binding proteins. Importantly the ChIP technique can also be used with antibodies that bind to TFs but in this case weaker binding to chromatin necessitates chemical cross-linking of the TF to chromatin (with formaldehyde) before DNA fragmentation as well as reversal of cross-linking before DNA isolation. Precipitated DNA following ChIP can be quantified using a bioanalyzer before library preparation to determine concentration and quality for ChIP sequencing. Generally, 20–50 ng/μl and 1 ng/μl of DNA in the size range 100–500 bp are required.

1.2 ChIP Sequencing with a Formaldehyde Cross-Linked TF

Comprehensive reviews of ChIP design [6], library preparation [7], and ChIP sequencing analysis using R [8] are available. Here we discuss ChIP of an epitope-tagged TF in human endothelial cells (HuVECs) and optimization steps and controls for ChIP

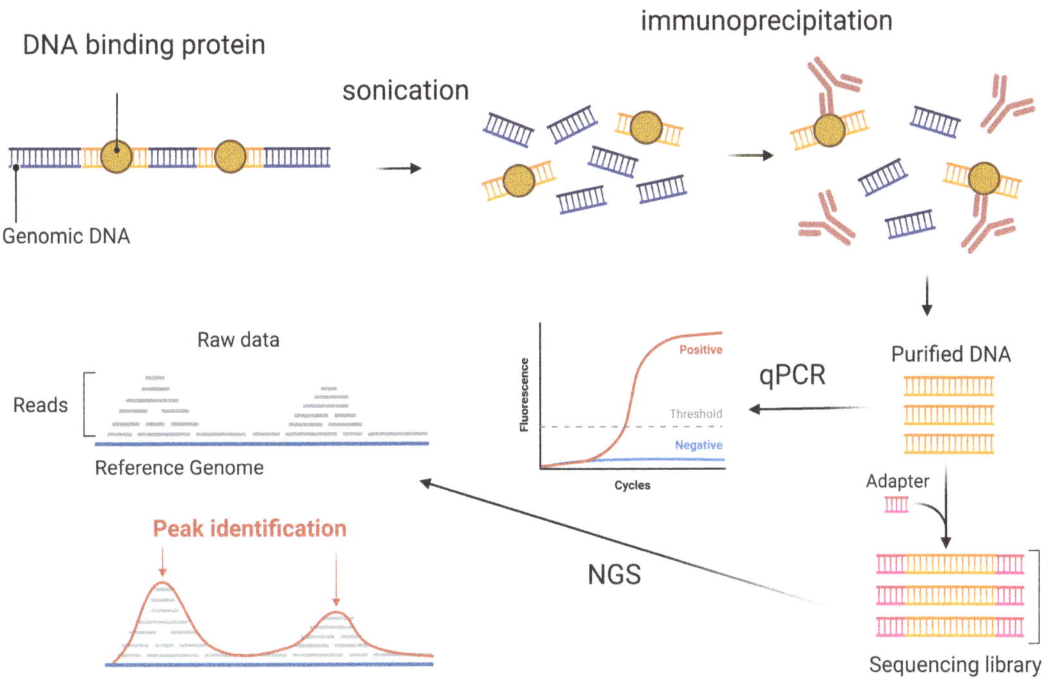

Fig. 1 An overview of ChIP followed by qPCR or whole genome sequencing. Transcription factors associated with chromatin are cross-linked to the chromatin and then following sonication of chromatin the cross-linked TFs are isolated by affinity purification using antibodies. DNA is isolated from the complexes and then known binding regions can be examined using qPCR or alternatively isolated DNA is ligated to primer adaptors and then amplified by PCR and a library of all DNA fragments associated with the TF is generated for Next Generation Sequencing (NGS). Fragments are sequenced and Reads are mapped to the reference genome allowing accumulation of read peaks at each position on the genome where the TF is associated

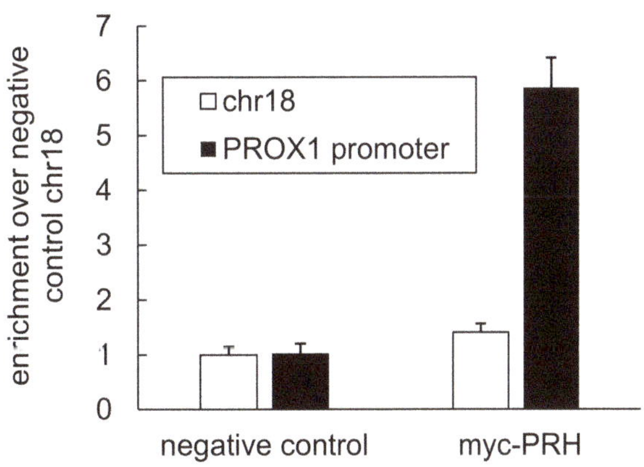

Fig. 2 Enrichment of TF binding at the PROX1 promoter. ChIP of myc-PRH binding at the PROX1 promoter compared to myc-PRH binding at chr18 (negative control) determined using qPCR

experiments prior to ChIP sequencing and qPCR. Epitope tagging is often used when a highly specific antibody against the TF that is

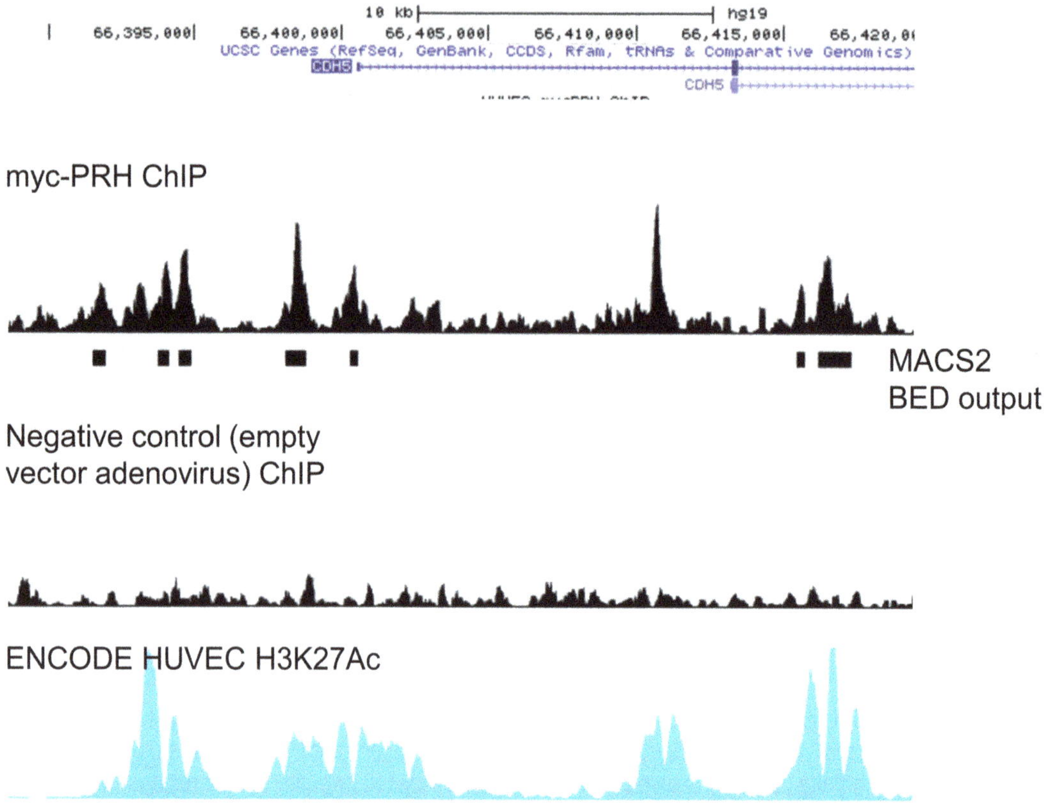

Fig. 3 Investigating TF binding genome-wide. Visualization of enrichment of myc-PRH binding at the CDH5 gene from a ChIP-seq experiment (UCSC genome browser). Called peaks (black bars) are shown in the CDH5 promoter region and also in the introns of the gene relative to the input DNA. The called peaks overlap with ENCODE HuVEC histone H3 K27 acetylation peaks indicating myc-PRH binding in regions of open accessible chromatin

useful for immunoprecipitation is not available [9, 10]. Common epitope tags include the hemagglutinin HA epitope, flag epitope, and the Myc 9E10-epitope. These are short peptides that are highly immunogenic and have well characterized antibodies raised against them that can be used for immunoprecipitation and ChIP. Although expression levels of tagged TFs need to be carefully compared to endogenous levels of the TF to avoid DNA binding artifacts from high expression levels, this is a commonly used method for ChIP in the absence of useful antibodies against the endogenous protein.

In this laboratory we have carried out ChIP sequencing for Myc-tagged PRH/HHEX from a variety of cell types [11]. HuVECs are infected with an adenoviral expression vector that expresses Myc-PRH in all cells (approximately five- to eight-fold above endogenous levels) or an empty adenoviral vector [12]. The ChIP protocol including formaldehyde cross-linking, immunoprecipitation with myc antibody, reverse cross-linking, and purification of DNA is outlined with consideration of different controls. To have confidence in the ChIP results it is of importance that proteins with known binding characteristics (positive controls) are used in this technique and tested for enrichment of known binding sequences using qPCR. Generally following ChIP for PRH and before using the ChIP DNA material for genome-wide sequencing we examine using qPCR whether the myc-epitope tagged PRH TF is associated with characterized DNA regulatory elements [11, 13]. It should be noted that sample to sample variation will occur and that DNA from the same sample should be analyzed using both techniques. Moreover, for reliable ChIP-seq it is advisable to repeat the experiment at least twice.

An example of ChIP followed by qPCR to confirm binding is shown (Fig. 2) and worked examples for calculation of ChIP as a percentage of input is shown in Subheading 3.5. Enrichment is calculated as the ratio of the ChIP percentage of input for a specific region (gene) to ChIP percentage of input for negative control regions (gene deserts on Chr18 [14]). We also describe bioinformatics processing steps required for analyzing raw ChIP data and show an example of the ChIP sequencing coverage track in UCSC for the called peaks for myc-PRH binding at the CDH5 gene in HuVECs (Fig. 3).

1.3 Chromatin Immunoprecipitation Sequencing

Chromatin immunoprecipitation sequencing (ChIP-seq) refers to next-generation sequencing (NGS) of all genomic DNA fragments isolated following ChIP and the reversal of the cross-links. As all of the precipitated DNA needs to be sequenced and is at very low concentration the DNA must be amplified. This is known as library preparation and although it can be performed in the laboratory using enzymes that are generally available, most often commercial kits are used or more frequently library preparation is carried out by sequencing providers because conditions for optimal library complexity are well-established.

For library preparation, the eluted DNA is prepared for ligation of adaptor sequences. T4 polynucleotide kinase, T4 DNA polymerase, and Klenow enzyme blunt end the fragments and 5′ phosphorylate the ends. The 3′ ends are also modified using Klenow (exo-) to allow A tailing. Then oligonucleotide adaptors for PCR are ligated onto the fragments. To allow for multiplex sequencing, adaptors containing barcodes are either ligated onto each sample or the adaptors containing the barcodes can be incorporated during subsequent PCR steps. The DNA fragments with adaptors ligated are size selected (~300 bp) by gel electrophoresis and isolation and elution of these DNA fragments removes excess unligated adaptors. This size selection allows optimal PCR amplification of the DNA library and is required for the NGS protocols. The DNA library consists of all the DNA sequences associated with chromatin features of interest and the PCR amplification step has to allow optimal amplification without introduction of PCR bias from repetitive or high GC content sequences. Thus the number of cycles to obtain the most complex library must be determined empirically.

The most useful control for ChIP-seq is the sequencing in parallel of the input chromatin. This is chromatin that has been fragmented but not subject to chromatin immunoprecipitation. This chromatin will include all the sequences in the genome. The increased representation (enrichment) of specific sequences in the ChIP experiment over input chromatin identifies the regions of the genome where specific chromatin features of interest are located.

1.4 ChIP Sequencing Depth

The fragments in the amplified library can be sequenced from one or both ends of the molecule. Sequencing from both ends is known as paired-end sequencing and this improves the mapping of the fragments to the genome. Single end reads (75–150 bp) are usually sufficient for analysis and 15–20 million reads are usually deemed sufficient. In some cases, increasing sequencing depth may lead to lower intensity signals being detected, although there are diminishing returns as this is ultimately limited by the number of unique fragments in the immunoprecipitated sample and the library complexity. After mapping all the reads to the genome, the bioinformatics software generates information about the number of aligned fragments for any particular region and the position of the centre of the ChIP peak, a genomic location with enrichment of aligned reads. The height of the peak corresponds to the amount of DNA purified by immunoprecipitation, averaged across all of the cells in the original biological sample. This depends on the fraction of cells containing the specific protein–DNA complex at that genomic region, but also antigen accessibility and cross-linking efficiency for that particular complex. Care must therefore be taken when comparing the heights of different peaks as a higher peak does not *necessarily* equate to more, or stronger, binding. In general TFs give discrete narrow peaks. Deeper sequencing (30–50 million reads) may be required for chromatin features that have much broader peaks such as histone modifications.

2 Materials

2.1 Cell Culture and Cross-Linking

- HuVECs (e.g., PromoCell or isolated from umbilical vein).
- Lonza EGM-2 medium.
- Methanol-free formaldehyde (e.g., Pierce 16% formaldehyde, methanol-free).
- Phosphate-buffered saline (PBS).
- M glycine in PBS.

2.2 Chromatin Isolation and Shearing

- Protease inhibitor cocktail (PIC) (e.g., Roche cOmplete).
- Water bath sonicator (e.g., Diagenode Bioruptor) and sonicator tubes.
- *Buffer A.*
  - 10 mM HEPES, pH 8.0.
  - 10 mM EDTA.
  - 0.5 mM EGTA.
  - 0.25% v/v Triton X-100 (*see* **Note 1**).
- *Buffer B.*
  - 10 mM HEPES, pH 8.0.
  - 200 mM NaCl.
  - 1 mM EDTA.
  - 0.5 mM EGTA.
  - 0.01% v/v Triton X-100.
- *ChIP buffer.*
  - 25 mM Tris-HCl, pH 8.0.
  - 150 mM NaCl.
  - 2 mM EDTA.
  - 1% v/v Triton X-100.
  - 0.25% w/v SDS.
- *ChIP dilution buffer.*
  - 25 mM Tris-HCl, pH 8.0.
  - 150 mM NaCl.
  - 2 mM EDTA.
  - 1% v/v Triton X-100.
  - 7.5% v/v Glycerol.

2.3 Immunoprecipitation

- Protein A/G magnetic beads (e.g., Pierce™ Protein A/G Magnetic).
- Beads or other protein A/protein G magnetic beads).
- Magnetic tube rack.
- 100 mM phosphate buffer, pH 8.0.
- 0.5% w/v bovine serum albumin (BSA) in 100 mM phosphate buffer, pH 8.0.

Antibody—to be determined by the user.

- *Wash buffer 1.*
  - 20 mM Tris-HCl, pH 8.0.
  - 150 mM NaCl.
  - 2 mM EDTA.
  - 1% v/v Triton X-100.
  - 0.1% w/v SDS.
- *Wash buffer 2.*
  - 20 mM Tris-HCl, pH 8.0.
  - 500 mM NaCl.
  - 2 mM EDTA.
  - 1% v/v Triton X-100.
  - 0.1% w/v SDS.
- *LiCl buffer.*
  - 10 mM Tris-HCl, pH 8.0.
  - 250 mM LiCl.
  - 1 mM EDTA.
  - 0.5% v/v NP-40 (*see* **Note 2**).
  - 0.5% sodium deoxycholate.
- *NaCl buffer.*
  - 10 mM Tris-HCl, pH 8.0.
  - 50 mM NaCl.
- *Elution buffer.*
  - 100 mM $NaHCO_3$.
  - 1% w/v SDS.

2.4 Cross-Link Reversal and DNA Isolation and Library Preparation

- Heat block or water bath.
- RNase A (10 mg/ml stock).
- Proteinase K (50 mg/ml stock).
- 5 M NaCl.
- 0.5 M EDTA.
- 80% v/v ethanol.
- *0.1× TE.*
  - 1 mM Tris-HCl, pH 8.0.
  - 0.1 mM EDTA.
- Solid phase reversible immobilization (SPRI) magnetic beads (e.g., Ampure XP, Axyprep).
- NGS library preparation kit (e.g., NEBNext Ultra II DNA Library Prep Kit for Illumina).

3 Methods

3.1 Cell Culture and Cross-Linking

- The number of cells required for a ChIP experiment will vary depending on a variety of factors including the expression level of the TF of interest, stability of the protein–DNA complex(es), antigen accessibility, and preservation during cross-linking, and antibody affinity. We find that a good starting point is 5×10^6 cells, but this should be optimized by the reader.
- Plate HuVECs at 5000 cells/cm^2 on gelatin-coated plasticware. Large vessels (e.g., T175 flasks or 150 mm dishes) may be needed to achieve the required cell number. For adenoviral protein overexpression, add virus at the required multiplicity of infection (MOI). For our myc-PRH/HHEX constructs we typically use an MOI of 50. Change medium the following day, and proceed with cross-linking 48 h after infection.
- Cross-linking can be done either in suspension (following enzymatic cell retrieval), or in situ on the culture vessel. We find that cross-linking in suspension leads to more reproducible sonication than cross-linking in situ. However, for the study of signaling pathways that respond quickly to changes in cell adhesion or cell–cell contacts, enzymatic retrieval of cells may not be appropriate.
- Before beginning cross-linking, pre-cool a centrifuge for 15 ml tubes to 4 °C.
- For suspension cross-linking, retrieve cells according to the standard protocol (with, e.g., trypsin, accutase, EDTA, etc.), and centrifuge in a 15 ml centrifuge tube. Resuspend cells in standard culture medium (including serum) at a concentration of 2×10^6 cells/ml. Add methanol-free formaldehyde to a final concentration of 1% (for some TFs this may need to be optimized, but we find that 1% is a good starting point). Incubate for 10 min at room temperature with constant agitation, either using a rotating mixer or by inverting the tube by hand.
- For in situ cross-linking, add methanol-free formaldehyde directly to the culture medium on the cells, to a final concentration of 1%. Incubate for 10 min at room temperature on a rocking platform set to approximately 20 cycles per minute.
- Quench formaldehyde by adding 1/4 volume of 2 M glycine in PBS (i.e., for a final concentration of 0.4 M). Centrifuge for 5 min at $300 \times g$, 4 °C in the pre-cooled centrifuge.
- Discard the supernatant and wash by gently resuspending the pellet in 10 mL of ice-cold PBS, then centrifuging for 5 min at $300 \times g$, 4 °C in the pre-cooled centrifuge. Repeat this wash once, for a total of two washes. During the second wash centrifugation step, add fresh protease inhibitor cocktail (PIC) to ice-cold buffer A and buffer B and keep on ice.

3.2 Chromatin Isolation and Shearing

- Lyse the cells to release nuclei by resuspending the pellet in ice-cold buffer A (with PIC) at a concentration of 2×10^6 cells/ml. Incubate for 10 min at 4 °C on a rotating mixer.
- Collect nuclei by centrifugation at 500 × *g* for 5 min at 4 °C. Discard the supernatant.
- Remove free soluble nuclear proteins by resuspending the pellet in ice-cold buffer B (with PIC) at a concentration of 2×10^6 cells/ml. Incubate for 10 min at 4 °C on a rotating mixer.
- Collect nuclei by centrifugation at 500 × *g* for 5 min at 4 °C. Discard the supernatant.

  Stopping point: If required, the isolated nuclear pellet can be snap-frozen and stored at −80 °C for several months.

  - *for time efficiency, Subheading 3.3 (binding of antibody to protein A/G beads) can be started at this point.*
- Add fresh PIC to ChIP Buffer. Resuspend the nuclear pellet in ChIP Buffer at a concentration of 3.33×10^7 cells/ml (i.e., 30 μl per 10^6 cells).
- Sonicate the chromatin in a water bath sonicator. Make sure to use the exact tubes and volumes recommended by the sonicator manufacturer. Use of standard Eppendorf tubes will lead to inefficient DNA shearing. If the sonicator does not have a dedicated cooling system, add ice to the water bath so that the total volume is approximately 25% ice/75% water (too little ice will lead to excessive heating and sample damage, whereas too much ice will lead to inefficient propagation of ultrasound). The ice may need to be replaced as sonication proceeds. The number of sonication cycles will need to be optimized for each cell type and cross-linking method used. On a Diagenode Bioruptor we find that 10–30 cycles of 30 s on/30 s off at high power (240 W) work for most cell types.
- For optimization of sonication, sonicate separate tubes for 10–40 cycles in steps of 5 s (i.e., 10, 15, 20,... *see* Fig. 4). If optimization is not required skip directly to Subheading 3.4 and proceed with reversal of cross-links and DNA purification. After DNA purification run sonicated DNA on a 1.5% agarose gel with a 100 bp ladder. Load 500 ng DNA per lane. "Optimal" sonication will depend to some extent on the DNA binding protein under study. For example, proteins which preferentially bind in heterochromatin (which is more sonication resistant than euchromatin) will require more sonication than proteins that bind to active promoters in euchromatin. A typical starting point is to aim for is fragments of 100–500 bp. Over sonication will result in inefficient PCR after DNA retrieval and under

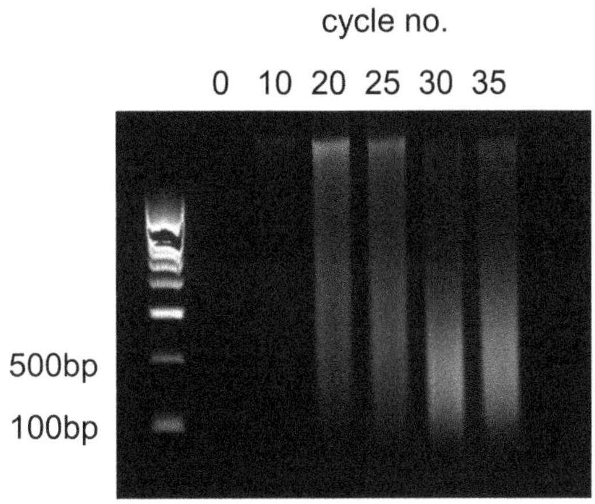

Fig. 4 Optimization of sonication. 1.5% agarose gel electrophoresis was used to visualize increased chromatin fragmentation with increasing number of sonication cycles and a DNA size marker. DNA stains most brightly (smear) between 100–500 bp at 30–35 cycles

sonication will not allow accurate identification of true ChIP peaks.

- Centrifuge at 16,000 × *g* for 10 min at 4 °C to remove any large debris.
- During this centrifugation step, add fresh PIC to ChIP Dilution Buffer.
- Collect the supernatant and dilute threefold with ChIP Dilution Buffer (i.e., add two parts ChIP Dilution buffer to one part sonicated chromatin).
- *Stopping point: If required, chromatin can be snap-frozen and stored at −80 °C for several weeks. Do not thaw more than once—if multiple samples will be required at different times, aliquot the chromatin before freezing.*

3.3 Immunoprecipitation

- Add protein A/G magnetic beads to low-binding Eppendorf-style tubes (e.g., Axygen Maxymum Recovery). Ensure that the binding capacity of the beads added is in small excess over the amount of antibody to be used. For example, we use Pierce Protein A/G Magnetic Beads, which have a binding capacity of around 55 μg IgG per mg of beads and are supplied at a bead concentration of 10 mg/ml. Therefore 10 μl of beads will have a total binding capacity of around 5.5 μg IgG.
- Wash the magnetic beads by adding 100 μl of 100 mM phosphate buffer (pH 8.0) followed by magnetic separation and removal of the supernatant. Repeat once for a total of two washes.

- Resuspend the beads in 10 μl of 0.5% w/v BSA in 100 mM phosphate buffer (pH 8.0). For all steps, ensure that the magnetic beads are thoroughly resuspended.
- The antibody to be used should be determined by the reader. Where possible ChIP-validated antibodies are preferred, however, especially for less well-studied proteins, this is not always possible. We recommend standard antibody validation techniques such as confirming loss/reduction of signal upon knockout/knockdown of the target protein. If an epitope-tagged exogenous protein is the target, non-tagged or empty vector controls can be used—we find that this is more robust than non-specific IgG controls, which tend to have much lower background than specific antibodies, potentially leading to the false conclusion that the protein of interest is "binding everywhere" in the genome.
- The amount of antibody to be used should also be optimized. 1–5 μg is a standard range for initial optimization experiments.
- Bind the antibody to the protein A/G beads by adding it to the beads, followed by incubation for 2 h at 4 °C with rotation.
- *If chromatin has been frozen*, thaw on ice followed by centrifugation at 16,000 × *g* for 10 min at 4 °C, and transfer the supernatant to a fresh tube.
- Remove an input sample (5% of chromatin volume) and add the remaining chromatin to the antibody-bound beads. Incubate for 3 h at 4 °C with rotation, and keep the input sample at 4 °C.
- Use the magnetic rack to separate the immune-complex-bound beads and remove the supernatant.
- Wash by adding 1 ml of wash buffer 1, incubate for 10 min at 4 °C with rotation, followed by magnetic separation.
- Repeat the step above using wash buffer 2 twice, LiCl buffer once, and NaCl buffer twice (for a total of 6 washes). Before separating the final NaCl wash, transfer to a fresh low-binding tube.
- Elute immune complexes by adding 50 μl of Elution Buffer. Incubate for 15 min at room temperature with shaking, then collect the eluate after magnetic separation.
- Repeat the step above. Pool the eluates for a total eluate volume of 100 μl.
- Add the elution buffer components to the input sample and top up to a total volume of 100 μl (e.g., to 25 μl of input, add 10 μl 10% SDS, 10 μl of $NaHCO_3$, and 55 μl molecular grade water; to 30 μl of input, add 10 μl 10% SDS, 10 μl of $NaHCO_3$, and 50 μl molecular grade water, etc.).

3.4 Cross-Link Reversal and DNA Isolation

- For each ChIP sample and input, remove RNA by adding 1 μl of RNase A from a 10 mg/mL stock, mix thoroughly using low-binding pipette tips, and incubate for 30 min at room temperature.
- Remove proteins and reverse cross-links by adding 4 μl of 5 M NaCl, 2 μl of 0.5 M EDTA, and 1 μl of 50 mg/mL Proteinase K, mixing thoroughly using low-binding pipette tips, and incubating overnight at 55 °C in a water bath.
- Prepare SPRI (solid phase reversible immobilization) beads by warming to room temperature and vortexing to ensure an even suspension.
- Bind DNA to beads by adding 1.8× reaction volume (i.e., to the 108 μl samples, add 194 μl beads). Incubate for 10 min at room temperature.
- Magnetically separate for 10 min. Remove and discard the supernatant, and wash twice with freshly made 80% ethanol.
- After removing the second ethanol wash, air-dry the bead pellet for 10–20 min until the shiny appearance of the beads starts to fade. The beads should appear dull, but not cracked—once cracks start to appear, the beads are over-dried and DNA elution will be poor. If using SPRI beads for the first time we recommend ethanol-washing and air-drying beads alone a few times to get used to the change in appearance, and to avoid loss of precious samples.
- Elute DNA by adding 50 μl of 0.1× TE and incubate for 5 min at room temperature. Magnetically separate the beads, and collect the supernatant containing the ChIP DNA fragments. Store at −20 °C.

3.5 qPCR Verification of Known Binding Regions

Pull down of specific genomic regions can be tested by qPCR. We typically use 5 μl of the eluted material per PCR reaction. Efficiency of all PCR primers should be calculated by constructing a standard curve of Ct against dilution factor, and the efficiencies should be taken into account when calculating enrichments, using the Pfaffl method. In our experience, normalization to a negative control (non-bound) region in the same sample is preferable to normalization to a negative control antibody, as non-specific IgGs often give a lower background signal than specific antibodies, leading to false positives. Enrichment can then be calculated according to the following equation:

$$\text{Enrichment} = \frac{E_a^{\text{Ct}_{a,\text{input}} - \text{Ct}_{a,\text{ChIP}}}}{E_-^{\text{Ct}_{-,\text{input}} - \text{Ct}_{-,\text{ChIP}}}} \quad (1)$$

where E and Ct are the primer efficiencies and threshold cycles, respectively, "a" refers to the putative TF binding site, and "–" to

Table 1
qPCR primer sequences for negative control regions of the human genome

Primer name	Sequence (5′-3′)
Human chr18 negative F	ACTCCCCTTTCATGCTTCTG
Human chr18 negative R	AGGTCCCAGGACATATCCATT

the negative control region. Non-bound regions should have an enrichment of approximately 1 and bound regions >1. In our datasets we find that regions that have strong ChIP-seq peaks are typically enriched approximately 5- to 20-fold by qPCR (*see* Fig. 2). For negative control regions, we typically use regions within "gene deserts," and in particular well-characterized regions on chr18 (*see* Table 1).

3.5.1 How to Calculate Enrichment as ChIP Percentage of Input Specific Region/Chip Percentage of Input Negative Control (CHr18)

The DNA obtained from the ChIP is used in qPCR for the negative control regions with Chr18 and with primers for the region of interest. In each case the Ct for the primers, that is, from the region of interest or control regions is compared to the adjusted Ct for the input chromatin control. This allows the concentration of the ChIP DNA to be calculated as a percentage of input at the region of interest and at negative control regions (*see* **Note 3**).

1. Adjustment of input Ct:

$$\text{Adjusted Ct}_{\text{input}} = \text{Ct}_{\text{input}} - \text{Log}_2(\text{dilution factor})$$
$$\text{Dilution factor} = 100/\text{input}\%$$

For example, 500 μl chromatin take 50 μl for "Input," add 5 μl to PCR reaction

$$5\ \mu\text{l of input, that is, } 5/500 \times 100/1 = 1\%\text{of}\text{``Input''}$$
$$\text{Dilution factor} = 100/1 = 100$$

Or if 500 μl chromatin take 50 μl for "Input," add 10 μl to PCR reaction

$$5\ \mu\text{l of input, that is, } 10/500 \times 100/1 = 2\%\text{of}\text{``Input''}$$
$$\text{Dilution factor} = 100/2 = 50.$$

2. ChIP enrichment for a gene as a *percentage* of input can be calculated as follows:

$$100 \times (\text{dilution factor for IP}) \times 2^{(\text{Adjusted Ct}_{\text{input}} - \text{CtChIP})}$$

For example, 500 μl chromatin take 450 μl for "ChIP," resuspend in 100 add 4 μl to PCR reaction

$$450 = 450/500 \times 100/1 = 90\%\text{of}\text{``Input''}$$

$$4/100 = 3.6\%\text{of}\text{``input''}\ (\text{i.e.}, 4/100 \times 90\% = 0.04 \times 90\% = 1)$$

$$\text{Dilution factor} = 100/3.6 = 27.8$$

$$\%\text{input} = 100 \times 27.8 \times 2^{(\text{Adjusted Ct}_{\text{input}} - \text{CtChIP})}$$

For example, 500 μl chromatin take 400 μl for "ChIP," resuspend in 400 add 5 μl to PCR reaction.

$$5/400\ 1\%\text{of}\text{``input''}\ (\text{i.e.}, 5/400 \times 80\% = 0.125 \times 80 = 1)$$

$$\text{Dilution factor} = 100/100 = 1$$

As mentioned above the same calculation is made using ChIP primers for gene desert negative controls. A ratio of the percentage input of the region of interest to the percentage of input for the negative control regions can be used to determine a ratio of enrichment. This should give the same enrichment as Eq. 1.

If IgG control is also included in the experiment, then enrichment compared to IgG control can instead be calculated. As mentioned above this is not the preferred method used in this laboratory due to the risk of false positives. If using an IgG control, the enrichment at all negative control regions should be 1 (i.e., no enrichment). If this is not the case, the IgG control is not suitable.

3.5.2 Example 1

500 μl chromatin

- 1% (5 μl) for input PCR (95 kept for input)
- 400 μl for "ChIP" resuspended in 400 add 5 μl to PCR reaction
  - 1% input Ct 20, IgG Ct40, Ab TF Ct21

$$\begin{aligned}\text{Adjusted Ct}_{\text{input}} &= \text{Ct}_{\text{input}} - \text{Log}_2(\text{dilution factor})\\ &= 20 - \text{Log}_2(100/1) = 20 - 6.64 = 13.36\end{aligned}$$

Percentage of input for IgG:

$$100 \times (\text{dilution factor for IP}) \times 2^{(\text{Adjusted Ct}_{\text{input}}\text{-CtChIP})}$$

$$100 \times 100 \times 2^{(13.36-40)} = 100 \times 100 \times 2^{(-26.64)} = 0.00009$$

Percentage of input for Ab TF:

$$100 \times (\text{dilution factor for IP}) \times 2^{(\text{Adjusted Ct}_{\text{input}} - \text{CtChIP})}$$

$$100 \times 100 \times 2^{(13.36-21)} = 100 \times 100 \times 2^{(-7.54)} = 53.7$$

3.5.3 Example 2

500 μl chromatin:

- 2% (10 μl) for input PCR (945 kept for input)
- 450 μl for "ChIP" resuspended in 200 add 5 μl to PCR reaction
- 2% input Ct 16, IgG Ct40, Ab TF Ct21

$$\begin{aligned}\text{Adjusted Ct}_{\text{input}} &= \text{Ct}_{\text{input}} - \text{Log}_2(\text{dilution factor}) \\ &= 20 - \text{Log}_2(100/2) = 16 - 5.64 = 10.36\end{aligned}$$

Percentage of input for IgG:

$$100 \times (\text{dilution factor for IP}) \times 2^{(\text{Adjusted Ct}_{\text{input}} - \text{CtChIP})}$$

$$\begin{aligned}100 \times (100/2.25) \times 2^{(10.36-40)} &= 100 \times 44.4 \times 2^{(-29.64)} \\ &= 0.000005\end{aligned}$$

Percentage of input for Ab TF:

$$100 \times 44.4 \times 2^{(\text{Adjusted Ct}_{\text{input}} - \text{CtChIP})}$$

$$100 \times 44.4 \times 2^{(10.36-21)} = 100 \times 44.4 \times 2^{(-10.64)} = 2.78$$

In the case of TFs that are not well characterised or TF that have been studied only in other tissues, it may be difficult to decide on appropriate "positive" regions. Given the decreasing costs of NGS, in such cases we would advocate a "sequence it and see" approach, rather than a brute force qPCR approach.

3.6 Library Production and Next-Generation Sequencing

Illumina sequencing platforms are most commonly used. Sequencing libraries can be prepared in-house using commercially available kits for adaptor ligation and amplification following kit protocols (e.g., NEBNext® Ultra™ II DNA Library Prep Kit for Illumina). To minimize the risk of PCR bias, the enrichment of several bound regions should be compared before and after library preparation and should be quantitatively similar. Many sequencing services will provide library preparation and in our experience this is a cost-effective approach unless sequencing a large number of samples.

For the HuVEC ChIP-seq we used read depth 40–50 million reads (2 × 100 bp pair end on Illumina HiSeq 2500) for mapping TF binding.

3.7 ChIP Sequencing Downstream Bioinformatics Processing

Generally, the sequencing data from the NGS sequence provider is delivered as a FASTQ file. The raw data can be processed using a large variety of bioinformatics tools but a common suite of tools for initial bioinformatics processing of ChIP experiments is the Galaxy webserver (usegalaxy.org). This is a useful solution if only one or two ChIP experiments are being performed, but larger experiments may require the use of command line tools or tools in the programming language R. Once FASTQ files are uploaded to Galaxy the following tools in Galaxy can be used.

1. FastQC is used to assess the quality of the sequencing data. Base call quality should be on average > 20 in PHRED units, where the PHRED score is $-10\log_{10}P$, with P as the probability of a miscalled base. There should be no overrepresented sequences (aside from possibly adaptor sequences), and average sequence length should match the length of sequence collected on the sequencer (e.g., 100 bp).
2. TrimGalore is used to remove adaptor sequences and trim low-quality bases. We use a PHRED cutoff of 20 for quality trimming.
3. BowTie is used to map sequencing reads to the genome, generating a .bam file that can be visualized in the UCSC genome browser (choose very sensitive local setting).
4. The FilterBam tool is used to remove duplicate reads and multi-mapped sequences. Choose IsDuplicates (No) and Mapquality >1. Reads aligned to poorly mapable regions (e.g., low-complexity repeat regions) should be removed by comparison to a reference "blacklist"—these are publicly available for many genomes [15]. We remove the blacklisted regions using the Subtract tool.
5. Peak calling is carried out on the filtered .bam file using MACS (model-based analysis of ChIP-seq) for single end sequencing or MACS2 for pair end sequencing and generates a Bed file. Coverage tracks for visualization in the UCSC genome browser can be generated using the bamCoverage tool.Option. To use control file—check Yes and specify that it is single or pair end and read sequence length, for example, 100 bp. The MACS2 algorithm determines enrichment above the background and the position of the peak [16]. The peak caller returns a statistical confidence for each peak in the form of a p value and an FDR-adjusted q value. The peak calling output in BED format can be simultaneously loaded into the UCSC browser to visually inspect the enriched regions detected by MACS2 (*see* Fig. 3).
6. Comparison to the input control or negative control chromatin is of great value to visually determine whether peaks are above background. Following peak assignment, genomic features close to the ChIP peaks can be identified using the ClosestBed tool which, along with a genome annotation, identifies the ChIP peaks near to desired genomic features (e.g., transcription start sites).

The sequences underlying ChIP peaks can be used for de novo identification of transcription factor binding motifs, using software such as HOMER or RSAT. The TF motifs are typically only a few base pairs in size (~8–40 bp), whereas the ChIP called peaks are typically 300 bp+ in size.

4 Notes

1. Triton X-100 is very viscous. To ensure pipetting accuracy, we first prepare a 10% v/v stock using a serological pipette and a pipette gun set to "slow."
2. NP-40 here refers to Nonidet P-40, which is also available as Igepal CA-630 (*octyl*phenoxypolyethoxyethanol), *not* tergitol NP-40 (*nonyl*phenoxypolyethoxyethanol).
3. The Ct of the input chromatin must be adjusted to reflect the volume of the input used in the qPCR reaction.

   [Note that instead of theoretical efficiency of 2 it is preferable to use the primer efficiency determined using serial dilution of genomic DNA, that is, $E^{(\text{Adjusted Ct}_{\text{input}} - \text{CtChIP})}$—but this is not essential.]

Acknowledgments

We are grateful to Dr. Ka Ying Lee for the data shown in Fig. 4. This work was funded by the UK Medical Research Council (MRC grant reference MR/N012615/1).

References

1. Jiang C, Pugh BF (2009) Nucleosome positioning and gene regulation: advances through genomics. Nat Rev Genet 10(3):161–172
2. Rothbart SB, Strahl BD (2014) Interpreting the language of histone and DNA modifications. Biochim Biophys Acta 1839(8):627–643
3. Catarino RR, Stark A (2018) Assessing sufficiency and necessity of enhancer activities for gene expression and the mechanisms of transcription activation. Genes Dev 32(3–4): 202–223
4. Gaston K, Jayaraman PS (2003) Transcriptional repression in eukaryotes: repressors and repression mechanisms. Cell Mol Life Sci 60(4):721–741
5. Visa N, Jordan-Pla A (2018) ChIP and ChIP-related techniques: expanding the fields of application and improving ChIP performance. Methods Mol Biol 1689:1–7
6. Jordan-Pla A, Visa N (2018) Considerations on experimental design and data analysis of chromatin immunoprecipitation experiments. Methods Mol Biol 1689:9–28
7. Head SR, Komori HK, LaMere SA, Whisenant T, Van Nieuwerburgh F, Salomon DR et al (2014) Library construction for next-generation sequencing: overviews and challenges. Biotechniques 56(2):61–64, 6, 8, passim
8. Kerdivel G, Boeva V (2021) Chromatin immunoprecipitation followed by next-generation sequencing (ChIP-Seq) analysis in ewing sarcoma. Methods Mol Biol 2226:265–284
9. Soufi A, Smith C, Clarke AR, Gaston K, Jayaraman PS (2006) Oligomerisation of the developmental regulator proline rich homeodomain (PRH/Hex) is mediated by a novel proline-rich dimerisation domain. J Mol Biol 358(4): 943–962
10. Marcolino E, Siddiqui YH, van den Bosch M, Poole AW, Jayaraman PS, Gaston K (2020) Blood platelets stimulate cancer extravasation through TGFbeta-mediated downregulation of PRH/HHEX. Oncogenesis 9(2):10
11. Kitchen P, Lee KY, Clark D, Lau N, Lertsuwan J, Sawasdichai A et al (2020) A Runaway PRH/HHEX-Notch3-Positive Feedback Loop Drives Cholangiocarcinoma and Determines Response to CDK4/6 Inhibition. Cancer Res 80(4):757–770
12. Gauvrit S, Villasenor A, Strilic B, Kitchen P, Collins MM, Marin-Juez R et al (2018)

HHEX is a transcriptional regulator of the VEGFC/FLT4/PROX1 signaling axis during vascular development. Nat Commun 9(1): 2704
13. Williams H, Jayaraman PS, Gaston K (2008) DNA wrapping and distortion by an oligomeric homeodomain protein. J Mol Biol 383(1):10–23
14. Nusbaum C, Zody MC, Borowsky ML, Kamal M, Kodira CD, Taylor TD et al (2005) DNA sequence and analysis of human chromosome 18. Nature 437(7058):551–555
15. Amemiya HM, Kundaje A, Boyle AP (2019) The ENCODE Blacklist: Identification of Problematic Regions of the Genome. Sci Rep 9(1):9354
16. Zhang Y, Liu T, Meyer CA, Eeckhoute J, Johnson DS, Bernstein BE et al (2008) Model-based analysis of ChIP-Seq (MACS). Genome Biol 9(9):R137

Chapter 21

In Vitro Co-culture of Fibroblast and Endothelial Cells to Assess Angiogenesis

Adam F. Odell and Aarren J. Mannion

Abstract

Angiogenesis relies on the spatial and temporal coordination of endothelial migration and proliferation to form new blood vessels. This occurs through synchronous activation of multiple downstream pathways which facilitate vascular development. Proangiogenic growth factors and supporting extracellular matrix allow the formation of capillary-like tubules, reminiscent of microvascular beds, in vitro. In this chapter, we describe a methodology for the establishment of vascular networks by co-culture of endothelial cells and fibroblasts to facilitate the study of tubulogenic and angiogenic potential. We detail the use of siRNA mediated knockdown to deplete target genes of interest, in either the endothelial or fibroblast cells, to allow the assessment of their role in angiogenesis. Finally, we detail how these vascular networks may be stained using immunofluorescence to allow quantification of angiogenic potential in vitro.

Key words Co-culture, Fibroblasts, Angiogenesis, HUVEC, VEGF, Tubulogenesis, Endothelial cells

1 Introduction

The formation of the circulatory system is shaped by the angiogenic coordination of endothelial cells during embryonic and postnatal development. The ability of endothelial cells to respond to chemotactic and microenvironmental cues through modulation to proliferation, migration, differentiation, metabolism, and remodeling of the extracellular matrix are just a few of the key requirements for angiogenic expansion [1, 2]. Angiogenesis refers to the establishment of new blood vessels from preexisting vasculature, which outside of disease settings is limited primarily to development. Presented with a facilitating environment and mitogenic signals, endothelial cells undergo a complex rearrangement to drive successful angiogenic growth. To date, numerous transgenic models (both global and endothelial specific) have been designed to manipulate vessel development and unraveling the role of numerous pathways in the angiogenic process. The coordination of these

Andrew V. Benest (ed.), *Angiogenesis: Methods and Protocols*, Methods in Molecular Biology, vol. 2441,
https://doi.org/10.1007/978-1-0716-2059-5_21,

multiple pathways, all indispensable for angiogenic growth, highlights the broad aspects of cellular orchestration required for unabated angiogenesis.

While sophisticated in vivo approaches provide physiologically relevant settings to delineate stages of angiogenesis in multiple organs, the ability to assess neovascularization in vitro allows similar experiments to be carried out with relative ease and in a reproducible manner. Furthermore, the use of such in vitro systems of angiogenesis greatly reduces the dependence on animal models.

Over the past two decades, many in vitro assays to determine the angiogenic potential of endothelial cells have been described and are discussed at length in the following reviews [3–5]. The aspects of angiogenesis that these systems test can differ greatly, as shorter-term assays are more focused toward migratory potential, whereas longer-term assays may be used to study lumen formation, branching, and proliferation rates. Therefore, the most suitable assay depends greatly on the aspect of angiogenesis the experimenter wishes to explore.

One example of this is the Matrigel cord formation assay which assesses endothelial cord or tubule formation by seeding endothelial cells to Matrigel coated plates [6]. The cords that form indicate the ability of endothelial cells to migrate, align, and form cell–cell contacts over a 12–24 h period. However, these primitive structures do not always develop lumens, which are seen in more developed vascular structures and are very cell density dependent. Other methods such as the sprouting assay can also be utilized to assess various aspects of angiogenesis but as with Matrigel cord assays, these methods only explore initial endothelial sprouting and migratory effects associated with the early stages of angiogenesis.

The angiogenic potential of endothelial cells can also be assayed by co-culture with fibroblasts, which stimulates the formation of a mature vascular network over a 7- to 10-day period [7]. The fibroblasts provide a source of vascular endothelial growth factor (VEGF-A) which potently stimulates angiogenesis [7]. Additionally, the extracellular matrix (ECM) deposition of fibroblasts also plays a major role in the development of endothelial tubules [8]. These conditions result in the formation of a tubular network through the elongation and branching of ECs to produce structures reminiscent of a primitive vascular network which embeds among the supporting layer of fibroblasts and the ECM they deposit [8]. Folkman and Haudenschild first documented the formation of in vitro lumenized capillary-like structures from isolated bovine and human endothelial cells [9]. Following this, Montesano et al. demonstrated that Swiss 3T3 fibroblasts embedded in collagen discs and co-cultured with endothelial cells resulted in the formation of capillary-like structures radiating toward the fibroblasts [10]. Bishop et al. proposed the direct co-culture of endothelial and fibroblasts and although several modifications to this original method have been adopted, the principal concept of a support cell type in the form of fibroblasts remains unchanged [7]. Several of these modifications include alterations to the time course of the assay, siRNA mediated knockdown of genes in endothelial cells, and changes to the source of fibroblasts, for example, foreskin versus dermal fibroblasts [8, 11–13]. Furthermore, it was also shown that the addition of ascorbate increases the rate of collagen deposition of the fibroblasts, which increased endothelial cell tubule density [8]. The supplement of mitogenic or angiogenic growth factors or small molecules to test for inhibition or stimulation of neovascularization has highlighted the screening potential of the assay. More recently, it was shown that lumenized tubules from these assays can be perfused to study endothelial permeability or trafficking of immune or cancer cells on a fabricated microfluidic device [14].

Here we describe an updated methodology to assess the angiogenic potential of endothelial cells, using a co-culture system which employs a monolayer of fibroblasts which facilitates tubule formation and angiogenic growth. Additionally, we provide a protocol detailing how siRNA mediated knockdown can be achieved in endothelial or fibroblast cells to assess the role of genes of interest in angiogenesis and tubulogenesis.

2 Materials

- Human umbilical vein endothelial cells (PromoCell).
- Endothelial cell basal medium (ECBM) (PromoCell) supplemented with 2% fetal calf serum (FCS) (v/v), 0.4% endothelial cell growth supplement, 0.1 ng/ml epidermal growth factor (recombinant human), 1 ng/ml basic fibroblast growth factor (recombinant human), 90 μg/ml heparin, and 1 μg/ml hydrocortisone.
- Primary and hTERT immortalized human dermal foreskin (HDFs) fibroblasts.
- DMEM (Sigma) supplemented with 10% (v/v) FCS (Biosera).
- 1× PBS (No calcium, no magnesium) (Sigma).
- 1× Trypsin (Sigma) or TrypLE™ Express Enzyme (1×), no phenol red (Thermo Fisher).
- 48-well or 96-well clear bottom tissue culture plate with black cassette for imaging (Perkin Elmer) (*see* **Note 1**).
- 0.2% gelatin (from porcine or fish skin) (Sigma) (w/v in 1× PBS) or human fibronectin (5 μg/ml) (Sigma).
- Recombinant human $VEGF_{165}$ (or VEGF-A) (PeproTech).
- Hemocytometer or automated cell counter.

- 4% PFA solution or ice-cold 70% ethanol for fixation.
- Anti-CD31-AF647 (WM-59; BioLegend), anti-CD144-AF488) (BV9; Santa Cruz Biotechnology).
- PBLEC staining solution (0.1% BSA, 1% Tween, 0.05% NaN_3, 1 mM $CaCl_2$, 1 mM $MgCl_2$, 0.1 mM $MnCl_2$).

2.1 siRNA Mediated Knockdown of Endothelial Targets and Assessment of Tubulogenesis

1. Lipofectamine RNAiMax (Invitrogen) transfection agent. DharmaFECT I or IV may also be used successfully on HUVECs.
2. Opti-MEM I Reduced Serum Medium, GlutaMAX Supplement (Gibco).
3. SMARTpool siRNA duplexes (GE Life Sciences) or Silencer select single siRNA duplexes (Ambion).
4. Scrambled control siRNA (ON-TARGETplus non-targeting control pool, Dharmacon).

3 Methods

3.1 Culturing Endothelial Cells

1. All volumes adjusted for T75 flask growth.
2. Blood endothelial cells are routinely grown on 0.2% porcine gelatin (w/v in 1× PBS) coated plates. Plates were coated for >2 h at 37 °C. For fibronectin coating (used for lymphatic endothelial cells), plates are coated overnight at 37 °C.
3. Matrix coated plates are aspirated, and plates are washed with 1× PBS before seeding of endothelial cells.
4. HUVEC endothelial cells are cultured in complete ECBM.
5. Harvest cells using 1× trypsin or TrypLE.
6. Remove culture medium, wash cells with 1× PBS, and incubate with 1 ml of 1× trypsin or TrypLE at 37 °C until cells began to detach.
7. Add 5 ml of 10% FCS-containing medium (DMEM) to dilute and inactivate the trypsin, disperse clumps by pipetting and collect cells in 15 ml tubes.
8. Pellet cells by centrifugation at 300 × g for 5 min at room temperature (RT) before passaging (1:4 for all endothelial cells and fibroblasts) or using in subsequent co-culture assays.

3.2 Culturing Fibroblasts

Human dermal fibroblasts isolated from foreskin are maintained in 10% FCS supplemented DMEM medium and split (1:4) once confluent. All indicated volumes are for passaging cells grown in T75 flasks.

1. Culture HDF fibroblasts in DMEM medium supplemented with 10% FCS and 400 μg/ml G418 (if maintaining hTERT selection during standard passaging).
2. Harvest cells using 1 ml 1× trypsin.
3. Aspirate culture medium and wash cells with 5 ml 1× PBS.
4. Remove excess 1× PBS and incubate with 1 ml of 1× trypsin at 37 °C until cells began to detach.
5. Once most of the cells have detached from the plate, resuspend in 5 ml 10% FCS-containing medium to dilute and inactivate the trypsin and pellet cells by centrifugation at 300 × *g* for 5 min at room temperature (RT) before passaging (1:4) or collecting to seed for subsequent use in the tubulogenesis assay.

3.3 Co-culture Tubulogenesis Assay

3.3.1 Day 1

1. Coat a 48-well or 96-well plate for approximately 2 h at 37 °C with 0.2% gelatin and then wash 1× with PBS and air dry in a laminar flow hood approximately 10–15 min before seeding HDF cell suspension.
2. Trypsinize HDF and resuspend in supplemented DMEM medium as described above.
3. Seed HDF at a density of 2×10^4 cells per well of a 48-well plate in 0.5 ml of supplemented ECBM:DMEM medium in a 1:1 ratio (or 1×10^4 cells per well of a 96-well plate in 0.2 ml of medium) and incubate overnight at 37 °C under 5.0% CO_2.

3.3.2 Day 2

1. Culture endothelial cells until confluent, as described in Subheading 3.1.
2. Harvest endothelial cells using 1 ml 1× trypsin or 1× TrpLE and inactivate by resuspending in FCS-containing medium.
3. Count endothelial cells using a cell counter or hemocytometer and resuspend at 1×10^4 cells/ml in ECBM medium.
4. Following incubation overnight at 37 °C under 5.0% CO_2, aspirate medium from the HDF monolayer.
5. Seed HUVEC cells at a density of 5×10^3 cells per well of HDFs in 0.5 ml of ECBM medium for 48-well plate (adjust to 2×10^3 cells per well for a 96-well plate).
6. Co-culture HDFs and the endothelial cells for 24 h before changing medium to ECMB for control wells or ECBM + VEGF-A (10 ng/ml) (PeproTech) to stimulate tubulogenesis (*see* **Note 2**).
7. Change control and VEGF-A-supplemented medium every 2 days for a period of 7–10 days.

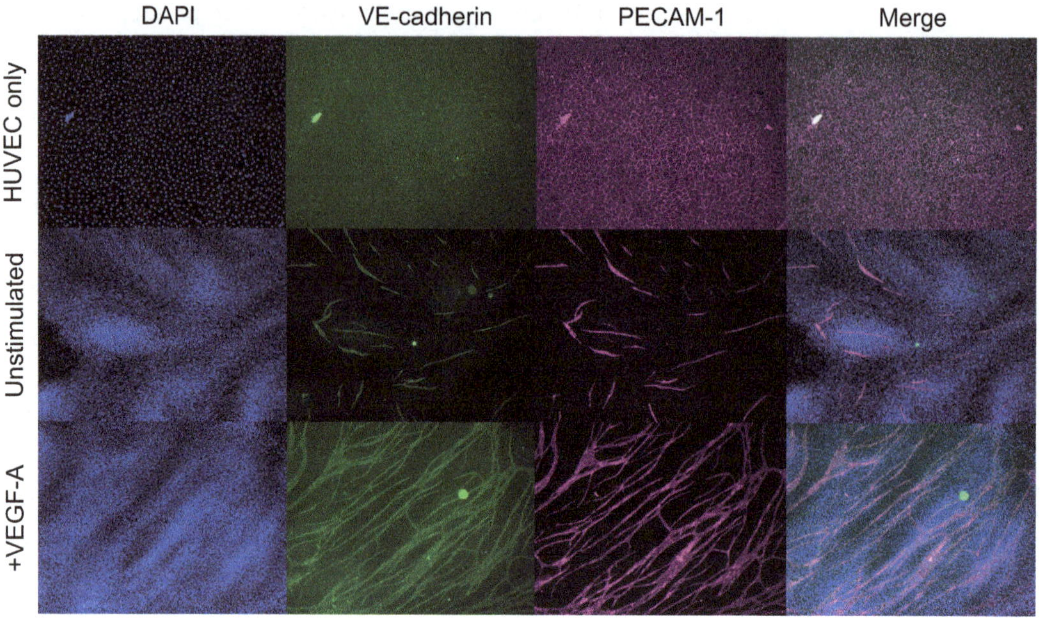

Fig. 1 Endothelial fibroblast co-culture assay. Immunofluorescent microscopy of indicated endothelial specific markers was carried out on HUVEC monolayers and co-cultures of HUVEC and fibroblasts. Co-cultures were either cultured in control medium (ECGM) or stimulated with VEGF-A (10 ng/ml) for 10 days before fixation, staining, and imaging

3.4 Fixation and Staining

1. Wash co-cultures 2× in PBS and fix in either 4% paraformaldehyde for 10 min or ice-cold 70% ethanol for 30 min at room temperature (*see* **Note 3**).
2. Wash wells 2× in PBS before staining with a fluorophore-conjugated mouse anti-CD31 antibody (PECAM-1; BioLegend) and anti-CD144 antibody (VE-cadherin; Santa Cruz Biotechnology) (both 1:400 dilution) in PBLEC staining solution (0.1% BSA, 1% Tween, 0.05% NaN_3, 1 mM $CaCl_2$, 1 mM $MgCl_2$, 0.1 mM $MaCl_2$). Cultures are stained for 1–2 h at RT or overnight at 4 °C with constant agitation (*see* **Note 4**).
3. Following antibody staining, wash 2× in PBS, and counterstain with DAPI or Hoechst for 5 min, wash 1× in PBS and image using either EVOS FL microscope or IncuCyte Live Cell Imager (Fig. 1).

3.5 siRNA Mediated Knockdown of Endothelial Targets and Assessment of Tubulogenesis

To investigate whether depletion of a gene of interest perturbs angiogenic potential, endothelial cells can be transfected with siRNA 24 h prior to seeding onto HDF cells [12, 13, 15]. This ensures efficient uptake of siRNA duplexes and substantial knockdown of targets in the initial stages of the tubulogenesis assay. While siRNA mediated depletion of target genes is only transient, it has been shown that the depletion of genes regulating aspects of angiogenesis can perturb endothelial tubulogenesis in this assay, even with only transient downregulation of target genes [13, 15]. Indeed, endothelial cells display sustained gene silencing using siRNA for up to 7 days post-transfection.

3.5.1 Day 1

1. Seed HDF as described in Subheading 3.3.
2. Wash HUVECs with 1× PBS, detach using 1× trypsin or 1× TrypLE, and count as above. Resuspend cells in Opti-MEM medium at 2–4 × 10^5/ml.
3. HUVECs are "reverse" transfected with 10 pmol of siRNA and 2 μl RNAiMAX (or DharmaFECT reagent) per well of a 12-well plate (HUVECs seeded at 1.5–2 × 10^5 cells/well) or 30 pmol siRNA and 5 μl RNAiMAX in a 6-well plate (3–4 × 10^5 cells/well).
4. Specifically, for a single well of a 6-well plate, 30 pmol siRNA duplexes are diluted in 200 μl of Opti-MEM medium and incubated at RT for 5 min.
5. At the same time 5 μl lipofectamine is diluted in 200 μl of Opti-MEM and incubated at RT for 5 min. siRNA and lipofectamine mixes are then combined within the well of a 6-well plate and gently mixed before incubating at RT for 20 min.
6. Following this, Opti-MEM suspended HUVECs are added to siRNA lipofectamine complexes at between 3–4 × 10^5 cells in 1 ml of Opti-MEM for each well of a 6-well plate.
7. Cells are incubated in this mixture for 4–6 h before transfection medium is carefully aspirated and replaced with supplemented ECBM culture medium.
8. The following day, endothelial cells are trypsinized following the protocol from Subheading 3.3.2 to seed siRNA treated cells to HDF monolayers for assessment of tubulogenesis.
9. For siRNA mediated knockdown in the HDF monolayer, *see* **Note 5**.

3.5.2 Analysis

Imaging and quantification can be performed using several different programs and tools in a manual, semiautomated, or fully automated fashion. Figure 2a displays images of HUVEC-HDF co-cultures obtained using an Essen IncuCyte Zoom live cell imager (Essen Bioscience) using 4× objective. Subsequent analysis of tubule length and the number of branch points within a network were performed using the Essen IncuCyte angiogenesis analysis software (or alternative software *see* **Note 6**) (Fig. 2b).

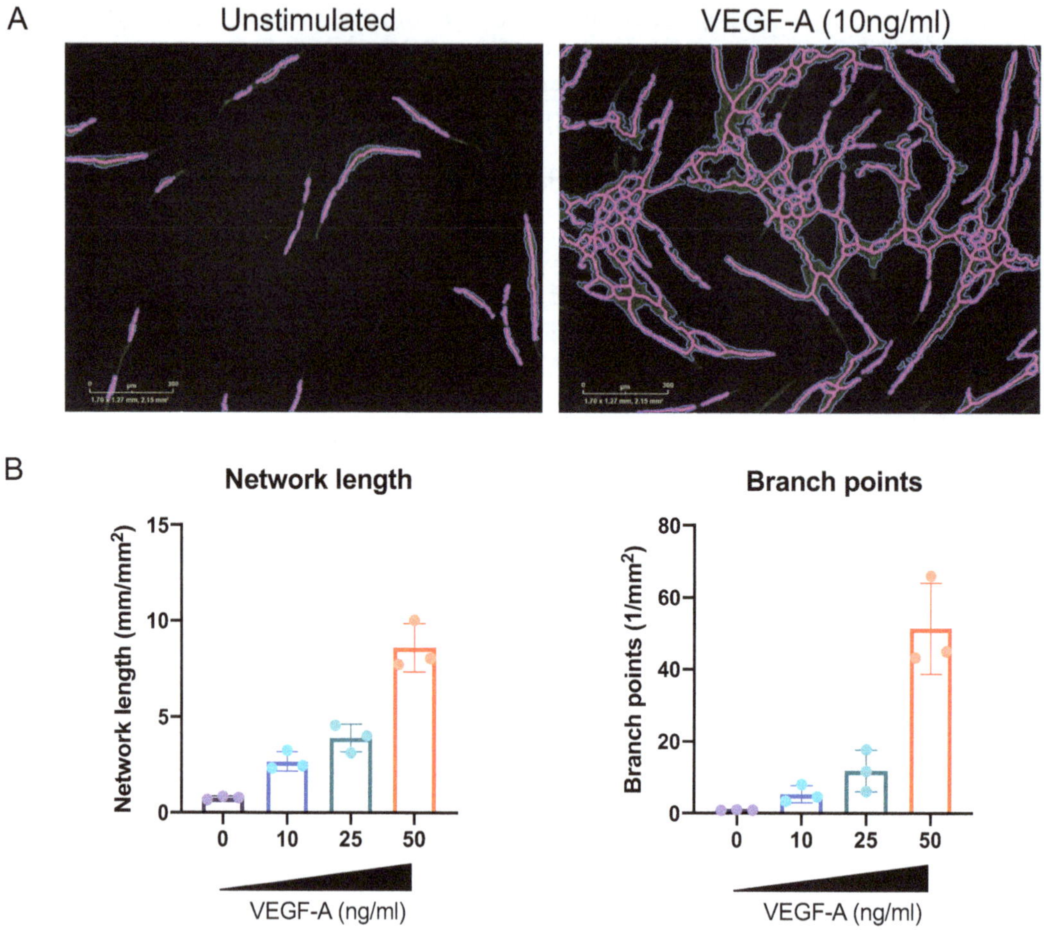

Fig. 2 Analysis and quantification of tubules. Co-cultures of HUVEC and fibroblasts were either cultured in control medium (ECGM) or stimulated with VEGF-A (10 ng/ml) for 10 days before fixation and staining with PECAM-1 (CD31). (**a**) Imaging was performed using Essen IncuCyte Zoom of both unstimulated and stimulated HUVEC-fibroblast co-cultures. (**b**) Quantification of tubules using Essen IncuCyte angiogenesis analysis software. Network length or branch points of tubule junctions under different VEGF-A stimulation conditions. Each data point represents the mean value from 1 random field of view. Error bars indicate ±SD

4 Notes

1. Normal transparent tissue culture plates can also be used for this assay but where possible, black-walled plates can help to reduce background when performing immunofluorescence.
2. VEGF-A supplementation can be delayed until later in the tubule growth period if desired but smaller networks will be formed. Additional small molecule inhibitors or activators can also be added at this stage to assess impact on tubulogenesis.
3. Tilt the plate to a 45° angle and wash gently to avoid any apical endothelial tubules from being dislodged.

4. Co-cultures can be stained with secondary antibodies to increase fluorescent signal of conjugated primary antibodies, donkey anti-mouse AF488, donkey anti-mouse AF555 or AF594 antibodies, and donkey anti-mouse APC, AF633, or AF647 antibodies may be used.
5. It is also possible to perturb gene expression in HDFs to assess their contribution to tubule formation. This would be performed during the steps described in Subheading 3.3.1 using 'reverse' siRNA transfection technique as outlined for HUVECs in Subheading 3.5.1. Thus, siRNA transfection of HDFs is conducted during the initial seeding of HDFs into 48-well or 96-well plates, before seeding of endothelial cells as in Subheading 3.3.2.
6. Image analysis of endothelial tubules can also be performed using ImageJ (Angiogenesis Analyzer plug-in), AngioTool64 (http://angiotool.nci.nih.gov) [16].

Acknowledgements

This work was supported by the University of Leeds, including a Faculty of Medicine and Health PhD Scholarship to A.J.M.

References

1. Adams RH, Alitalo K (2007) Molecular regulation of angiogenesis and lymphangiogenesis. Nat Rev Mol Cell Biol 8:464–478
2. Carmeliet P (2003) Angiogenesis in health and disease. Nat Med 9:653–660
3. Staton CA, Stribbling SM, Tazzyman S, Hughes R, Brown NJ, Lewis CE (2004) Current methods for assaying angiogenesis in vitro and in vivo. Int J Exp Pathol 85:233–248
4. Goodwin AM (2007) In vitro assays of angiogenesis for assessment of angiogenic and anti-angiogenic agents. Microvasc Res 74:172–183
5. Stryker ZI, Rajabi M, Davis PJ, Mousa SA (2019) Evaluation of angiogenesis assays. Biomedicine 7:1–13
6. Brown RM, Meah CJ, Heath VL, Styles IB, Bicknell R (2016) Tube-forming assays. In: Martin SG, Hewett PW (eds) Angiogenesis protocols. Springer, New York, NY, pp 149–157
7. Bishop ET, Bell GT, Bloor S, Broom IJ, Hendry NF, Wheatley DN (1999) An in vitro model of angiogenesis: basic features. Angiogenesis 3:335–344
8. Sorrell JM, Baber MA, Caplan AI (2007) A self-assembled fibroblast-endothelial cell co-culture system that supports in vitro vasculogenesis by both human umbilical vein endothelial cells and human dermal microvascular endothelial cells. Cells Tissues Organs 186:157–168
9. Folkman J, Haudenschild C (1980) Angiogenesis in vitro. Nature 288:551–556
10. Montesano R, Pepper MS, Orci L (1993) Paracrine induction of angiogenesis in vitro by Swiss 3T3 fibroblasts. J Cell Sci 105 (Pt 4):1013–1024
11. Hetheridge C, Mavria G, Mellor H (2011) Uses of the in vitro endothelial–fibroblast organotypic co-culture assay in angiogenesis research. Biochem Soc Trans 39:1597–1600
12. Mavria G, Vercoulen Y, Yeo M, Paterson H, Karasarides M, Marais R, Bird D, Marshall CJ (2006) ERK-MAPK signaling opposes Rho-kinase to promote endothelial cell survival and sprouting during angiogenesis. Cancer Cell 9:33–44
13. Latham AM, Odell AF, Mughal NA, Issitt T, Ulyatt C, Walker JH, Homer-Vanniasinkam S, Ponnambalam S (2012) A biphasic endothelial stress-survival mechanism regulates the cellular response to vascular endothelial growth factor A. Exp Cell Res 318:2297–2311

14. Jeon JS, Bersini S, Gilardi M, Dubini G, Charest JL, Moretti M, Kamm RD (2015) Human 3D vascularized organotypic microfluidic assays to study breast cancer cell extravasation. Proc Natl Acad Sci U S A 112:214–219
15. Fearnley GW, Abdul-Zani I, Latham AM, Hollstein MC, Ladbury JE, Wheatcroft SB, Odell AF, Ponnambalam S (2019) Tpl2 is required for VEGF-A-stimulated signal transduction and endothelial cell function. Biol Open 8
16. Zudaire E, Gambardella L, Kurcz C, Vermeren S (2011) A computational tool for quantitative analysis of vascular networks. PLoS One 6: e27385

Chapter 22

Quantification of Leukocyte–Endothelial Cell Interaction During In Vitro Models of Thrombosis at Arterial and Venous Shear Rates

Joshua H. Bourne

Abstract

Upon injury, stable thrombi formation requires the recruitment of platelets, leukocytes, and various clotting factors, to provide sufficient inhibition of hemostasis. Classical models of thrombosis involve either ex vivo isolation of platelets and subsequent quantification of aggregation through light transmission aggregometry or in vivo murine intravital thrombosis models (laser injury, ferric chloride, or rose Bengal). Flow adhesion models allow for accurate quantification of the contribution of cell-types to thrombi formation. Here, we describe the use of flow chambers to flow human blood over activated endothelial cells to observe leukocyte-endothelial adhesion at arterial and venous shear rates.

Key words Flow adhesion, Endothelium, Leukocyte, Thrombosis, Thromboinflammation

1 Introduction

Venous thrombosis, alongside major complications such as thromboembolisms, is a global concern. While antiplatelet drugs such a warfarin are effective in preventing clot formation, excess bleeding accompanies issues with treatment [1]. Leukocytes have been identified as key contributors to venous thrombosis [2, 3]. In physiology, neutrophils are classically seen to roll along the vessel endothelium. Endothelial cells express 2 carbohydrate-binding c-type lectins: P-selectin (CD62P) and E-selectin (CD62E). Selectins recognize a host of glycan epitopes on a lactosamine backbone—the determining selectin-binding determinant is a terminal tetrasaccharide, sialyl Lewis X (sLe^x) [4]. Human circulating leukocytes express sLe^x within protein scaffolds including P-selectin glycoprotein ligand-1 (PSGL-1) [5], CD43 [6], and CD44 [7]. sLe^x-expressing leukocytes form weak, transient bonds with the endothelium through selectins, as they are forced along the

Andrew V. Benest (ed.), *Angiogenesis: Methods and Protocols*, Methods in Molecular Biology, vol. 2441,
https://doi.org/10.1007/978-1-0716-2059-5_22,

vessel through a force generated by the flow of erythrocytes; as old bonds break, new ones are formed. Upon activation as a result of infection-, lesion- or sterile-induced inflammation, leukocyte receptor lymphocyte function-associated antigen-1 (LFA-1) undergoes a conformational change to form strong bonds with selectins expressed on the endothelium [8]. Simultaneously, platelet–leukocyte aggregates form, which promotes leukocyte adhesion to the endothelium through von Willebrand factor (VWF) strands.

While in vivo thrombosis assays are effective, it is difficult to identify the contribution of a specific cell-type to a phenotype while also refining and reducing the use of animals. Concurrently, in vitro assays using isolated platelets and leukocytes do not accurately represent the complex inflammatory microenvironment and interaction between the endothelium and blood. Here, we demonstrate an ex vivo assay identifying a method to assess leukocyte–endothelial interaction effectively, and concisely, by recreating the endothelial environment—thus providing a platform to administer drugs in a "blood vessel" with ease.

2 Materials

Prepare all solutions using sterile, ultrapure water (18 MW-cm at 25 °C) and analytical grade reagents, unless otherwise stated. Prepare all reagents on the day of use.

2.1 Blood Collection

1. Ethical approval for blood donation granted by the relevant internal ethical committee.
2. Venous blood collected from consenting, healthy, drug-free volunteers.
3. 3.2% trisodium citrate BD Vacutainers (Becton Dickinson, Oxford, UK).
4. PPACK dihydrochloride, thrombin inhibitor (Sigma, Germany).

2.2 Endothelial Culture

1. Human umbilical vein endothelial cells (HUVECs) (PromoCell, Germany).
2. Endothelial cell growth media 2 (PromoCell, Germany).
3. Sterile phosphate buffered saline (PBS).
4. Sterile trypsin-EDTA solution.
5. T75 EasYFlask tissue culture surface flask with filtered cap (Thermo Fisher, UK).
6. Gelatin type B, 2% H_2O, tissue culture grade (Sigma, Germany).

7. Cell culture grade laminar flow hood.
8. Cell culture grade humidified incubator (5% CO_2).

2.3 Preparation for Flow Adhesion

1. Phosphate buffered saline (PBS) pH 7.
2. 2% bovine serum albumin (BSA) in PBS.
3. Recombinant human TNFa (1 μg/ml stock).
4. DiOC6(3), (3,3′-dihexloxacarbocyanine iodide) (Thermo Fisher, UK).
5. Water bath 37 °C.

2.4 Flow Adhesion Assay

1. μ-Slide VI 0.4, 6-well flow chamber (Ibidi, Germany).
2. Silicone tubing, inner diameter 1 mm, outer diameter 3 mm.
3. Elbow Luer connector (×2) (Ibidi, Germany).
4. Luer connector male (Ibidi, Germany).
5. 20 ml Fisherbrand plastic syringe (internal diameter 19.13 mm)—this determines the flow rate and is important to modify if different syringes are used.
6. Standard Infuse/Withdraw PHD 22/2000 syringe pump (Harvard Apparatus, USA).
7. Formalin.

2.5 Microscopy Acquisition

1. EVOS M5000 Imaging system—digital inverted desktop microscope with multi-color fluorescence and transmitted light imaging, with 20× magnification.
2. EVOS FL Auto 2 Imaging System Software.

3 Methods

Perform all procedures at room temperature, unless otherwise specified. Blood handling should be treated with necessary personal protective equipment.

3.1 Endothelial Cell Culture

For the following experiments, human umbilical vein endothelial cells (HUVECs) were used to model vasculature. Other organ-specific endothelial cells may require particular culture methodology.

1. Preheat cell media, PBS, and trypsin-EDTA to 37 °C prior to use.
2. Culture endothelial cells in a humidified incubator at 5% CO_2 and 37 °C in 10 ml of endothelial cell growth media within a T75 tissue culture flask.

3. Cells should be cultured to 80% confluence and passaged using 3 ml 0.25% trypsin-EDTA solution, following a PBS wash.
4. Wash the flask with 3 ml cell media, and collect the now 6 ml total, and centrifuge the cells at 500 × *g* for 5 min.
5. Resuspend the resulting pellet in 10 ml fresh cell media, and add 2.5 ml of this mixture to a new T75 flask, containing 7.5 ml media.
6. Endothelial cells should be split to a maximum of passage 6 (*see* **Note 1**).

3.2 Flow Chamber Preparation

1. Remove the Ibidi flow chamber from the sterile packaging within a laminar flow hood.
2. Preheat 2% gelatin solution to 37 °C, or until the solution is runny, and no longer gelatinous.
3. Prepare 0.5% gelatin solution by diluting 250 μl 2% gelatin with 750 μl PBS and vertexing for 5 min.
4. Open the Ibidi chamber by lifting the plastic, transparent lid, and tilt the chamber at 45°, so that all openings are facing forwards.
5. Add the 0.5% gelatin solution to the 6 openings which are at the top of the chamber—approximately 100 μl/channel. Absolutely avoid bubbles in the chamber (*see* **Note 2**).
6. Incubate for 1 h at room temperature in a sterile atmosphere.

3.3 Endothelial Seeding in Flow Chamber

1. Wash the flow chamber containing gelatin thrice with PBS (*see* **Note 3**).
2. Split a confluent T75 flask of endothelial cells (as previously described in Subheading 3.1), but resuspend the cell pellet in 1 ml media, to create a dense cell mixture.
3. Add 100 μl of the cell mixture to each well of the flow chamber, again absolutely avoiding bubbles—add the remaining volume of the cell mixture to a new culture flask.
4. Incubate the endothelial cell-containing flow chamber at 37 °C for 1 h in a humidified incubator.
5. After 1 h, check to see if the endothelial cells have spread to form a monolayer across the bottom of the flow chamber.
6. Wash each flow chamber thrice with clean endothelial media. Cells obtained from washing can be added to another fresh T75 cell culture flask and cultured as normal.
7. Incubate the flow chamber overnight in a humidified incubator.

3.4 Preparation for Flow Adhesion Assay

1. The next morning, confirm the endothelial cells have formed a complete monolayer with very few gaps, then wash thrice with endothelial cell media.
2. Dilute stock TNFa (1 μg/ml) to 10 ng/ml in 1 ml endothelial cell media (*see* **Note 4**).
3. Add 100 μl of TNFa (10 ng/ml) to each well of the flow chamber and incubate for at least 4 h (*see* **Note 5**).
4. After at least 4 h, wash each chamber thrice with endothelial cell media, and block in sterile-filtered 2%BSA-PBS for 1 h.

3.5 Fluid Dynamics and Flow Setup

1. Calculate the desired flow rate, depending on the shear stress of the artery or vein which is being modeled using the following equation:

$$T = \eta \cdot 176.1 \cdot \phi$$

T: Flow rate (ml/min); η: Dynamical viscosity (dyn·s/cm^2); ϕ: Shear stress (dyn/cm^2)

2. Attach 2 lengths of silicon tubes to the relevant elbow Luer elbow connectors, and then 2 of them to a male Luer connector.
3. Attach the male Luer connector to a 20 ml plastic syringe, held in a Standard Infuse/Withdraw PHD 22/2000 syringe pump (*see* **Note 6**).
4. Place the flow chamber on a benchtop microscope, and secure in place.
5. Fill each well of the flow chamber with PBS, ensuring there are no bubbles.
6. Connect the Luer elbow connecters into the flow chamber, which is now ready for perfusion (Fig. 1).

3.6 Blood Preparation and Perfusion

1. Draw blood from the vein of a healthy volunteer on citrate and use within 2 h of acquisition.
2. Aliquot a volume of blood (dependent on the desired flow rate) to a clean, sterile falcon tube.
3. Add fluorescent membrane dye, DioC6 (1 μM), to the blood, and incubate at 37 °C for 10 min (*see* **Note 7**).
4. During this period, find focus on the microscope of the uppermost point of the endothelial cells.
5. Using PBS in a flacon tube, begin infusion on the syringe pump to flow PBS over the endothelial cells at the same flow rate as the modeled vessel for 5 min to align endothelial cells in the direction of flow.

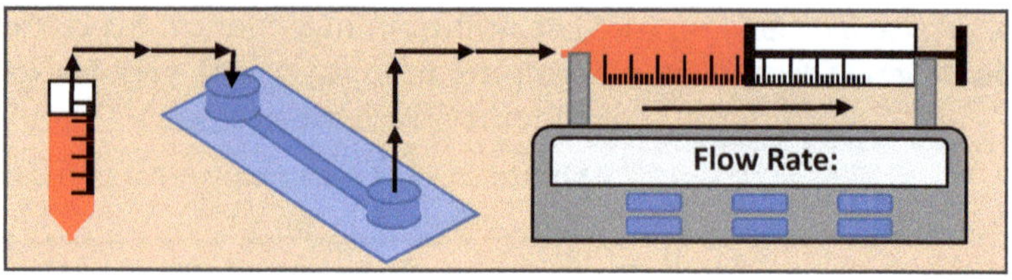

Fig. 1 Assembly of the flow adhesion system using the Ibidi μ-Slide VI 0.4, 6-well flow chamber, and a Standard Infuse/Withdraw PHD 22/2000 syringe pump to perfuse blood over a monolayer of activated endothelial cells to model thrombosis

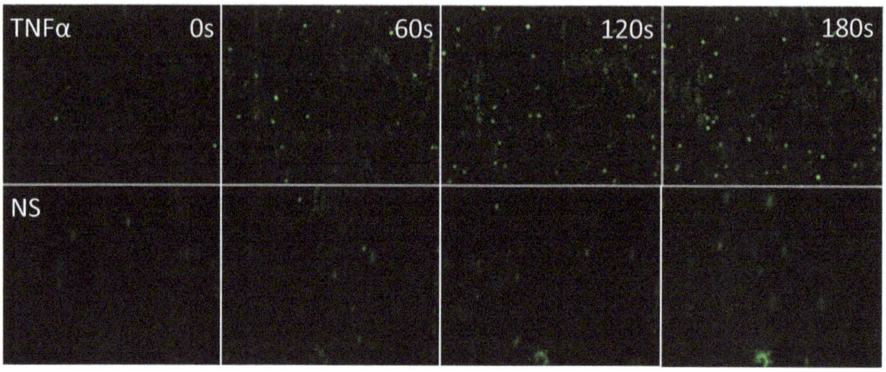

Fig. 2 Image acquisition of thrombi formation and leukocyte adhesion over 3 min, flown over TNFa, or non-stimulated (NS) endothelial cells using DioC6-stained, PPACK-treated human blood (*see* **Note 10**)

6. To avoid bubbles, pinch the silicon tube when transferring it from each solution.
7. Upon completion of the 10-min incubation for DioC6, add 40 μM PPACK to the blood (*see* **Note 8**) and begin flowing.
8. Begin recording for 3 min (Fig. 2).
9. Switch the silicon tube from the blood to PBS, and flow PBS over the endothelial cells until erythrocytes have cleared. Take brightfield images once cleared.
10. Upon completion, formalin-fix each chamber for 10 min, being careful not to disrupt any endothelial–leukocyte complexes (*see* **Note 9**).
11. Repeat for each flow chamber.

3.7 Flow Analysis and Quantification

1. Convert the acquired videos to tiff images.
2. Using ImageJ, convert tiff images to 8-bit.
3. Threshold the image to generate a quantifiable mask (Fig. 3).
4. Quantify area fraction to determine thrombus surface coverage.

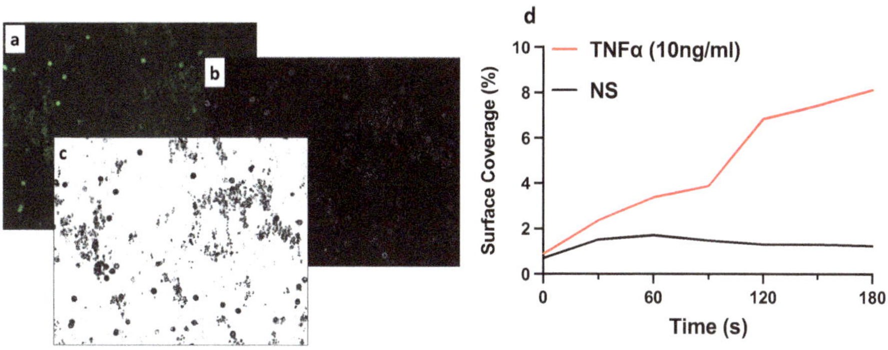

Fig. 3 Image analysis of thrombi formation and leukocyte adhesion over TNFa-activated, or non-stimulated endothelial cells over 3 min. (**a**) The tiff image, (**b**) conversion to 8-bit, and (**c**) threshold mask was used to generate the (**d**) surface coverage at intervals throughout the time course

3.8 Leukocyte Quantification

Leukocyte quantification may occur through multiple ways:

1. Immunofluorescent staining of myeloid cells ($CD11b^+$), neutrophils ($Ly6G^+$), or monocytes ($Ly6C^+$).
2. Counting the number of leukocytes in a field of view multiple time points by eye based on size using the DioC6 dye.
3. Counting the number of cells per field of view, based on cell morphology, using abrightfield microscope (Fig. 4).

4 Notes

1. Passage number is dependent on the origin of the endothelial cell. Adhesion receptors, such as ICAM and VCAM, have a reduction in presentation as passages increase—this is important to comprehend when quantifying data across passages.
2. Bubbles can be avoided by wedging a pipette tip into the chamber opening and slowly expulsing the liquid. A smooth flow can be confirmed by gently rocking the chamber back and forth, observing the transfer of liquid from one side of the chamber to the other.
3. To ensure good washing technique throughout, designate one "in" hole, and one "out" hole of the chamber, and stay consistent with this throughout each wash.
4. Due to the unstable nature of recombinant proteins, store stock concentrations at a high concentration, in small volumes, and at −80 °C. Keep on ice when using them to administer endothelial cells.
5. Adhesion receptor upregulation, such as ICAM and VCAM, is a transcriptional event, which takes at least 4 h [9]. Further incubation may present more receptors, but may also disrupt

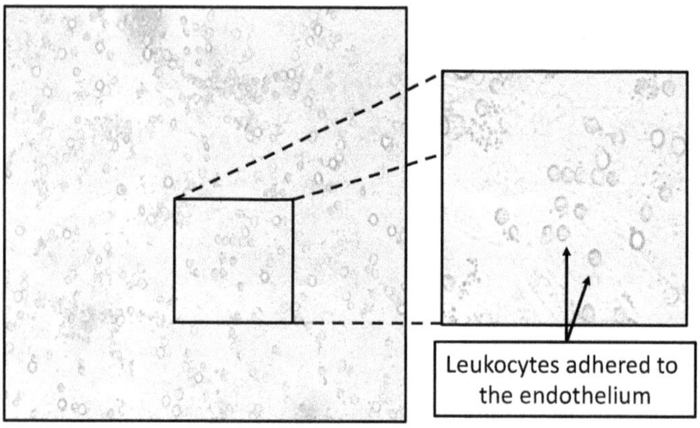

Fig. 4 Quantification of leukocyte adhesion over TNFa-activated endothelial cells after 3 min of blood perfusion using brightfield microscopy

endothelial integrity—remain consistent throughout activation incubations.

6. Syringe size and model are vital. The internal diameter is used to calculate flow rate on the syringe pumps and varies dependent on volume and model.
7. While DioC6 is administered in this model at this time point, other fluorescent markers or inhibitors may be given here, dependent on necessary incubation times.
8. PPACK must be kept on ice and in the dark until the point of treatment. It is used to remove thrombin and prevent the polymerization of fibrinogen to fibrin [10].
9. A rapid addition of formalin to the flow chamber can easily remove leukocytes from the endothelium, therefore add the fixative in a "dropwise" manner.
10. An unstimulated control confirms successful blocking. Platelets bound to VWF strands, generated by shear stress, are still visible in unstimulated controls. However, a lack of adhesion receptors induces their embolism.

Acknowledgments

This work was supported by a college-funded studentship from the University of Birmingham and the Centre of Membrane Proteins and Receptors.

References

1. Pirmohamed M (2006) Warfarin: almost 60 years old and still causing problems. Br J Clin Pharmacol 62(5):509–511. https://doi.org/10.1111/j.1365-2125.2006.02806.x
2. Swystun LL, Liaw PC (2016) The role of leukocytes in thrombosis. Blood 128(6):753–762. https://doi.org/10.1182/blood-2016-05-718114
3. Rayes J, Bourne JH, Brill A, Watson SP (2020) The dual role of platelet-innate immune cell interactions in thrombo-inflammation. Res Pract Thromb Haemost 4(1):23–35. https://doi.org/10.1002/rth2.12266
4. Munro JM, Lo SK, Corless C, Robertson MJ, Lee NC, Barnhill RL et al (1992) Expression of sialyl-Lewis X, an E-selectin ligand, in inflammation, immune processes, and lymphoid tissues. Am J Pathol 141(6):1397–1408
5. Huo Y, Xia L (2009) P-selectin glycoprotein ligand-1 plays a crucial role in the selective recruitment of leukocytes into the atherosclerotic arterial wall. Trends Cardiovasc Med 19(4):140–145. https://doi.org/10.1016/j.tcm.2009.07.006
6. Woodman RC, Johnston B, Hickey MJ, Teoh D, Reinhardt P, Poon BY et al (1998) The functional paradox of CD43 in leukocyte recruitment: a study using CD43-deficient mice. J Exp Med 188(11):2181–2186. https://doi.org/10.1084/jem.188.11.2181
7. McDonald B, Kubes P (2015) Interactions between CD44 and hyaluronan in leukocyte trafficking. Front Immunol 6:68. https://doi.org/10.3389/fimmu.2015.00068
8. Shamri R, Grabovsky V, Gauguet JM, Feigelson S, Manevich E, Kolanus W et al (2005) Lymphocyte arrest requires instantaneous induction of an extended LFA-1 conformation mediated by endothelium-bound chemokines. Nat Immunol 6(5):497–506. https://doi.org/10.1038/ni1194
9. Lawson C, Ainsworth M, Yacoub M, Rose M (1999) Ligation of ICAM-1 on endothelial cells leads to expression of VCAM-1 via a nuclear factor-kappaB-independent mechanism. J Immunol 162(5):2990–2996
10. Kattula S, Byrnes JR, Wolberg AS (2017) Fibrinogen and Fibrin in Hemostasis and Thrombosis. Arterioscler Thromb Vasc Biol 37(3):e13–e21. https://doi.org/10.1161/ATVBAHA.117.308564

Chapter 23

Isolation of Cardiac Endothelial Cells for Transcriptomic Analysis of the Zebrafish and Mouse Heart

Ziwen Li, Katherine M. Ross Stewart, Finnius A. Bruton, Martin A. Denvir, and Mairi Brittan

Abstract

Isolation of high quality cardiac endothelial cells is a prerequisite for successful bulk and single cell sequencing for RNA (scRNA-seq). We describe a protocol using both enzymatic and mechanical dissociation and fluorescence-activated cell sorting (FACS) to isolate endothelial cells from larval and adult zebrafish hearts and from healthy and ischemic adult mouse hearts. Endothelial cells with high viability and purity can be obtained using this method for downstream transcriptional analyses applications.

Key words Cardiac endothelial cells, Enzymatic dissociation, Mechanical dissociation, FACS, Single cell transcriptional analysis, Endocardial cells

1 Introduction

Tissue dissociation is a process where individual cells are released from tissues via enzymatic digestion and mechanical forces [1–3]. The goal is to obtain highly viable cells of interest with minimal perturbation of their normal functions, which is essential for downstream transcriptomics analyses. Enzymes such as collagenases and dispase can break down the extracellular matrix to release cells into suspension. DNase-I can prevent cell aggregation by degrading free-DNA released from lysed cells [4]. The obtained cell suspension can be purified via centrifugation, magnetic activated cell sorting (MACS), or fluorescence-activated cell sorting (FACS) [4, 5]. Though centrifugation and MACS are rapid and inexpensive they are prone to contamination from unwanted cell types and lack effective methods for removing cell doublets/aggregates, which are more thoroughly controlled for in FACS [6]. Here we demonstrate that high quality cardiac endothelial cells can be isolated from freshly harvested larval zebrafish, adult zebrafish, and adult mouse

Andrew V. Benest (ed.), *Angiogenesis: Methods and Protocols*, Methods in Molecular Biology, vol. 2441,
https://doi.org/10.1007/978-1-0716-2059-5_23,

hearts through enzymatic-mechanical dissociation and FACS. These protocols have been used to isolate viable and pure endothelial cell populations for the generation of scRNA-seq data for studies of endothelial heterogeneity and regenerative vascular responses in the healthy and ischemic mouse heart [7].

Cell-type specific reporter lines can be exploited in order to isolate endothelial cells for RNA studies or cell culture. A number of transgenic zebrafish lines use vascular endothelial cell-specific promoters to drive expression of fluorescent proteins. One of the most widely used of these transgenic lines is *Tg(fli1a:EGFP)*^{y1} [8], where endothelial-specific expression of enhanced green fluorescent protein (EGFP) allows straightforward selection of hearts under a basic fluorescence microscope and can later be used during FACS to distinguish and isolate endothelial cells from other cell types. RNA sequencing at the single cell level is gaining popularity in zebrafish studies, though there are still relatively few publicly available scRNA-seq datasets that are focused on the analyses of the zebrafish heart [9–11].

2 Materials

Prepare and store all reagents at room temperature (unless indicated otherwise). Diligently follow all waste disposal regulations when disposing waste materials, paying special attention to biological waste from transgenic animals. Prepare all consumables including scalpels, cell culture Petri dishes, gentleMACS™ C tubes (Miltenyi Biotec), conical centrifuge tubes, cell strainers (Corning), and FACS tubes beforehand in accordance with the number of heart samples to be processed. It is essential to carry out the procedures efficiently to maximize cell viability.

2.1 Adult Mouse Heart Tissue Collection

Prepare all solutions for mouse tissue collection in a laminar flow cabinet the day before tissue harvest, filtered through 0.22 μm syringe filters (Merck Millipore), and stored at 4 °C.

1. Tissue collection solution: RPMI 1640 medium with 10% fetal bovine serum (FBS) and 1× GlutaMAX Supplement. For each mouse heart, prepare 4 ml tissue collection solution in a 15 ml conical centrifuge tube. Mix 3.56 ml RPMI 1640 medium (Thermo Fisher Scientific) with 0.4 ml FBS (Thermo Fisher Scientific) and 40 μl GlutaMAX Supplement (100×, Thermo Fisher Scientific).
2. 70% EtOH: In a spray bottle add 700 ml distilled H_2O and 300 ml 100% EtOH.
3. Surgery scissors and forceps (WORLD PRECISION INSTRU MENTS).
4. Dissecting board and pins.
5. Ice bucket filled with ice.

2.2 Mouse Heart Tissue Dissociation

1. Tissue digestion buffer: For each mouse heart, prepare 5 ml tissue digestion buffer in a 50 ml conical centrifuge tube (*see* **Note 1**). Mix 4.45 ml RPMI 1640 medium with 0.5 ml FBS and 50 μl GlutaMAX Supplement (100×). Dissolve 4.25 mg Collagenase V (Sigma-Aldrich), 6.25 mg Collagenase D (Roche), 10 mg Dispase (Gibco), and 300 μg DNase-I (Roche) in the solution. Filter the solution using a 0.22 μm filter into a sterile conical centrifuge tube.
2. FACS buffer: Prepare in bulk. Open a new bottle of 500 ml 1×DPBS (no calcium, no magnesium, Thermo Fisher Scientific) and transfer 11 ml buffer to a 15 ml conical centrifuge tube. Add 10 ml FBS and 1 ml 0.5 M EDTA (Sigma-Aldrich) to the bottle and mix well (*see* **Note 2**).
3. RNaseZAP™ (Thermo Fisher Scientific).
4. gentleMACS™ Dissociator (Miltenyi Biotec).
5. Refrigerated benchtop centrifuge accommodating 15 ml and 50 ml conical centrifuge tubes.
6. 37 °C water bath.
7. Vortex mixer.
8. Disposable scalpels No 10 (Swan Morton).
9. gentleMACS™ C tube (Miltenyi Biotec).
10. 40 μm cell strainer (Corning).
11. Disposable 90 mm Petri dishes (Thermo Fisher Scientific).
12. An ice bucket filled with ice.

2.3 Adult Zebrafish Heart Tissue Collection

1. 70% EtOH: In a spray bottle add 700 ml distilled H_2O and 300 ml 100% EtOH.
2. Dissection kit: 2 pairs of Dumont 5 Biology Tweezers (0.05 × 0.01 mm tip), 1 pair of blunt anatomical forceps, 1 pair of microdissection spring scissors (10.5 cm, 8 mm blades), a dissection sponge (1.5 × 5 × 3 cm) with a single center groove cut using a scalpel (0.5 × 2.5 cm), and disposable scalpels No 10 (Swan Morton).
3. An ice bucket filled with ice.
4. Excision buffer; 10% FBS in PBS: Into a clean, labeled 15 ml Falcon tube add 10 ml 1× PBS and 1 ml FBS. Mix well by inverting. Store at 4 °C until use, then keep on ice.
5. 1 M Tricaine methanesulfonate (MS-222) in 1× PBS: Add 48 ml of H_2O to 200 mg of Tricaine powder and 2 ml of 1 M Tris base (pH 9) to prepare 50 ml of stock solution. Adjust to pH 7.
6. Weighing boat.

2.4 Larval Zebrafish Heart Tissue Collection

1. 70% EtOH: In a spray bottle, add 700 ml distilled H_2O and 300 ml 100% EtOH.
2. 1 M Tricaine methanesulfonate (MS-222) in 1× PBS: Add 48 ml of H_2O to 200 mg of tricaine powder and 2 ml of 1 M Tris base (pH 9) to prepare 50 ml of stock solution. Adjust to pH 7.
3. Excision buffer (10% FBS in Leibovitz's L-15 medium): Into a clean, labeled 15 ml Falcon tube, add 9 ml 1× PBS and 1 ml FBS. Mix well by inverting. Store at −20 °C until use then keep on ice.
4. Plastic 5 ml syringe with 19-gauge needle.
5. Petri dishes (30 mm and 100 mm).
6. Plastic Pasteur pipettes.

2.5 Adult/Larval Zebrafish Heart Tissue Digestion

1. Digestion cocktail [9] (5 mM $CaCl_2$,10 mg/ml B. Licheniformis protease, 125 U/mL DNase-I in 1× PBS): In a clean 15 ml Falcon tube, add 12.5 μl 1 M $CaCl_2$ (Thermo Fisher Scientific) to 237.5 μl PBS to make a 50 mM 10× stock, which should be stored at room temperature. Weigh 25 mg 7 U/ml Bacillus Licheniformis protease (Sigma-Aldrich) and transfer to a clean 1 ml microcentrifuge tube. Add PBS to a volume of 250 μl to make a 100 mg/ml 10× BLP stock. Store at −20 °C. Add 1.56 mg 1250 U/ml DNase-I (Sigma-Aldrich) to a 5 ml tube and add PBS to a final volume of 2.5 ml to make a 1250 U/ml 10× DNase-I stock. Store at 4 °C. In a fresh 1 ml microcentrifuge tube, add 10 μl of the 10× DNase-I stock, 10 μl of the 10× $CaCl_2$ stock, and 10 μl of the 10× BLP stock to 70 μl PBS to make up 100 μl of the digestion cocktail per sample—this should be made fresh each time and kept on ice.
2. FACS buffer (2% FBS in PBS): Into a clean, labeled 15 ml Falcon tube add 10 ml 1× PBS and 2 ml excision buffer (10% FBS in PBS). Mix well by inverting. Store at −20 °C until use then keep on ice.
3. 40 μm cell strainer.
4. 15 ml Falcon tube.

2.6 Fluorescence-Activated Cell Sorting

1. FACS buffer (*see* Subheading 2.2, **item 2** for mouse and Subheading 2.5, **item 2** for zebrafish FACS buffer compositions).
2. Post FACS cell collection buffer: For each mouse heart, prepare 500 μl cell collection buffer in 1.5 ml Eppendorf tubes. Dissolve 0.005 g bovine serum albumin in 0.5 ml 1×DPBS (no calcium, no magnesium).
3. FACSAria™ II Cell Sorter (BD).
4. 5 ml round-bottom polystyrene tubes for FACS (Thermo Fisher Scientific).

5. Alexa Fluor® 647 anti-mouse CD31 Antibody (BioLegend).
6. DAPI solution (Thermo Fisher Scientific).

3 Methods

Tissues and cells should be maintained at temperatures around 4 °C during collection, single cell suspension preparation (except for mouse tissue dissociation at 37 °C), staining and sorting, to achieve high cell viability and to mitigate gene expression changes caused by the procedures.

3.1 Preparation for Mouse Heart

1. Warm up the tissue digestion buffer in 37 °C water bath (*see* **Note 3**).
2. Turn on the refrigerated benchtop centrifuge and cool down to 4 °C (*see* **Note 4**).
3. Put premade FACS buffer and post FACS cell collection buffer on ice.
4. Decontaminate working surface with RNase decontamination solution RNaseZAP™ (*see* **Note 5**).
5. Lay out Petri dishes (labeled if processing multiple samples) and scalpels on the working surface according to the number of hearts to be processed (*see* **Note 6**).

3.2 Mouse Sacrifice and Heart Tissue Collection

1. Cull the mouse rapidly and humanely by cervical dislocation.
2. Place the mouse on a dissecting board with ventral-side up and pin down the four limbs.
3. Wet the fur with 70% EtOH (*see* **Note** 7).
4. Use tough-cut-scissors and forceps to cut the skin from mid abdominal region, through the thoracic area to the throat and carefully separate the skin from the underlying musculature to expose both the thoracic and abdominal musculature.
5. Use forceps to hold the xiphoid process and make an incision immediately below it through the abdominal muscle wall. Extend the incision to the midaxillary lines on both sides.
6. Raise the xiphoid process to expose the diaphragm and carefully cut the diaphragm along its insertions on the rib cage.
7. Open the rib cage by cutting through the dorsal third of ribs on both sides, then recline the rib cage and sternum towards the head with hemostatic forceps to hold down.
8. Carefully cut the great vessels and dissect the surrounding connective tissues and remove the heart (*see* **Note 8**).
9. Gently tap off excess blood on clean paper towel (*see* **Note 9**) before transferring into ice-cold tissue collection solution.

3.3 Mouse Heart Tissue Dissociation and Single Cell Suspension Preparation

1. Place the harvested heart on a 90 mm Petri dish and remove excess collection solution using a sterile transfer pipette.
2. Carefully remove the fat tissue and extract ventricles with forceps and scalpels.
3. Finely mince the ventricles with scalpels and transfer to a gentleMACS™ C tube (*see* **Note 10**).
4. Add 5 ml pre-warmed tissue digestion buffer to the gentleMACS™ C tube.
5. Incubate the minced ventricles in digestion buffer in 37 °C water bath for 20 min under continuous agitation and vortex the sample for 10 s every 5 min.
6. Place the gentleMACS™ C tube on the gentleMACS™ dissociator [3] and homogenize the tissue using pre-programmed setting "heart01" (*see* **Note 11**).
7. Add 6 ml ice-cold FACS buffer to the digest immediately after homogenization.
8. Pass the digest through a 40 μm cell strainer and collect the cell suspension into ice-cold 50 ml conical centrifuge tube.
9. Rinse the strainer with another 2 ml of FACS buffer and collect the suspension into the same tube.
10. Centrifuge cell suspension at 500 × *g* for 5 min at 4 °C.
11. Slowly decant the supernatant.
12. Resuspend the cell pellet in 1 ml Red Blood Cell Lysing Buffer Hybri-Max™ (Sigma-Aldrich) and mix for exactly 1 min (*see* **Note 12**).
13. Add 30 ml cold FACS buffer to dilute the lysis buffer to stop the reaction.
14. Centrifuge the cell suspension at 500 × *g* for 5 min at 4 °C.
15. Carefully remove the supernatant and resuspend in 500 μl ice-cold FACS buffer.

3.4 Adult Zebrafish Heart Tissue Collection

1. Prepare the heart collection bench: Sterilize the dissection area and tools with 70% EtOH. Fill a 30 mm Petri dish with cold excision buffer. Place on ice ready to collect hearts. If you are extracting RNA, spray the area liberally with RNaseZAP™.
2. Euthanize adult *Tg(fli1a:EGFP)*^{y1} zebrafish (aged 6–12 months) by immersion in 1 M tricaine (MS-222), until loss of righting reflexes. Use the blunt anatomical forceps to move the fish from the tricaine onto the benchtop, placing it on some paper towel. Using the scalpel, confirm death by cervical dislocation or according to local animal guidelines.
3. Place the euthanized fish ventral-side up in the groove on the dissecting sponge, and place under a dissecting microscope (*see* **Note 13**).

4. Using a pair of Dumont 5 Biology Tweezers, gently lift one operculum and pull to one side, try to avoid tearing the tissue; at the same time, use the microdissection scissors to cut through the triangular piece of tissue between the two opercula.
5. Using the second pair of Dumont 5 Biology Tweezers, pull the triangular piece of tissue backwards. The heart, which often continues to beat after cervical dislocation, should rise to the surface of the wound (*see* **Note 14**).
6. Gently pick the heart up by the paler-colored bulbous arteriosus (BA), being careful to avoid damaging the myocardium (*see* **Note 15**).
7. Carefully excise the heart with minimal manipulation by tugging the BA until the heart comes free and quickly place it into the excision buffer.
8. Once all hearts have been extracted, use the Pasteur pipette to transfer them to a small weighing boat (*see* **Note 16**) and manually mince the tissue with a sterile scalpel. Collect the tissue into a 15 ml falcon tube (*see* **Note 17**).
9. Centrifuge the sample at 350 × *g* for 5 min. Carefully discard the supernatant (*see* **Note 18**).
10. Resuspend the pellet in 100 μl ice-cold protease solution.

3.5 Larval Zebrafish Heart Tissue Collection

Protocol adapted from [12].

1. Prepare the heart collection bench: Sterilize the dissection area with 70% EtOH. Fill a 30 mm Petri dish with cold excision buffer. Place on ice ready to collect hearts. Combine a 5 ml syringe and a 19-gauge needle. If you are extracting RNA, spray the area liberally with RNaseZAP™.
2. Euthanize larval zebrafish (<5 days post-fertilization) by adding 1 M tricaine (MS-222) to the water in a drip-wise fashion until the larvae cease movement (*see* **Note 19**).
3. Using a plastic Pasteur pipette, collect approximately 40 larvae and place in a 2 ml microcentrifuge tube, draining as much water as possible with the Pasteur pipette (*see* **Note 20**). Label the tube "shearing."
4. Add 500 μl ice-cold excision buffer to resuspend the larvae.
5. Shear the larvae by pipetting up and down 10 times with the 19-gauge needle (*see* **Note 21**).
6. Use the plastic Pasteur pipette to transfer the solution from the shearing tube into the lid of a 100 mm Petri dish for inspection under the fluorescent microscope (*see* **Note 22**). Using a 20 μl pipette, manually select the fluorescent hearts drawing them individually into the pipette and transferring into the 30 mm Petri dish containing excision buffer, on ice.

7. Some hearts will remain in the bodies of the larvae, so once all free-floating hearts are retrieved place the solution back into the 2 ml shearing microcentrifuge tube for another round of shearing (*see* **Note 23**). Add additional ice-cold excision buffer as necessary.
8. Once all hearts are collected in the 3 ml Petri dish, under the fluorescent microscope, use a Pasteur pipette to collect them (*see* **Note 24**) and place hearts into a fresh microcentrifuge tube. Keep on ice.
9. Centrifuge the sample at 350 × *g* for 5 min at 4 °C. Carefully discard the supernatant.
10. Resuspend the pellet in 100 μl ice-cold protease solution.

3.6 Adult/Larval Zebrafish Heart Dissociation and Single Cell Suspension Preparation

1. Keeping everything on ice at all times, use a 200 μl pipette to triturate the tissue in the protease solution for 30 s every minute for 10 min.
2. Centrifuge samples at 350 × *g* for 5 min at 4 °C.
3. Discard the supernatant being careful not to dislodge the pellet.
4. Resuspend pellet in 500 μl ice-cold FACS buffer.
5. Place a 40 μm cell strainer on a sterile 15 ml Falcon tube. Pass the whole sample through the cell strainer.
6. Wash the filter with 500 μl ice-cold FACS buffer.
7. Keep on ice and take to flow cytometry.

3.7 Mouse: Fluorescence-Activated Cell Sorting

1. Load 50 μl cell suspension into a 15 ml conical centrifuge tube as unstained control for FACS and store at 4 °C (*see* **Note 25**).

   Steps 2–5 are not required for transgenic zebrafish tissue, which can be isolated based on endogenous fluorescence.
2. Add AF647 anti-mouse CD31 antibody (5 μg/ml, BioLegend) to the 450 μl cell suspension in 50 ml conical centrifuge tube and incubate in dark for 30 min at 4 °C (*see* **Note 26**).
3. Coat FACS tubes for both the control and stained samples using 1 ml neat FBS and place the coated tubes on ice while waiting for the stain.
4. Add 500 μl ice-cold FACS buffer to the stained cells and centrifuge the cell suspension at 500 × *g* for 5 min at 4 °C.
5. Resuspend the stained cells in 500 μl ice-cold FACS buffer.
6. Add 450 μl ice-cold FACS buffer to the 15 ml conical centrifuge tube with the control sample and mix well.
7. Load both stained and unstained samples into the coated FACS tubes and proceed with FACS (Fig. 1).

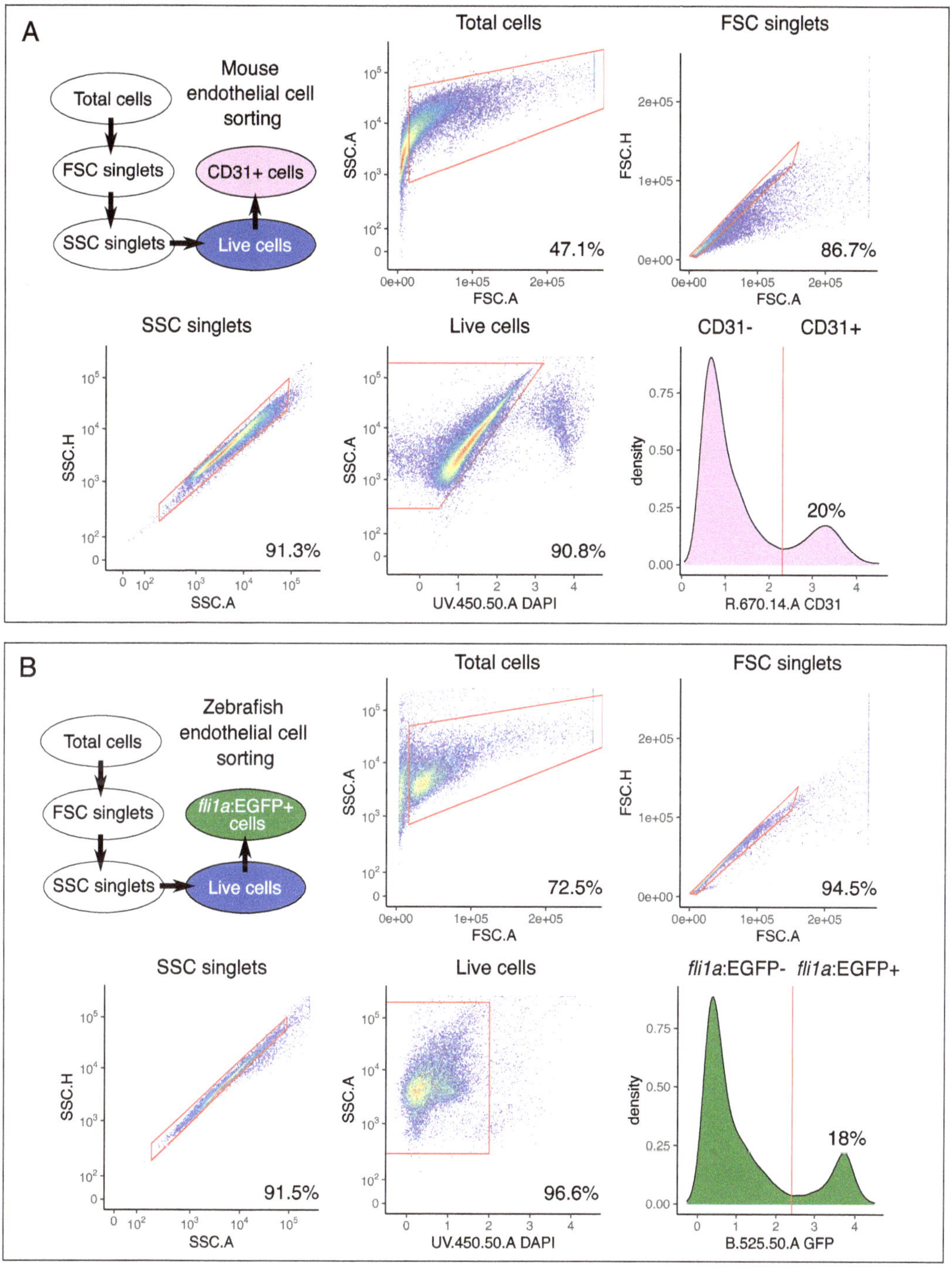

Fig. 1 Endothelial cell enrichment via fluorescence-activated cell sorting (FACS) using mouse or zebrafish hearts. Hierarchical gating strategies and flow cytometry plots are shown for endothelial cell purification using an AF647 conjugated anti-CD31 antibody in mouse (**a**) and using endogenous GFP expression in *Tg(fli1a: EGFP)*^{y1} zebrafish (**b**). Total dissociated cells are gated based on side scatter area (SSC-A) versus forward scatter area (FSC-A), doublets are then excluded by positive selection of cells in a singlet gate in plots of forward scatter area (FSC-A) versus height (FSC-H) and then SSC-A versus SSC-H. Viable DAPI-negative cells

8. Add DAPI to the samples just before the sort for viability check (*see* **Note 27**).
9. Viable CD31 positive cells (mouse) or GFP+ cells (zebrafish) are collected in 500 μl ice-cold post FACS cell collection buffer in a 1.5 ml Eppendorf tube and stored on ice for downstream transcriptomics analysis (*see* **Notes 28** and **29**).

4 Notes

1. When making the tissue digestion buffer, weigh out the collagenases, dispase, and DNase in 1.5 ml Eppendorf tubes, dissolve in RPMI 1640 medium, respectively, and then mix into the final solution.
2. EDTA is a chelating agent sequestering divalent metal ions that are involved in integrin mediated cell adhesion. Including EDTA (1–5 mM) in the FACS buffer can limit cell adhesion and aggregation.
3. Ensure the tissue digestion buffer is warmed up to 37 °C as it is the optimal temperature at which enzymes work at maximum speed and efficiency.
4. Minimizing the processing time is key to high cell viability. Make sure to coordinate the use of equipment in a multi-occupancy lab and complete the preparations before going to the animal facility for tissue collection.
5. RNase decontamination solution should be applied regularly to the working surface, equipment, and gloves to avoid RNase contamination.
6. Label the Petri dishes, conical centrifuge tubes, and gentleMACS™ tubes during experiment setup when processing multiple samples, to save time and reduce errors.
7. Wetting the fur with 70% EtOH helps to minimize contamination of hairs and other allergens.
8. Act quickly when collecting the heart to ensure high cell viability. Ideally, the heart still beats after the excision.
9. Removing excess blood after excision can help avoid prolonged red blood cell lysis treatment, minimizing cell stress and improving cell viability.
10. Mince the ventricles into fine tissue pieces ($\leq 0.5 \times 0.5 \times 0.5$ mm^3) is to increase the surface area for more efficient enzymatic digestion. To dislodge tissues/cells

Fig. 1 (continued) are then selected and endothelial cells isolated based on CD31 expression in mouse (**a**) or EGFP expression in zebrafish (**b**)

attached to the Petri dish, add 0.1–0.3 ml tissue collection buffer, gently scrap the tissues/cells with the sterile transfer pipette, and transfer the suspension to the gentleMACS™ C tube.

11. Tightly cap the gentleMACS™ C tube and avoid twisting the tube when removing from the gentleMACS™ dissociator, as the screw cap may become loose, risking losing cell suspension.
12. If lysis is incomplete, **steps 12–15** may be repeated, but prolonged exposure to RBC lysis buffer will affect the viability of endothelial cells.
13. It is important to work quickly, but carefully through the following steps. You do not want to tear any tissue or damage the heart, equally, if you take too long, the heart may stop beating, which will make extraction much more difficult and may compromise the cell viability.
14. The heart may be encased in the silvery pericardium. If so, gently tear the tissue to reveal the beating heart.
15. This can be a little difficult as the heart is slippery and does not naturally orientate that way. If possible, it is also acceptable to use the forceps to pinch the small bit of tissue in between the BA and the ventricle.
16. Try to transfer as little liquid as possible during this step, as the excision buffer here acts as a wash step, removing tissue detritus and red blood cells.
17. If desired, the weighing boat can be washed with excision buffer to be sure that all tissue is transferred to the falcon tube. If this is done, then be sure to balance the centrifuge correctly in **step 9**.
18. If the pellet becomes dislodged, centrifuge the tissue again.
19. Be aware of local laws regarding animal anesthesia and euthanasia. In the UK, larvae sub 5 days post-fertilization (dpf) do not require confirmation of death under the schedule 1 laws that apply to post 5 dpf fish.
20. Though it is not essential to remove all the water from the shearing tube, try to remove as much as possible as the excision buffer is designed to maximize viability, and thus the researcher should aim to avoid diluting it with fish-water where possible.
21. The larvae are to be dissociated by shear force, so the up and down motions should be relatively quick, though the researcher should be careful not to overflow the shearing tube on the expiration. The excision buffer may begin to froth.
22. Many of the hearts, which will be visible under the fluorescence channel, will have come loose from the larvae.

23. Typically shearing will need to be repeated 3× to retrieve all hearts. Yield should be 50–70%.
24. Be careful to ensure that there are no eyes transferred as they have high endogenous fluorescence.
25. If multiple samples are processed at the same time, 50 μl cell suspension will be taken from each sample and mixed to create the unstained control for FACS.
26. Other antibodies can be added to the FACS panel to simultaneously collect endothelial subpopulations and to reduce contamination of unwanted cell types, e.g., using a podoplanin antibody can help further select lymphatic endothelial cells and using a CD45 antibody can exclude macrophages and other inflammatory or hematopoietic lineages.
27. High cell viability is a prerequisite for obtaining good transcriptomic data. We recommend >90% as the minimum viability required. Cell viability can also be assessed using the trypan blue staining exclusion assay with a hemocytometer or automated cell counter such as TC20™ (Bio-Rad).
28. If the protocol is used to generate cell suspensions for scRNA-seq, we suggest sorting at least 100,000 cells *per* mouse sample. We have observed reduced viability during cell sorting; thus, we recommend to stop the sort once the target cell number is reached, especially when multiple samples are processed at the same time.
29. Our experience using this method has yielded approximately 40,000 EGFP$^+$ endothelial cells from 15 adult *Tg(fli1a:EGFP)^{y1}* ventricles and approximately 2000 cells from 15 larval hearts (Ross Stewart, *unpublished data*). At present, the number of cells used for zebrafish scRNA-seq studies in published literature is variable, and it should be noted that methods will have different requirements with regard to cell number and quality control. For example, Honkoop et al., [9] used SOrting and Robot-assisted Transcriptome SEQuencing (SORTseq) to analyze 768 FACS-isolated *citrine*-expressing cardiomyocytes from cryoinjured adult (4–18 months of age, males and females, TL strain) *TgBAC(nppa:mCitrine)* hearts. Cao et al. [10] used FACS isolation and a Fluidigm C1 chip to analyze 39 single *EGFP*-expressing epicardial cells and a pooled sample of 5000 *EGFP*-expressing epicardial cells from uninjured adult hearts (6 months of age, Ekkwill strain) *Tg(tcf21:nucEGFP)*. Koth et al. [11] used the Chromium 10× Genomics platform to sequence 12,000–12,500 FACS-isolated *citrine*-expressing cells per group from 20 cryoinjured adult (age not stated, KCL strain) *Tg(BAC-runx1P2:Citrine;kdrl:Hsa.HRAS-mCherry)* hearts.

References

1. Luo S, Truong AH, Makino A (2016) Isolation of mouse coronary endothelial cells. J Vis Exp (113). https://doi.org/10.3791/53985
2. van Beijnum JR, Rousch M, Castermans K, van der Linden E, Griffioen AW (2008) Isolation of endothelial cells from fresh tissues. Nat Protoc 3(6):1085–1091. https://doi.org/10.1038/nprot.2008.71
3. Jungblut M, Oeltze K, Zehnter I, Hasselmann D, Bosio A (2009) Standardized preparation of single-cell suspensions from mouse lung tissue using the gentleMACS dissociator. J Vis Exp (29). https://doi.org/10.3791/1266
4. Reichard A, Asosingh K (2019) Best practices for preparing a single cell suspension from solid tissues for flow cytometry. Cytometry A 95(2):219–226. https://doi.org/10.1002/cyto.a.23690
5. Manconi F, Markham R, Fraser IS (2000) Culturing endothelial cells of microvascular origin. Methods Cell Sci 22(2–3):89–99. https://doi.org/10.1023/a:1009895723488
6. Cossarizza A, Chang HD, Radbruch A, Acs A, Adam D, Adam-Klages S et al (2019) Guidelines for the use of flow cytometry and cell sorting in immunological studies (second edition). Eur J Immunol 49(10):1457–1973. https://doi.org/10.1002/eji.201970107
7. Li Z, Solomonidis EG, Meloni M, Taylor RS, Duffin R, Dobie R et al (2019) Single-cell transcriptome analyses reveal novel targets modulating cardiac neovascularization by resident endothelial cells following myocardial infarction. Eur Heart J 40(30):2507–2520. https://doi.org/10.1093/eurheartj/ehz305
8. Lawson ND, Weinstein BM (2002) In vivo imaging of embryonic vascular development using transgenic zebrafish. Dev Biol 248(2):307–318. https://doi.org/10.1006/dbio.2002.0711
9. Honkoop H, de Bakker DE, Aharonov A, Kruse F, Shakked A, Nguyen PD et al (2019) Single-cell analysis uncovers that metabolic reprogramming by ErbB2 signaling is essential for cardiomyocyte proliferation in the regenerating heart. eLife 8. https://doi.org/10.7554/eLife.50163
10. Cao J, Navis A, Cox BD, Dickson AL, Gemberling M, Karra R et al (2016) Single epicardial cell transcriptome sequencing identifies Caveolin 1 as an essential factor in zebrafish heart regeneration. Development 143(2):232–243. https://doi.org/10.1242/dev.130534
11. Koth J, Wang X, Killen AC, Stockdale WT, Potts HG, Jefferson A et al (2020) Runx1 promotes scar deposition and inhibits myocardial proliferation and survival during zebrafish heart regeneration. Development 147(8):dev186569. https://doi.org/10.1242/dev.186569
12. Burns CG, MacRae CA (2006) Purification of hearts from zebrafish embryos. BioTechniques 40(3):274, 6, 8 passim

Chapter 24

The Evaluation of Neovessel Angiogenesis Behavior at Tissue Interfaces

Hannah A. Strobel and James B. Hoying

Abstract

Angiogenesis, the formation of new vessel elements from existing vessels, is important in homeostasis and tissue repair. Dysfunctional angiogenesis can contribute to numerous pathologies, including cancer, ischemia, and chronic wounds. In many instances, growing vessels must navigate along or across tissue-associated boundaries and interfaces tissue interfaces. To understand this dynamic, we developed a new model for studying angiogenesis at tissue interfaces utilizing intact microvessel fragments isolated from adipose tissue. Isolated microvessels retain their native structural and cellular complexity. When embedded in a 3D matrix, microvessels, sprout, grow, and connect to form a neovasculature. Here, we discuss and describe methodology for one application of our microvessel-based angiogenesis model, studying neovessel behavior at tissue interfaces.

Key words Angiogenesis, Microvessel, Interface, Vascularization, Neovessel

1 Introduction

During angiogenesis, new vessels (neovessels) sprout and elongate from existing parent microvessels. A neovessel consists of a dynamic arrangement of a guiding tip cell, stalk cells, and phalanx cells in an immature vessel structure. The neovessel advances as an intact structure, often changing directions and branching as it navigates through the stromal space [1]. Our isolated microvessel system recapitulates this neovessel dynamic as intact fragments of microvessels serve as the parent microvessels for angiogenesis [2]. Fragments of microvessels, isolated from adipose tissue, provide a more mimetic model for studying angiogenesis than single cell or co-culture models (Fig. 1). Microvessels retain their native structure, and contain a complete spectrum of cell types, including endothelial cells, mesenchymal stem cells, macrophages, pericytes, and others. When embedded in a native matrix, neovessels sprout from parent vessels, grow, and interconnect with each other forming a

Andrew V. Benest (ed.), *Angiogenesis: Methods and Protocols*, Methods in Molecular Biology, vol. 2441,
https://doi.org/10.1007/978-1-0716-2059-5_24,

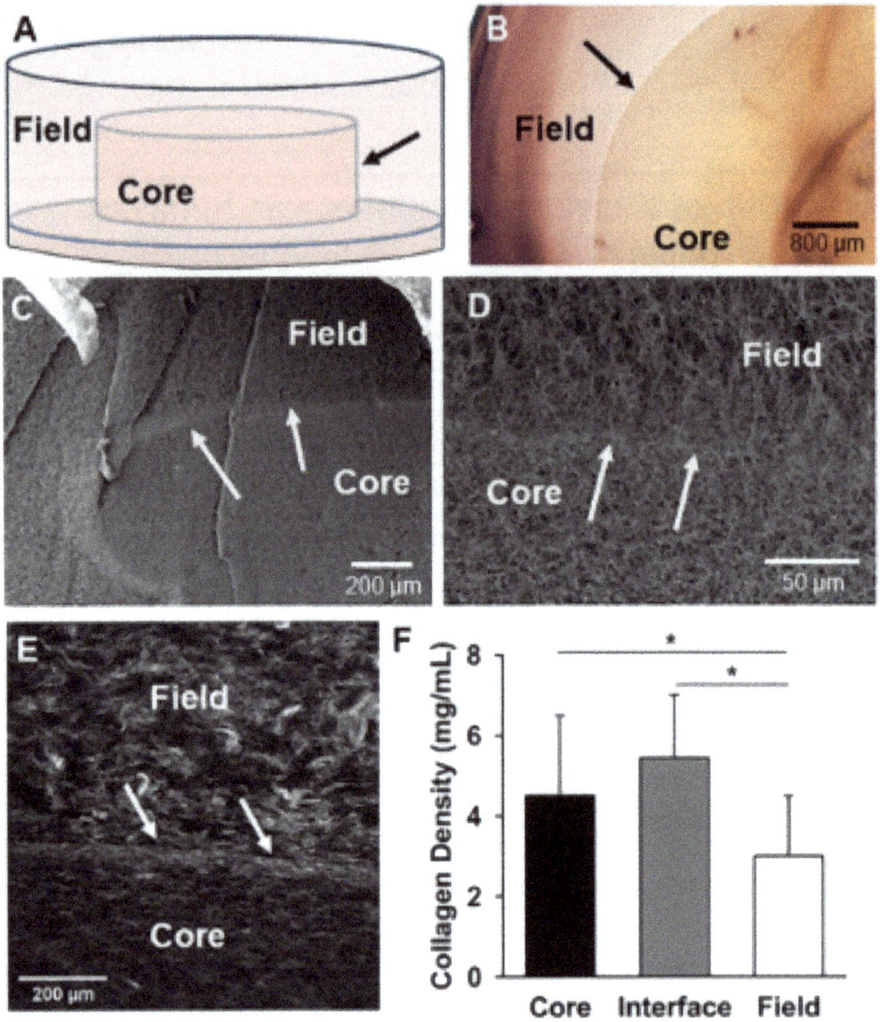

Fig. 1 The core-in-field (CIF) tissue boundary model. (**a**) Schematic of the CIF model showing the preformed "core" sitting on a thin bed of gelled collagen surrounded by an additional "field" of gelled collagen. (**b**) Top view of a phase microscopy image of a microvessel-free and cell-free CIF construct. (**c**) SEM image of a cross-section view of a CIF construct. (**d**) Higher-magnification of the interface and peri-interface region of a CIF construct. (**e**) Second harmonic generation image of the native collagen fibril structure comprising the interface and adjacent regions. (**f**) Fibril densities of the three regions in the CIF construct based on a validated method of measurement from SHG images. Bars are mean $\pm$ SD, $N = 15$, one-way ANOVA with Holm–Sidak post-hoc analysis. $^*P < 0.05$. In all cases, arrows indicate the interface between the core and field. (Used with permission from [3])

neovasculature [2, 4, 5]. When vessels are implanted, they rapidly inosculate with host vasculature [6–8]. We have used microvessels in preclinical studies to prevent the progression of ischemic lesions following myocardial infarction [9] and to improve biomaterial biocompatibility [10]. Furthermore, we have performed a number of experimental investigations investigating stromal cell and vascular precursor dynamics [8, 11, 12], angiogenesis-tissue biomechanics [5, 13, 14], imaging modalities to assess neovascular behavior [4, 15], and post-angiogenesis microvascular maturation and patterning [16–18]. We routinely use this MV system as an in vitro experimental assay platform to evaluate angiogenic factors [2, 17, 19], identify putative angiogenic agents [20, 21], evaluate microvascular instability [22], determine MMP-related angiogenic activity [19], define matrix dynamics during angiogenesis [4, 5, 13], and vascularize tissue models [11, 12, 23].

Interfaces are present between and within different compartments of a tissue and tissue beds. Such interfaces are also present following implantation and transplantation of tissues and tissue elements. Consisting of matrix structures and/or cellular layers, these interfaces serve to define microanatomical and functional regions (e.g. between muscle groups or lobules of the liver) or barriers to isolate and segregate. During angiogenesis, growing neovessels must navigate through not only the tissue space but also along and/or across these interfaces to fully vascularize the tissue. Very little is known about neovessel behavior at such interfaces, much less the mechanisms mediating this behavior.

Based on our experience with manipulating native matrices and using the isolated microvessel platform, we developed an experimental, in vitro model of tissue interfaces. This model consists of two compartments of adjacent type I collagen, which either or both compartments may or may not contain cells and/or microvessels [3]. For our angiogenesis modeling, a "core" compartment comprised of collagen containing microvessels is first formed, which is then surrounded by a second collagen "field" compartment that may or may not contain other cell types (Fig. 2). Due to how the model is established, an interface of dense collagen forms between the two compartments acting as a potential native, structural barrier. To form the core, microvessels are mixed with collagen and seeded in a single well of a 96 well plate. After gelling, this collagen plug is removed to the center of a well of a 48 well plate. Then, fresh collagen is flooded over the gelled core and allowed to gel thereby forming the field and an interface between the two. When constructs are fabricated in this way, the interface between the core and the field consists of a higher density of collagen fibrils that exhibit some alignment than either tissue compartment [3]. Angiogenic neovessels originating from the parent microvessels in the core grow toward the interface, but are deflected and continue to grow parallel to the interface (Fig. 3). Interestingly, when stromal cells were also included in the construct, either in the core or the field, neovessels crossed the interface and invaded the field region [3]. We are continuing to use this system to understand the mechanisms controlling neovessel crossing to gain additional insight into tissue vascularization.

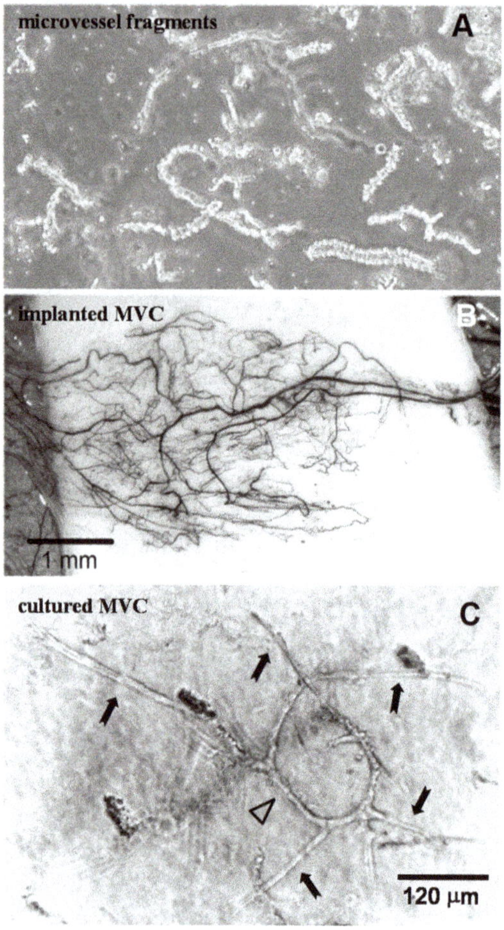

Fig. 2 (**a**) Phase contrast image of isolated microvessel fragments. (**b**) Vascular cast of the microcirculation derived by microvessels. (**c**) Sprouting angiogenesis (arrows) from a parent microvessel (arrow) in a 3D culture

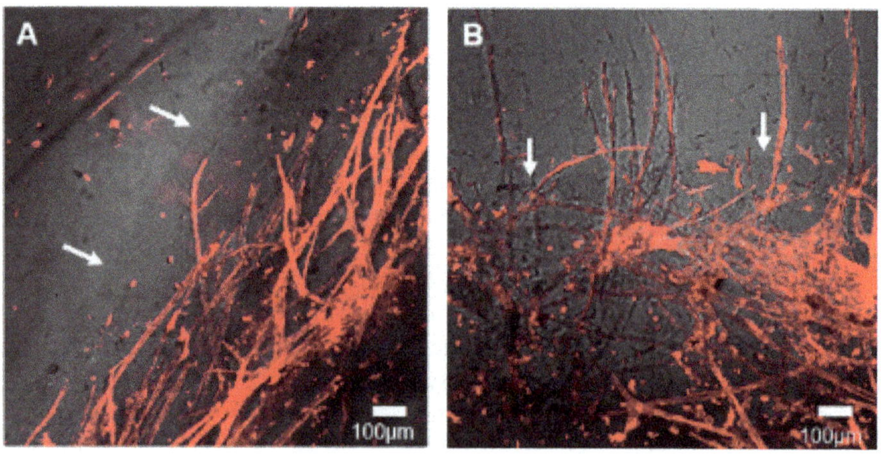

Fig. 3 Without stromal cells, microvessels are deflected at the interface (**a**). When stromal cells are included in either the core or field region, neovessels readily cross the interface (**b**). Lectin stain (red = endothelial cells). White arrows point to the interface. (Used with permission from [3])

2 Materials

All materials used must sterile and manipulated using aseptic technique.

- Thawing medium; DMEM, 10% FBS.
- Culture medium (for rat vessels); DMEM low glucose (Gibco), 20% FBS, 1% penicillin-streptomycin, 1% amphotericin B.
- Culture medium (for human vessels); RPMI (Corning), B27 supplement (Gibco), 50 ng/ml vascular endothelial growth factor (VEGF, Peprotech).
- Collagen solution (*see* preparation instructions below); 4× DMEM, Ultrapure water, 1 M Sodium Hydroxide (filter sterilized), High concentration rat tail collagen (Corning).
- Transfer pipette.
- Mini-scoopula.
- Cotton swabs.
- Pipettes of various sizes (20 μl, 100 μl, 1000 μl) and filter tips.
- 48 and 96 well cell culture plates.
- 37 °C humidified, CO_2-controlled incubator.
- Rodent, adipose microvessels isolated according to previously published methods [3] OR human, adipose microvessels (Advanced Solutions Life Sciences).

3 Methods

1. Thaw microvessels into thawing medium by quickly moving them from liquid nitrogen to a 37 °C water bath. Swirl vials continuously (being careful not to submerge the cap) until only a small amount ice remains, then transfer quickly to thawing medium with a micropipette. Centrifuge at 400 × *g* for 4 min to pellet microvessels (*see* **Note 1**).
2. While microvessels are spinning down, make collagen at a concentration of 3 mg/ml. Make enough so that you will have 200 μl per sample, plus ~200 μl extra to compensate for tube and pipette wall adhesion (*see* **Notes 2–4**).
3. Calculate the volume (V) of reagents needed to dilute your stock collagen to the desired concentration using the following equations:

$$\text{V Stock Collagen} = \frac{(\text{Desired Collagen Concentration}) * (\text{Desired V Collagen})}{\text{Concentration of Stock Collagen}}$$

$$\text{V } 4 \times \text{DMEM} = (\text{Desired V Diluted Collagen}) \times (0.25)$$

$$\text{V DI Water} = \text{Desired V Collagen} - (\text{V } 4 \times \text{DMEM} + \text{V Stock Collagen})$$

4. Add all the reagents together in a 15 ml centrifuge tube (on ice). The collagen stock should be the last reagent added.
5. Multiply the volume of stock collagen by 0.023 to obtain the amount of NaOH needed to neutralize the solution. Add the NaOH and observe the mix turn red/reddish pink (reflecting pH 7.4). Magenta means it is too basic (add back 1 M HCl) and orange means it is still too acidic (add more NaOH). *Do not add NaOH or HCl once the microvessels are in the collagen.
6. Aspirate the supernatant above the microvessel pellet, so that ~50 μl culture medium remains. Resuspend the microvessels with a micropipette and place the tube on ice for 1–2 min (*see* **Note 5**).
7. Add the collagen solution to the microvessels. Add enough so that you will have 100 μl per well, plus ~10% extra (some collagen will adhere to the tube). Microvessels should be used at a concentration of 60 k/ml for rat microvessels, and 100 k/ml for human.
8. Add 100 μl of the collagen/microvessel suspension to each well of a 96 well plate. Add 100 μl of the remaining collagen (without microvessels) to an equal number of wells in a 48 well plate, making sure the bottom of the well is completely covered. Remove any air bubbles. This is to collagen coat the wells where the cores will later be placed, which helps prevent contraction during culture.
9. Incubate 30 min at 37 °C to allow the collagen to gel.
10. After about 25 min, begin prepping for the next step. Prepare enough collagen to add 250 μl per well according to **step 2** (*see* **Note 6**).
11. Use the sterile mini-scoopula to gently separate the collagen cores from the 96 well plate, by running it around the edge of each well (Fig. 4).
12. Tilt the 96 well plate and rest it on a pipette tip box or something similar (Fig. 4).
13. Use a sterile transfer pipette to carefully remove the cores. Squeeze the transfer pipette slightly, then position the tip on the middle of the top of the core (Fig. 5). To pick up the core, slowly release pressure on the pipette to apply gentle suction,

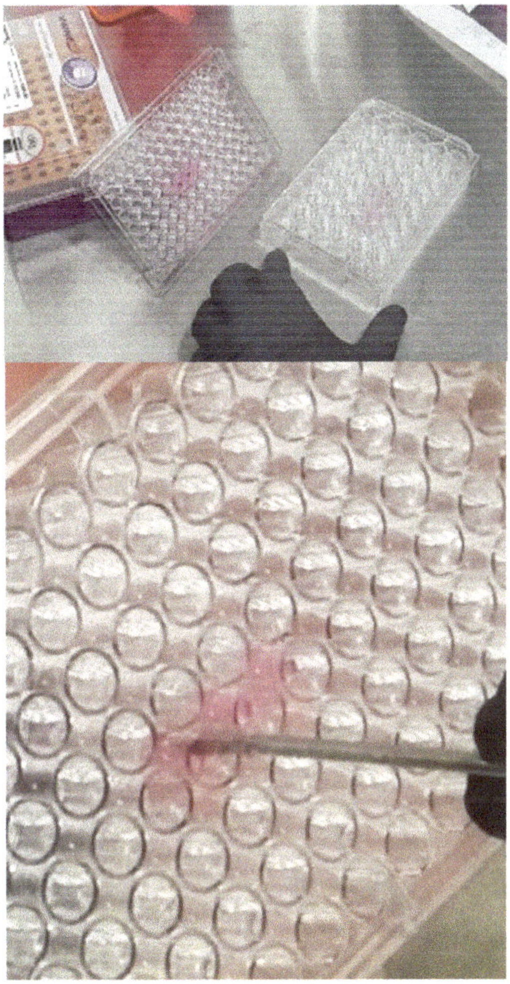

Fig. 4 Tilt the plate onto a pipette box (top). Then use the mini-scoopula to separate the cores from the well edges

while making a very slight motion toward the bottom side of the well.

14. Release the core into the wells of the 48 well plate with the collagen-coated bottom. If the core sticks to the side of the well, immediately push it down onto the collagen surface or it will begin to dry and stick to the plastic.
15. After cores have been transferred, use sterile cotton swabs to gently remove excess liquid from the well and around the core. Be careful not to touch the core, as it will stick to the cotton swab.
16. Add 250 μl collagen per well (prepared in **step 7**) over the core to form the field.

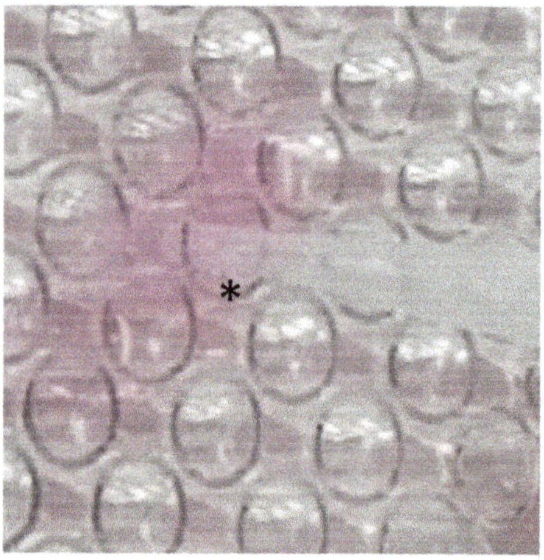

Fig. 5 Use transfer pipette to gently remove cores from wells. Use slight downward motion toward bottom edge of well plate while applying suction

17. Immediately after adding the field, tap the dish to help center and flatten out the cores.
    (a) If any cores are still touching the side of the well, push them toward the middle with the mini-scoopula, taking care not to touch the bottom collagen coating in the well. If this is not done within 1–2 min after adding the field collagen, stop trying to center the cores. You will do more damage by moving them after this point, as the collagen is starting to set.
    (b) Try to aspirate any remaining bubbles in the construct using a micropipette, but take care not to remove collagen.
18. Incubate the constructs at 37 °C for 30 min. Add 350 μl culture medium per well. Medium should be changed every 4 days. More frequent medium changes may slow microvessel growth (*see* **Note 7**).

4 Notes

1. If cells are to be included in the cores with the microvessels, they should be thawed and combined with microvessels before spinning down.
2. Undiluted, collagen stocks can be stored at 4 °C for 6 months. Longer storage times can result in poor neovessel growth.
3. All components of the collagen mix should be kept on ice throughout the preparation.
4. Once prepared, use the collagen within 30 min.

5. If the 50 μl microvessel suspension will be more than 25% of your total volume after adding collagen, increase the working collagen concentration accordingly so that the final concentration will be approximately 3 mg/ml. For example, if making 100 μl collagen/microvessel suspension, make 6 mg/ml collagen instead of 3 mg/ml. In this case, the microvessel suspension (~50 μl) will be combined with 50 μl of collagen for a final concentration of 3 mg/ml collagen.
6. If cells are to be included in the field region, thaw them at this time and centrifuge.
7. Microvessels will begin to sprout after 3–5 days in culture and will continue to grow and form neovascular networks throughout culture. We have observed that microvessels, when plated in the core region with no other cells or vessels in the field, grow toward the interface but are largely deflected upon reaching it and do not cross. When stromal cells are included in either the core or field, vessels will instead cross the interface into the field (Fig. 3), although it is possible to see this when vessels are initially plated at a very high density as well [3]. We are continuing to investigate the biological processes that cause microvessels to cross a tissue interface. Such knowledge may prove beneficial in applications such as implant vascularization, tissue vascularization and repair, and understanding microvascular-related pathologies.

References

1. Rauff A et al (2019) Imaging the dynamic interaction between sprouting microvessels and the extracellular matrix. Front Physiol 10: 1011
2. Hoying JB, Boswell CA, Williams SK (1996) Angiogenic potential of microvessel fragments established in three-dimensional collagen gels. In Vitro Cell Dev Biol Anim 32(7):409–419
3. Strobel HA et al (2020) Stromal cells promote neovascular invasion across tissue interfaces. Front Physiol 11:1026
4. Kirkpatrick ND et al (2007) Live imaging of collagen remodeling during angiogenesis. Am J Physiol Heart Circ Physiol 292(6): H3198–H3206
5. Krishnan L et al (2007) Interaction of angiogenic microvessels with the extracellular matrix. Am J Physiol Heart Circ Physiol 293(6):H3650–H3658
6. Shepherd BR et al (2004) Rapid perfusion and network remodeling in a microvascular construct after implantation. Arterioscler Thromb Vasc Biol 24(5):898–904
7. Nunes SS et al (2010) Implanted microvessels progress through distinct neovascularization phenotypes. Microvasc Res 79(1):10–20
8. Nunes SS et al (2010) Angiogenic potential of microvessel fragments is independent of the tissue of origin and can be influenced by the cellular composition of the implants. Microcirculation 17(7):557–567
9. Shepherd BR, Hoying JB, Williams SK (2007) Microvascular transplantation after acute myocardial infarction. Tissue Eng 13(12):2871–2879
10. Gruionu G et al (2010) Encapsulation of ePTFE in prevascularized collagen leads to peri-implant vascularization with reduced inflammation. J Biomed Mater Res A 95(3):811–818
11. Rhoads RP et al (2009) Satellite cell-mediated angiogenesis in vitro coincides with a functional hypoxia-inducible factor pathway. Am J Physiol Cell Physiol 296(6):C1321–C1328
12. Hiscox AM et al (2008) An islet-stabilizing implant constructed using a preformed vasculature. Tissue Eng Part A 14(3):433–440

13. Krishnan L et al (2008) Effect of mechanical boundary conditions on orientation of angiogenic microvessels. Cardiovasc Res 78(2):324–332
14. Krishnan L et al (2004) Design and application of a test system for viscoelastic characterization of collagen gels. Tissue Eng 10(1–2):241–252
15. Kirkpatrick ND et al (2006) In vitro model for endogenous optical signatures of collagen. J Biomed Opt 11(5):054021
16. Chang CC et al (2012) Determinants of microvascular network topologies in implanted neovasculatures. Arterioscler Thromb Vasc Biol 32 (1):5–14
17. Chang CC et al (2010) Angiogenesis in a microvascular construct for transplantation depends on the method of chamber circulation. Tissue Eng Part A 16(3):795–805
18. Chang CC, Hoying JB (2006) Directed three-dimensional growth of microvascular cells and isolated microvessel fragments. Cell Transplant 15(6):533–540
19. Vartanian KB et al (2006) The non-proteolytically active thrombin peptide TP508 stimulates angiogenic sprouting. J Cell Physiol 206(1):175–180
20. Carter WB, Ward MD (2001) Parathyroid-produced angiopoietin-2 modulates angiogenic response. Surgery 130(6):1019–1027
21. Carter WB et al (2000) Parathyroid-induced angiogenesis is VEGF-dependent. Surgery 128(3):458–464
22. Carter WB (2001) HER2 signaling—induced microvessel dismantling. Surgery 130(2):382–387
23. Strobel HA, Gerton T, Hoying JB (2021) Vascularized adipocyte organoid model using isolated human microvessel fragments. Biofabrication 13:035022

Chapter 25

Endothelial Cell Fibrin Gel Angiogenesis Bead Assay

Eva M. Clavane, Hannah A. Taylor, Richard M. Cubbon, and Paul J. Meakin

Abstract

The fibrin gel angiogenesis bead assay provides a controlled in vitro setting for observing endothelial angiogenic sprouting in response to modified variables. Endothelial cells are coated onto microcarriers and embedded into a fibrin clot containing necessary growth factors. Following a 24-h incubation, endothelial sprouts are imaged using a light microscope. This method is useful for rapidly and affordably investigating the effects of genetic or chemical manipulation to endothelial function.

Key words Angiogenesis, Fibrin matrix, Cytodex® 3 beads, Endothelial cells, Sprouting

1 Introduction

The in vitro fibrin gel angiogenesis bead assay is widely used to gain an understanding of the basic biology of blood vessel growth in health and disease [1]. It allows researchers to rapidly and affordably determine whether potential angiogenesis-manipulating compounds are worth pursuing in more expensive in vivo experimentation. Moreover, endothelial cells which have been genetically manipulated can also be exploited in this assay to provide further insights.

The assay involves attaching endothelial cells, for example, human umbilical vein endothelial cells (HUVECs) and pulmonary endothelial cells (PECs), to a microcarrier such as Cytodex® 3 beads (Fig. 1). The collagen coated bead simulates the extracellular matrix which is vital in supporting and organizing endothelial sprouting [2]. The HUVEC-coated Cytodex® 3 beads are suspended in a fibrin gel matrix, formed by reacting fibrinogen with thrombin, which is conducive to endothelial assembly. Fibrin is the provisional matrix involved in wound healing and possesses pro-angiogenic features; thus, being exploited extensively in in vitro angiogenesis assays [1]. Supplemented growth factors including vascular endothelial growth factor (VEGF), fibroblast

Andrew V. Benest (ed.), *Angiogenesis: Methods and Protocols*, Methods in Molecular Biology, vol. 2441,
https://doi.org/10.1007/978-1-0716-2059-5_25,

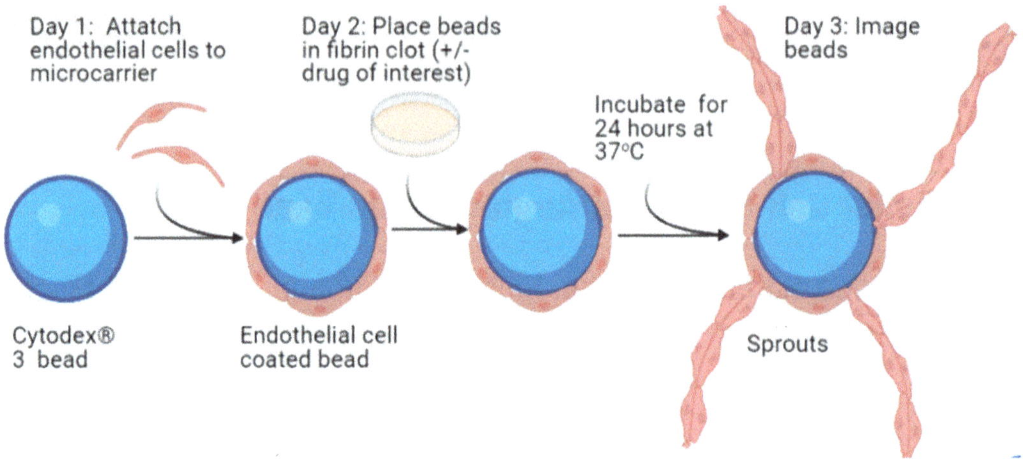

Fig. 1 Stages of the fibrin gel angiogenesis bead assay. Day 1 consists of attaching endothelial cells to a microcarrier. Day 2 involves embedding the coated beads in a fibrin clot consisting of growth factors and the compound of interest. Finally, following a 24-h incubation at 37 °C, the beads are imaged on day 3

growth factors (FGF), and aprotinin initiate endothelial cells to acquire a pro-angiogenic phenotype and encourage three-dimensional sprouting.

Impressively, the assay recapitulates early stages of endothelial sprout formation, with nascent vessels displaying intercellular lumens bordered by polarized endothelial cells [3]. This gives the assay an advantage over the minimalistic two-dimensional cell migration assay, which focuses primarily on the movement of endothelial cells and not their sprout-forming attributes [4].

Previous versions of the fibrin gel angiogenesis bead assay include co-culturing endothelial cells with other cell types such as pericytes and perivascular fibroblasts [5, 6]. Although these methods have proven useful in investigating cellular interactions relating to angiogenesis, using isolated endothelial cells allows researchers to specifically focus on endothelial cell physiology.

2 Materials

2.1 Endothelial Cell Preparation

1. Human umbilical vein endothelial cells (HUVECs).
2. Endothelial Cell Culture Medium 2 (ECGM2).
3. Trypsin-EDTA 0.25%.
4. Phosphate buffered saline (PBS).
5. Hemocytometer.

2.2 Bead Preparation

1. Cytodex® 3 microcarrier beads.
2. 5 mL FACs tubes.
3. 0.6 mL, 0.5 mL, 15 mL tubes.
4. T75 and T25 culture flasks.

2.3 Fibrin Clot

1. Fibrinogen Type 1 (from bovine plasma).
2. Human VEGF-A165.
3. Human basic FGF (FGF-2).
4. Aprotinin (from bovine lung).
5. Thrombin (from bovine plasma).
6. Filter sterilizing cap (0.2 μm acrodisc) and syringe.
7. 24-well plate (sterile and uncoated).

3 Methods

3.1 Endothelial Cell Preparation

1. Seed 7.5×10^5 HUVECs in a T75 flask with ECGM2 and incubate at 37 °C (*see* **Notes 1–3**).
2. Once confluent, aspirate medium from cells and wash with 6 mL of warm PBS (*see* **Note 4**).
3. Aspirate off the PBS and harvest the cells by incubating them with 1 mL trypsin-EDTA 0.25% for approximately 3 min at 37 °C (*see* **Note 5**).
4. Once the HUVECs have lifted off the base of the flask, add 8 mL of ECGM2 to neutralize the trypsin-EDTA (*see* **Notes 6** and 7).
5. Transfer the 9 mL cell suspension to a 15 mL tube.
6. Centrifuge the cells for 5 min at 500 RCF to pellet them.
7. Aspirate off the medium and resuspend the pellet in 4 mL of ECGM2 (*see* **Note 8**).

3.2 Cell Counting

1. Prepare a 1:1 mix of cell suspension and trypan blue by adding 10 μL of each to a 0.6 mL tube and mixing well.
2. Add 10 μL of the mix to each grid on an hemocytometer slide.
3. Under a light microscope, accurately count the number of live cells in each grid (*see* **Note 9**).
4. Multiply the average number from both grids by 2 to account for the trypan blue dilution.
5. Multiply this value by 10^4 to attain the number of cells per mL.
6. 500,000 cells are required for each condition of the assay. Therefore, to obtain the dilution factor, divide the number of cells in 1 mL by 500,000 (*see* **Note 10**).
7. Divide 1 mL by the dilution factor to achieve the amount of cell suspension containing 500,000 cells.
8. Make a 1 mL mix of 500,000 cells and medium in preparation for adding to the beads in Subheading 3.3 (*see* **Note 11**).

3.3 Attaching Endothelial Cells to Cytodex® 3 Beads

1. Prepare Cytodex® 3 beads according to manufacturer's instructions (*see* **Note 12**).
2. Place 21 μL of resuspended Cytodex® 3 beads in a 1.5 mL tube (*see* **Notes 13** and **14**).
3. Wash the beads in 50 μL of PBS, waiting for a minimum of 5 min for the beads to re-settle before carefully aspirating off the supernatant.
4. Resuspend the beads in 1 mL ECGM2 and remove the supernatant once the beads have settled.
5. Add 1 mL of cell suspension to the beads in 1.5 mL tube, mix gently and transfer the bead/cell mix to a FACs tube.
6. Add an additional 500 μL medium to the FACs tube and flick the tube numerous times to resuspend them before incubating the beads at 37 °C for a minimum of 4 h (*see* **Note 15**).
7. Agitate the beads with a flicking action every 20 min for the duration of the 4 h (*see* **Note 16**).
8. After 4 h, transfer the 1.5 mL cell/bead mix from the FACs tube into a T25 flask using a pipette.
9. Add an additional 4 mL of ECGM2 and leave overnight at 37 °C.

3.4 Preparation of a Fibrinogen Mix

1. Dissolve 2 mg/mL Fibrinogen Type 1 in PBS and heat in a 37 °C water bath for 30 min (*see* **Note 17**).
2. Once fully dissolved, filter sterilize the fibrinogen using a sterile syringe and cap.
3. Prepare a sufficient quantity of fibrinogen mix for each condition containing 0.15 U/mL aprotinin, 5 ng/mL VEGF, and 5 ng/mL FGF-20 (*see* **Notes 18** and **19**).

3.5 Placing Endothelial-Coated Beads into a Fibrin Clot

1. Very gently, dislodge the HUVEC-coated beads from the T25 flask using a Stripette and transfer the HUVEC-coated beads into a new 15 mL tube (*see* **Note 20**).
2. Rinse the T25 with an additional 4 mL of ECGM2 and add this to the tube.
3. Leave the beads for 15 min, allowing them to pellet, then aspirate off the excess medium.
4. Resuspend the pellet very gently in 1 mL ECGM2 and transfer the beads into a 1.5 mL tube before doing three 1 mL washes with ECGM2 (*see* **Note 21**).
5. Once washes are complete, resuspend the beads in 2.5 mL of fibrinogen mix previously prepared in Subheading 3.4.
6. Reverse pipette 12.5 μL of 50 U/mL thrombin in the center of a well of a 24-well plate (*see* **Note 22**).

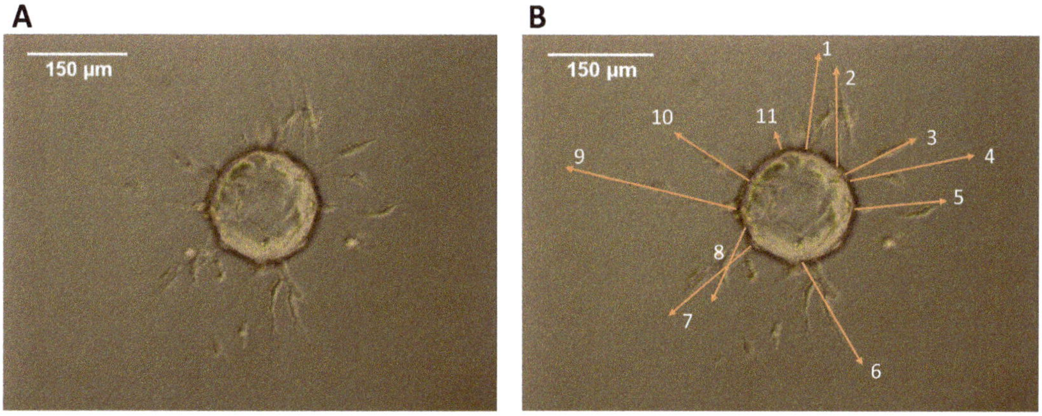

Fig. 2 Imaging and analysis of sprouted beads. (**a**) For each image, a single bead must be central and in focus. (**b**) A straight-line tool on ImageJ is used to draw a line from each sprout that protrudes from the bead; starting from the edge of the bead and finishing at the longest branch of the sprout

7. Slowly add 500 µL of the fibrinogen/bead mix to the thrombin, ensuring that it is being adequately mixed to allow dispersion of the beads (*see* **Notes 23** and **24**).
8. Plate four wells for each condition to allow for a sufficient number of repeats.
9. Leave the plate undisturbed for 5 min at room temperature before transferring it to 37 °C for 10–15 min, allowing the clot to solidify (*see* **Note 25**).
10. Once solid, carefully add 1 mL of ECGM2 drop-wise onto the clot using a p1000 pipette (*see* **Note 26**).
11. Place the plate at 37 °C for 24 h.

3.6 Imaging Beads

1. Following the incubation, image 10 sprouted beads per well using a light microscope equipped with a camera.
2. On the imaging software, burn a scale onto an image so that it can be used later in analysis.
3. Analyze the beads using the straight-line tool on the software ImageJ (Fig. 2) (*see* **Notes 27** and **28**).
4. Average the number and length of beads per well (*see* **Note 29**).

4 Notes

1. Use HUVECs up to passage number 4 (P4) to ensure that they are able to be manipulated throughout the assay. Older cells lose their endothelial properties and may find it more difficult to coat the Cytodex® 3 beads and sprout effectively.

2. A confluent T75 flask should produce approximately three to four million HUVECs. Each condition for the bead assay requires 500,000 cells. Therefore, it will provide a sufficient number of cells for a minimum of six conditions.
3. Primary isolated PECs have also been successfully used at this concentration in this assay.
4. Ensure that the medium and PBS are warm before adding to the HUVECs; this will prevent damaging the cells.
5. Once lifted from the flask, cells produce a curled composition under the microscope.
6. If the Trypsin-EDTA is left on for longer than necessary, without being neutralized by ECGM2, it will begin to damage or kill the cells.
7. Use this medium to harvest the more stubborn cells from the side of the flasks by pipetting up and down numerous times.
8. A 4 mL suspension provides a manageable number of cells to count under the light microscope.
9. Live cells are bright white, whereas dead cells are a dull blue color.
10. 500,000 cells are an optimal number of cells for attaching to the number of Cytodex® 3 beads required in each condition.
11. It is useful to make a master-mix of cell suspension and medium with 500,000 cell per mL. Therefore, extra suspension can be made to account for any pipetting inaccuracies.
12. This step may be required in advance. For example, some Cytodex® 3 beads require swelling in PBS and autoclaving.
13. The assay requires 400 cells per bead. Therefore, the number of beads added at this stage may vary depending on the density of the Cytodex® 3 beads. However, if the beads are a standard particle size of >175 μm, then 21 μL of beads will be optimal.
14. Resuspend beads sufficiently by swirling the bottle before transferring them to a 1.5 mL tube.
15. Use this 500 μL to rinse out the remaining beads/cells in the 1.5 mL tube.
16. Encourage a "whirlpool" type motion in the center of the tube to ensure that the cells and beads have all been resuspended sufficiently, optimizing adherence.
17. Each well requires 500 μL fibrinogen mix and each condition requires four wells for repeats.
18. To save time, it is useful to make the fibrinogen mix during the bead washes mentioned in Subheading 3.5.
19. Add the compound/ drug of interest to the fibrinogen mix as the HUVEC-coated beads will be submerged in this.

20. Ensure very delicate handling of the beads to prevent removing the attached cells.
21. Ensure that you allow 5 min between each wash to enable the beads to settle to prevent loss.
22. Keep thrombin on ice and do not pipette more than four wells of thrombin at any one time to ensure optimal activity and clotting.
23. Take approximately 5 s to slowly expel the mix while stirring it in circular motions. However, do not take too long as the clot will start to solidify.
24. Prevent scratching the bottom of the well as this will be seen when imaging the beads.
25. Be extremely careful when transferring the plate as the clots are fragile and can dislodge. Dislodged clots are problematic as they are extremely difficult to image.
26. Do this step extremely slowly and carefully to prevent dislodging the clots.
27. Trace over each sprout with the straight-line tool, ensuring the length overlays the longest branch of the sprout.
28. Exclude counting beads that are in close proximity to the side of the well or sprouts from other beads. Only one bead should be present in each image.
29. Control beads will have approximately 7–11 sprouts with a length of 70–140 μm.

Acknowledgments

This work was supported by the British Microcirculation and Vascular Biology Society and the British Heart Foundation.

References

1. Morin KT, Tranquillo RT (2013) In vitro models of angiogenesis and vasculogenesis in fibrin gel. Exp Cell Res 319(16):2409–2417
2. Davis GE, Senger DR (2005) Endothelial extracellular matrix: biosynthesis, remodeling, and functions during vascular morphogenesis and neovessel stabilization. Circ Res 97(11):1093–1107
3. Nakatsu MN, Davis J, Hughes CC (2007) Optimized fibrin gel bead assay for the study of angiogenesis. J Vis Exp 3:186
4. Jonkman JEN et al (2014) An introduction to the wound healing assay using live-cell microscopy. Cell Adhes Migr 8(5):440–451
5. Nehls V, Drenckhahn D (1995) A microcarrier-based cocultivation system for the investigation of factors and cells involved in angiogenesis in three-dimensional fibrin matrices in vitro. Histochem Cell Biol 104(6):459–466
6. Unterleuthner D et al (2017) An optimized 3D coculture assay for preclinical testing of pro- and antiangiogenic drugs. SLAS Discov 22(5):602–613

Chapter 26

Live Cell Imaging and Analysis of Cancer-Cell Transmigration Through Endothelial Monolayers

Aarren J. Mannion

Abstract

During metastasis, a subset of cancer cells will break away from the primary tumor and invade into the surrounding tissue. Cancer cells which are able to breach the endothelium and enter the blood stream are then transported in the circulation to new target organs where they may seed as a distant metastasis. In order to invade this new organ, the cancer cells must bind to and traverse the vascular wall, a process known as transendothelial migration (TEM) or extravasation. This chapter describes an in vitro approach to automated live cell imaging and analysis of TEM in order to accurately quantify these kinetics and aid the researcher in dissecting the mechanisms of tumor–endothelial interactions during this phase of metastasis.

Key words Transendothelial migration, Metastasis, Extravasation, Endothelium, HUVEC, Live cell imaging

1 Introduction

Adhesion and subsequent extravasation require close interactions between the cancer cell and the endothelial cells, with active contribution from both cell types to facilitate transendothelial migration (TEM) [1]. Extravasation and TEM refer to the movement of cells through an endothelial barrier to exit the blood stream and enter a tissue. Metastatic cells which intravasate into the circulation must adhere to the endothelium in order extravasate and colonize secondary tissues. As with leukocytes, cancer cells have been observed undergoing rolling, tethering, firm adhesion, and transmigration in vitro (Fig. 1) [2]. However, tethering and rolling of cancer cells have rarely been observed in vivo although intraluminal crawling of cancer cells prior to extravasation from the vasculature has been shown in a zebrafish model of adhesion and extravasation [3].

Transendothelial migration or transmigration refers to the step in the metastatic cascade where cancer cells penetrate through the

Andrew V. Benest (ed.), *Angiogenesis: Methods and Protocols*, Methods in Molecular Biology, vol. 2441, https://doi.org/10.1007/978-1-0716-2059-5_26,

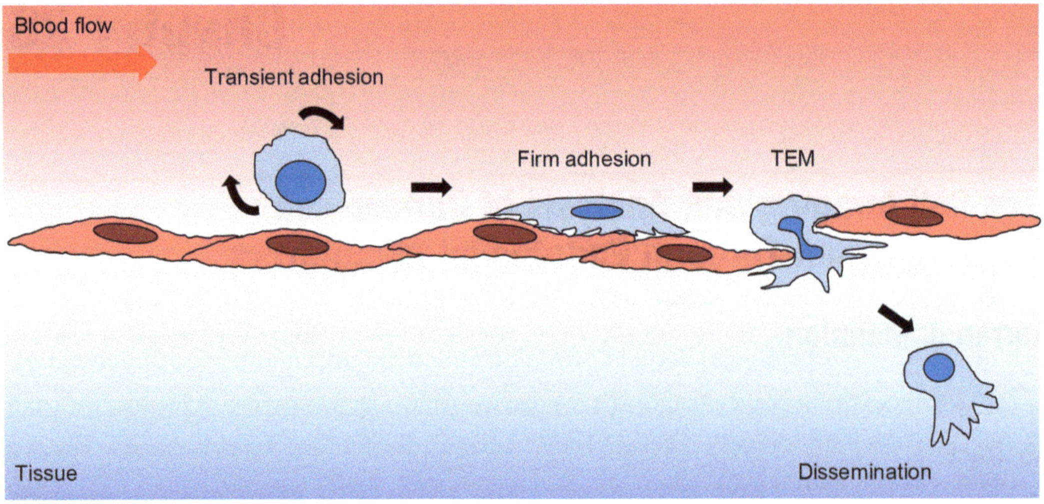

Fig. 1 Cancer cell adhesion and transendothelial migration. Once in the vasculature, cancer cells circulate until they adhere to the vascular wall. Initial transient adhesion reduces the velocity of the circulating cancer cells and engagement of multiple adhesion molecules eventually leads to the firm adhesion of the cancer cell to the endothelial monolayer. Cancer cells extravasate out of the blood flow by transendothelial migration (TEM) and disseminate into the surrounding parenchyma

endothelium lining blood vessels into surrounding tissue. As with adhesion, there is considerable overlap between the mechanisms of leukocyte adhesion and TEM with that of cancer cells [4]. The success rate for cancer cells in transmigrating and forming metastases is very inefficient (<0.01%) [5, 6], which has been postulated to be due to a lack of multiple integrins required for adhesion and TEM [4]. Once a circulating cancer cell has adhered to the endothelium, a series of events occur which allow the bound cells to traverse the endothelial monolayer. Most importantly, the uncoupling of multiple endothelial cell-cell junctional complexes is required for migrating cancer cells to penetrate the endothelium and colonize the surrounding tissue. TEM can be split into two pathways; (1) paracellular TEM directs migrating cells to the cell–cell junctions where the uncoupling of adhesion molecules allows the migrating cell to pass through the endothelium and (2) transcellular TEM which involves the movement of migrating cells through the endothelial cells cytoplasm by the reorganization of the actin cytoskeleton [1, 7].

The assessment of cancer cell metastasis has previously placed a heavy emphasis on the use of in vivo models. There are great benefits to in vivo analysis of transmigration, however, the associated difficulty of establishing such techniques is significant, and while the importance of in vivo systems to assess tumor progression are unlikely to diminish, there are obvious advantages to in vitro methodologies in metastatic research. For transmigration in particular, some progress has been made in modelling in vivo vascular conditions using microfluidic devices incorporating endothelial and metastatic cells to explore their interactions [8–10]. These platforms offer a highly adaptable, physiologically relevant setting (under conditions of flow), to answer complex questions regarding cancer cell extravasation which would otherwise require in vivo techniques such as intravital imaging. Other 3D migration assays such as modified Boyden chambers are frequently used to study transendothelial migration. While these assays are thought of as a gold standard in the assessment of migration and TEM, real-time imaging can be cumbersome and often relies on endpoint analysis for quantification. Alternatively, impedance assays such as the XCelligence (Agilent) allow the real-time assessment of transmigration by measurement of endothelial monolayer conductance [11].

The use of 2D systems incorporating endothelial and cancer cells offers a manipulable and reproducible platform to assess the transmigration of cancer cells. Live cell imaging can provide a large amount of information regarding TEM of tumor cells (such as migration kinetics, spreading efficiency, and endothelial junction integrity) with an increasing number of quantifiable parameters as the resolution and sophistication of microscopes improves. The ability to image transendothelial migration and quantify the number of cancer cells which have undergone this process was recently exemplified by Hilfenhaust et al., where it was demonstrated that compounds which inhibited cancer cell TEM could be screened for using a high throughput imaging platform [12]. Earlier documentation of in vitro transmigration imaging was elegantly described by Reymond et al., and showed a role for Rho GTPases in the cancer cell TEM using confocal microscopy [13]. The recent modification made by Hilfenhaus and colleagues indicates that the quantification of cell area of fluorescently labelled tumor cells which are able to spread and displace the endothelium is indicative of TEM efficiency. Importantly, validation of these in vitro models shows a parallel effect in vivo, suggesting that such systems can be used to predict in vivo effects on transmigration.

This chapter describes a methodology to allow the experimenter to perform imaging and quantitative analysis of TEM of cancer cells on a medium throughput basis using the Incucyte Zoom imaging platform. Thus, allowing the unbiased real-time capture of cancer cell transmigration kinetics.

2 Materials

1. Human umbilical vein endothelial cells (HUVEC, PromoCell).
2. Endothelial cell basal medium (ECBM) (PromoCell) supplemented with 2% fetal calf serum (FCS) (v/v), 0.4% endothelial cell growth supplement, 0.1 ng/ml epidermal growth factor

(recombinant human), 1 ng/ml basic fibroblast growth factor (recombinant human), 90 μg/ml heparin, and 1 μg/ml hydrocortisone.

3. Cells in which you wish to assess transmigration. Here the MDA-MB-231 triple negative breast cancer cell line is used because of their documented metastatic and invasive capacity. MDA-MB-231 (MDA-231) was isolated from the pleural effusion of a patient with metastatic breast cancer [14] and was purchased from the European Collection of Cell Cultures (ECACC).
4. RPMI-1640 (Sigma) supplemented with 10% (v/v) (FCS) (Biosera) (Sigma).
5. 1× PBS (No calcium, no magnesium) (Sigma).
6. 96-well clear bottom with black cassette for imaging (Perkin Elmer).
7. 1× Trypsin (Sigma).
8. Appropriate medium for associated cell lines.
9. 0.2% gelatin (Sigma) (w/v in 1× PBS) or human fibronectin (5 μg/ml).
10. Hemocytometer or automated cell counter.
11. Cell tracker green-CMFDA (0.4 μM) (Invitrogen).
12. Incucyte Zoom S3 (Essen bioscience), 20× objective.

3 Methods

3.1 Culturing Endothelial Cells

1. HUVEC endothelial cells are cultured in ECBM.
2. Harvest cells using 1× trypsin.
3. Remove culture medium, wash cells with 1× PBS, and incubate with 1 ml of 1× trypsin at 37 °C until cells began to detach.
4. Add FCS-containing medium to dilute and inactivate the trypsin and pellet cells by centrifugation at 300 × *g* for 5 min at room temperature (RT) before passaging (1:5 for MDA-231 cancer cell lines and 1:4 for all endothelial cells) or using in subsequent assays.
5. Endothelial cells are grown on 0.2% gelatin (w/v in 1× PBS) coated plates. Plates were coated for >2 h at 37 °C. For fibronectin coating, plates were coated overnight at 37 °C.
6. Matrix coated plates are aspirated, and plates washed with 1× PBS before seeding of endothelial cells.
7. Endothelial cells are grown to confluency before passaging every 2–3 days.

3.2 Establishing Endothelial Monolayers

1. Culture endothelial cells until confluent, as described above.
2. Harvest endothelial cells using 1 ml 1× trypsin and inactivate by resuspending in FCS-containing medium.
3. Count endothelial cells using a cell counter or hemocytometer and resuspend at 1×10^5 cells/ml.
4. Seed endothelial cells in 0.2% gelatin-coated 96-well plates at a density of 1×10^4 cells/well in 100 μl to achieve confluent monolayers in 24–48 h (*see* **Note 1**).
5. Set up the experiment so that there are at least, triplicates per condition.

3.3 Cell Tracker Green (CTG) Labeling of Cells for TEM Imaging

Once endothelial monolayers have reached confluency, cancer cells are cocultured with the endothelial monolayers and subsequent adhesion and transmigration is imaged using live microscopy.

1. MDA-231 cells are detached using 1× trypsin and resuspended in 10% FCS supplemented RPMI medium.
2. Count cells using a hemocytometer or Countess II FL Automated Cell Counter and pellet 1×10^5 cells by centrifugation at 300 × *g* for 5 min.
3. Remove the supernatant, resuspend the cell pellets in 1 ml of serum-free RPMI medium containing 0.4 μM cell tracker green-CMFDA, and incubate for 30 min at 37 °C (*see* **Note 2**).
4. Pellet the cells by centrifugation at 300 × *g* for 5 min and discard the supernatant. Wash cells by resuspending in 1 ml 10% FCS supplemented RPMI medium and pelleting cells once more.
5. Cells are then resuspended at 2×10^5 cells/ml in 10% FCS supplemented RPMI medium.
6. Seed 1×10^4 CTG-labelled cancer cells per well, in 50 μl volume, onto confluent endothelial monolayers, bringing the total volume of each well to 150 μl and place the plate in the Incucyte imaging tray.

3.4 Live Cell Imaging Using the Incucyte Zoom

1. Select the option for the correct vessel that has been used for the experiment (96-well plate format) and ensure that both the "green" and "bright-field" options have been selected to acquire images in both channels.
2. When prompted, select multiple regions of the well for subsequent image acquisition. Four regions per well were selected across three triplicate wells per condition. Once repeated in three independent experiments, this ensures 36 images will have been analyzed in total when collating the analyzed data.

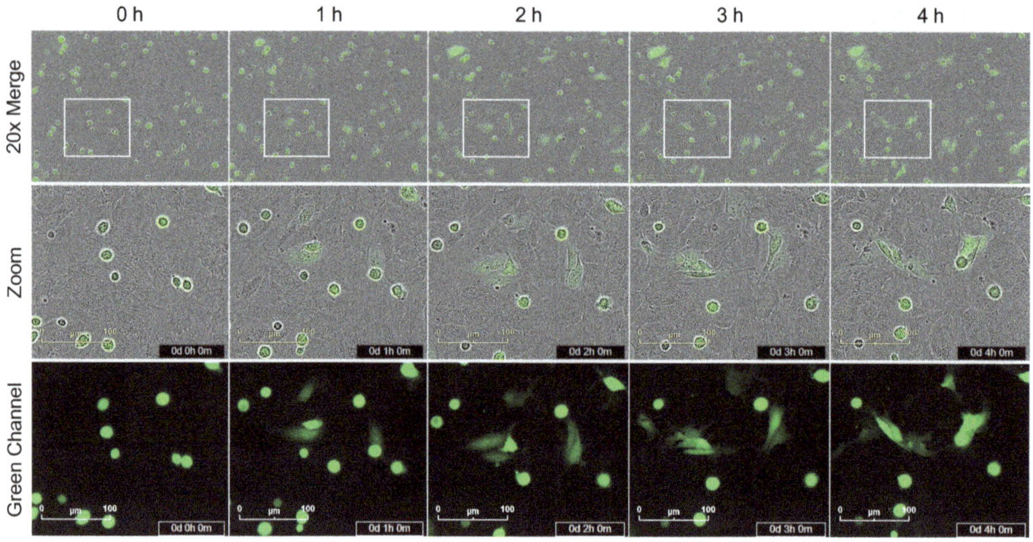

Fig. 2 Live cell imaging of MDA-231 transmigration through HUVEC endothelial monolayers. HUVEC cells were seeded to fibronectin coated 96-well plates until confluent (24 h) before CTG-labelled MDA-231 cells were seeded to EC monolayers and imaged for 4 h using the Incucyte Zoom live cell imager. Images were captured using the 20× objective and exported as a merged, blown up zoom panel, and single green channel TIF file. Scale bar: 100 μm

3. Depending on the number of conditions, images can be acquired up to every minute for individual fields of view. However, it is important to consider the temperature of the instrument, as imaging at this frequency can result in an increase in temperature which may be detrimental to the imaging unit. A warning note will appear if this happens and can be remedied by increasing the time between acquisitions, to allow the instrument to cool adequately before imaging commences.
4. Captured images can be exported as merged or individual channels as shown in Fig. 2.

3.5 Image Analysis

Although the interface of the software has changed significantly over recent years, the principles of the image analysis described in this method remain relevant. The analysis described below was carried out using the v2020B version of the Incucyte Zoom analysis software. However, previous versions of the software (v2011A) have also successfully been used to analyze similar data [15]. This description of image analysis was performed using version v2020B and data generated in Fig. 2 was generated using version v2020B.

The limitations of this assay and analysis are that migration through the monolayer is only inferred and there is no discrimination between those spreading on-top of, and in between the EC cell junctions. However, as the masking does not allow for individual selection of specific cells in the analysis, this strategy allows an unbiased methodology of determining cancer cell spreading on or between endothelial cells.

1. For image analysis, the green imaging channel capturing CTG-labelled MDA-231 cells was subject to thresholding using the mask function within the Incucyte analysis software.
2. To start the analysis software, select the "launch analysis" setting within the Incucyte software and create a new analysis definition using the "basic analyzer" function.
3. A threshold based on cell size with a radius of 100 μm was applied. Further filters were applied with a minimal area of 150 μm and a maximum eccentricity of 0.7 (where 0.0 represents a perfect circle). This analysis mask allows discrimination between MDA-231 cells which remain on the apical surface of the EC monolayer and have not undergone TEM, and those which have begun to intercalate and spread, which is indicative of TEM.
4. To preview the analysis mask, select a number of images from different timepoints and select "preview all." This applies the current settings of your analysis definition to these images and allows you to make adjustments if you can see that cells are not being correctly selected by the analysis mask. Select a broad range of timepoints, incorporating images from the start, middle and end of the time course and from a number of different wells, to ensure that the analysis mask is accurately selecting rounded non-transmigrated cells. Figure 2 shows images from time 0 h and 3 h indicating the cells which have been detected by the analysis mask.
5. This mask is created in the "green" channel to allow the thresholding to apply to only the CTG labelled MDA-231 cells (Fig. 3a).
6. This analysis definition is used to analyze the cell number of non-transmigrated cells of each individual image captured at each time point.
7. To calculate the percentage of cells which have undergone TEM at a given timepoint, subtract the number of cells counted by the analysis definition at the selected timepoint from the number of cells counted at time zero, at the initiation of the assay. This allows determination of the percentage of CTG cells which have undergone TEM when we consider that no cells had undergone TEM at the initial stage of the assay.

Results following analysis can then be exported in an excel file format either as values for each individual field of view or grouped to create an average for a whole well or across multiple wells. In Fig. 2, four individual fields of view from three technical replicates of $n = 3$ biological replicates were analyzed to give 36 data points

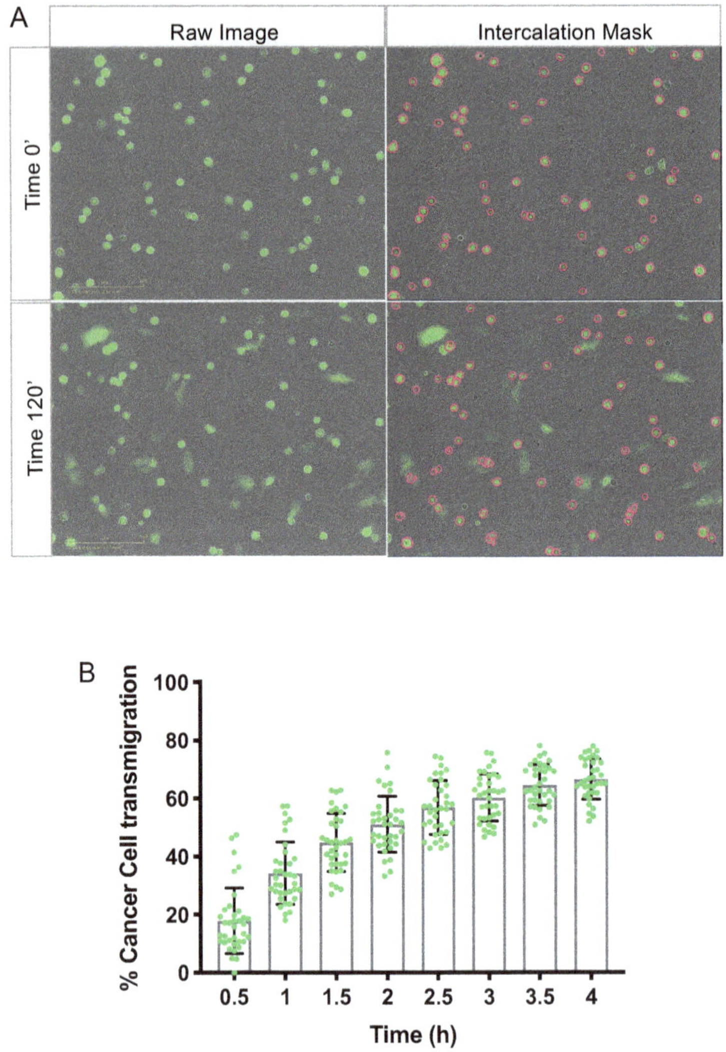

Fig. 3 Image analysis of cancer cell transmigration. (**a**) Example merged images of CTG-labelled MDA-231 cancer cells at 0 and 120 min seeded to confluent HUVEC monolayers. Intercalation mask (purple mask) to discriminate between MDA cells which have undergone TEM was created using the Incucyte basic analyzer software function and was used to quantify the number of MDA cells which had undergone TEM. Scale bar: 200 μm. (**b**) Quantification of MDA-231 cells which have undergone spreading when cocultured with HUVEC confluent monolayers. MDA spreading was calculated as percentage of the total number of cells within a field of view using the intercalation mask displayed in panel **a**. $n = 3$ biological replicates, with three technical replicates and four fields of view imaged per well. Each data point represents one field of view. Bar graph depict means ± SD

shown for each time point. The percentage of cancer cell TEM was calculated by subtracting the number of non-transmigrated cells from the total number of cells detected by the analysis mask at time 0.

For example, if 50 CTG-labelled cells in a field of view within the time 0 image are captured by the counting mask, this is marked as 0% of the cells undergoing TEM. As the assay progresses, and a number of cancer cells undergo TEM, the number of rounded cells decreases. There may be now only 25 cells which are captured by the analysis mask (which highlights those cells which are still rounded). This indicates that, subtracted from the analysis of time 0 image, 50% of the cancer cells within this field of view have deviated from a rounded morphology and have likely undergone spreading and TEM.

4 Notes

1. It is important that a confluent monolayer of endothelial cells is achieved, otherwise, the cancer cells in the subsequent steps of the assay will begin to spread on patchy areas of sub-confluent endothelial coverage, thus giving inaccurate results. However, if there are areas where there is not 100% confluent monolayer growth, these images can be discounted at a later stage when analyzing/quantifying.
2. Stand falcon tube in incubator while staining with cell tracker green-CMFDA.

Acknowledgments

This work was supported by the University of Leeds, including a Faculty of Medicine and Health PhD Scholarship to A.J.M.

References

1. Reymond N, d'Água BB, Ridley AJ (2013) Crossing the endothelial barrier during metastasis. Nat Rev Cancer 13:858–870
2. Ley K, Laudanna C, Cybulsky MI, Nourshargh S (2007) Getting to the site of inflammation: the leukocyte adhesion cascade updated. Nat Rev Immunol 7:678–689
3. Follain G, Osmani N, Azevedo AS, Allio G, Mercier L, Karreman MA, Solecki G, Garcia Leòn MJ, Lefebvre O, Fekonja N, Hille C, Chabannes V, Dollé G, Metivet T, Der Hovsepian F, Prudhomme C, Pichot A, Paul N, Carapito R, Bahram S, Ruthensteiner B, Kemmling A, Siemonsen S, Schneider T, Fiehler J, Glatzel M, Winkler F, Schwab Y, Pantel K, Harlepp S, Goetz JG (2018) Hemodynamic forces tune the arrest, adhesion, and extravasation of circulating tumor cells. Dev Cell 45:33–52.e12
4. Madsen CD, Sahai E (2010) Cancer dissemination—lessons from leukocytes. Dev Cell 19: 13–26
5. Chambers AF, Groom AC, MacDonald IC (2002) Dissemination and growth of cancer cells in metastatic sites. Nat Rev Cancer 2: 563–572
6. Fidler IJ (1970) Metastasis: quantitative analysis of distribution and fate of tumor emboli labeled with 125 I-5-iodo-2′-deoxyuridine. J Natl Cancer Inst 45:773–782
7. Mamdouh Z, Mikhailov A, Muller WA (2009) Transcellular migration of leukocytes is mediated by the endothelial lateral border recycling compartment. J Exp Med 206: 2795–2808
8. Hynes WF, Pepona M, Robertson C, Alvarado J, Dubbin K, Triplett M, Adorno JJ,

Randles A, Moya ML (2020) Examining metastatic behavior within 3D bioprinted vasculature for the validation of a 3D computational flow model. Sci Adv 6:eabb3308
9. Offeddu GS, Hajal C, Foley CR, Wan Z, Ibrahim L, Coughlin MF, Kamm RD (2021) The cancer glycocalyx mediates intravascular adhesion and extravasation during metastatic dissemination. Commun Biol 4:255
10. Coughlin MF, Kamm RD (2020) The use of microfluidic platforms to probe the mechanism of cancer cell extravasation. Adv Healthc Mater 9:e1901410
11. Rahim S, Üren A (2011) A real-time electrical impedance based technique to measure invasion of endothelial cell monolayer by cancer cells. J Vis Exp 2792
12. Hilfenhaus G, Mompeón A, Freshman J, Prajapati DP, Hernandez G, Freitas VM, Ma F, Langenbacher AD, Mirkov S, Song D, Cho B-K, Goo YA, Pellegrini M, Chen J-N, Damoiseaux R, Iruela-Arispe ML (2021) A high-content screen identifies drugs that restrict tumor cell extravasation across the endothelial barrier. Cancer Res 81:619–633
13. Reymond N, Im JH, Garg R, Vega FM, Borda d'Agua B, Riou P, Cox S, Valderrama F, Muschel RJ, Ridley AJ (2012) Cdc42 promotes transendothelial migration of cancer cells through β1 integrin. J Cell Biol 199: 653–668
14. Cailleau R, Olivé M, Cruciger QV (1978) Long-term human breast carcinoma cell lines of metastatic origin: preliminary characterization. In Vitro 14:911–915
15. Mannion AJ, Odell AF, Taylor A, Jones PF, Cook GP (2019) CD99 regulates cancer cell transendothelial migration and endothelial cell function via CDC42 and actin remodelling. bioRxiv:760934

Chapter 27

Differentiation and Characterization of Human Pluripotent Stem Cell-Derived Cardiac Endothelial Cells for In Vitro Applications

Kavita Raniga, Nguyen T. N. Vo, and Chris Denning

Abstract

Various protocols have been developed to generate endothelial cells for disease modeling, angiogenesis, vascular regeneration, and drug screening. These protocols often require cell sorting, as most differentiation strategies result in a heterogenous population of endothelial cells (ECs). For any given model system, one important consideration is choosing the appropriate EC subtype, as different EC populations have unique molecular signatures.

Herein, we describe a protocol for cardiac EC differentiation and a protocol for endothelial cell characterization. This protocol is aimed at investigating differentiation efficiency by measuring endothelial lineage markers, CD31, VE-Cadherin, and VEGFR2 by flow cytometry. Collectively, these protocols comprise the tools required to generate cardiac ECs efficiently and reproducibly from different hPSC lines without the need for cell sorting. Our protocol adds to the panel of hPSCs for cardiac EC differentiation and addresses reproducibility concerns of hPSC-based experiments. The approaches described are also applicable for complex model generation where multiple cardiovascular cell types are involved and may assist in optimizing differentiations for different cell lineages, including cardiomyocytes, cardiac endothelial cells, and cardiac fibroblasts.

Key words Cardiac endothelial cells, Stem cell differentiation, Cardiac mesoderm, Flow cytometry, VEGF

1 Introduction

Endothelial cells (ECs) are indispensable for sustaining vascular homeostasis and regulate many critical physiological processes including, thrombosis [1], platelet adhesion [2], immune and inflammatory responses [3], and inter-organ communication [4, 5]. Dysfunctional ECs are connected to a large variety of vascular pathologies such as atherosclerosis [6], hypertension [7], cardiomyopathy [8], retinopathy [9], and diabetes mellitus [10]. Although primary human endothelial cells have provided

Andrew V. Benest (ed.), *Angiogenesis: Methods and Protocols*, Methods in Molecular Biology, vol. 2441,
https://doi.org/10.1007/978-1-0716-2059-5_27,

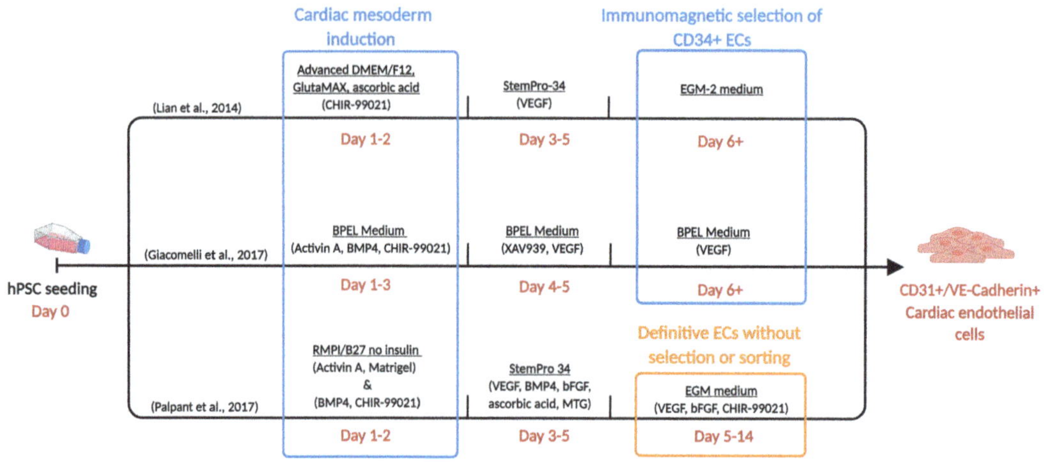

Fig. 1 Protocol comparison for the generation of cardiac endothelial cells from hPSCs. The schematic highlights differentiation protocols toward the endothelial cell fate. Differentiation begins with cardiac mesoderm induction via the modulation of activin/nodal and BMP pathways, followed by further directed differentiation with vascular endothelial growth factor (VEGF). This graphical abstract was created with https:/BioRender.com

important insights into endothelial dysfunction and disease states, these cells are often difficult to obtain, are of inconsistent genetic and disease status, and have a limited replicative capacity. Alternatively, human pluripotent stem cells (hPSCs) have high self-renewal capacity while maintaining the ability to differentiate into any cell type. As such, hPSCs are an attractive source for generating ECs, and multiple techniques of generating subtypes of endothelial cells have been published in the last decade, reviewed in [11].

Protocols allowing for the differentiation of hPSCs toward the cardiac lineage offer the opportunity to develop complex models of cardiovascular disease in vitro and more predictive cardiovascular toxicity screening. The ability to generate scalable quantities of definitive cardiac ECs would complement well-established hPSC-derived cardiomyocyte platforms, allowing for the simultaneous study of both cell types. Figure 1 outlines three protocols for cardiac endothelial cell differentiation [12–14]. Similarities between protocols include cardiac mesoderm induction via temporal modulation of Wnt/β-catenin signaling using small-molecule GSK3 inhibitors (CHIR-99021), followed by further differentiation to endothelial cells using high concentrations of VEGF (*see* Fig. 1). Palpant et al. [14] showed BMP4 and Activin A modulation can give rise to hemogenic and cardiogenic mesodermal precursors and both progenitor populations can be matured into >85% CD31 + and >90% VE-cadherin+/VEGFR2+ definitive ECs without the need for cell sorted as seen with the other two protocols.

In this chapter, we describe in detail the generation of hPSC-derived cardiac ECs adapting from the protocol published by

Palpant et al. [14]. We demonstrate the reproducibility of this protocol and outline methods for protocol optimization and the characterization of ECs.

All methods described here have been robustly reproduced in two hPSC lines: in-house REBL-PAT and commercial CGT-RCiB-10 (Cell Therapy Catapult).

2 Materials

2.1 Cells

Two human induced pluripotent stem cell lines were used with this protocol: in-house REBL-PAT (passage 25–50, derived from a skin punch biopsy from a male subject) and the commercial CGT-RCiB-10 (isolated from peripheral blood of a female subject, Cell Therapy Catapult).

2.2 hPSC Culture and Maintenance

1. Matrigel-coated T25 flasks: Matrigel (Corning) is stored at −80 °C. Place the matrigel on ice in a 4 °C refrigerator to thaw overnight before using. Dilute Matrigel in Dulbecco's Modified Eagle Medium (DMEM) at 1:100 dilution. Add 5 ml medium in a T25 flask and incubate at room temperature (RT) for 30 min and store at 4 °C.
2. Accutase Cell Dissociation Reagent (Accutase, LifeTechnologies).
3. Ca^{2+}/Mg^{2+}-free Dulbecco's Phosphate Buffer Saline (D-PBS, Gibco).
4. Essential 8 medium (E8, Gibco): The medium is prepared by warming the medium and gently thawing the supplement. Once thawed, the supplement is added to a 500 ml medium bottle in the laminar flow cell culture hood. The complete medium can be stored at 4 °C for up to 1 month and is prewarmed to 37 °C prior to use in routine cell culture.
5. E8 medium supplemented with 10 μM Y-27632 (Rock Inhibitor, Tocris).

2.3 hPSC Endothelial Cell Differentiation

1. Matrigel-coated T25 flasks (Thermo Scientific Nunc cell culture-treated flasks or plates).
2. Accutase Cell Dissociation Solution.
3. D-PBS.
4. Pre-conditioning medium: RPMI basal medium containing B27 minus insulin (Gibco) and supplemented with 100 ng/ml Activin A (LifeTechnologies) and Matrigel (1:60).
5. RPMI basal medium containing B27 minus insulin, and supplemented with 5 ng/ml BMP4 (R&D Research) and 1 μM CHIR-99021 (Cayman Chemicals).

6. StemPro™34-Serum Free Medium (Gibco) containing supplement provided, 2 mM L-Glutamine (Gibco), 50 μg/ml ascorbic acid (Sigma-Aldrich), 10 ng/ml BMP4, 5 ng/ml bFGF (Peprotech), 300 ng/ml VEGF (Peprotech), and 4×10^{-4} 1-Thioglycerol (MTG, Sigma-Aldrich).
7. StemPro™34-Serum Free Medium containing supplement provided 10 μM Y-27632.
8. Endothelial cell growth medium BulletKit (EGM, Lonza) supplemented with 20 ng/ml VEGF, 20 ng/ml human bFGF, and 1 μM CHIR-99021. The kit contains human EGF, hydrocortisone, gentamicin (GA-1000), fetal bovine serum, bovine brain serum, and ascorbic acid. The medium is prepared by warming EGM medium and thawing the kit components. Once thawed, the individual components are added to 500 ml bottle. Store at 4 °C and use within 2 weeks. This medium is used for cell collection, and endothelial cell maintenance and expansion.

2.4 Flow Cytometry Preparation of hPSC-Derived Endothelial Cells

1. 1% BSA in PBS (PBA, ice-cold, sterile filter).
2. Flow cytometry tubes or V-bottom 96-well plates.

2.5 Cryopreservation of Cardiac Endothelial Cells

1. CryoStor CS10 (Stem Cell Technologies).
2. Mr. Frosty™ Freezing Container.

2.6 Equipment and Supply

1. CEDEX HiRes counter (Roche) or equivalent.
2. Inverted light microscope.
3. Tissue culture-treated 96-well plates and flasks.
4. Flow cytometer.

3 Methods

All cell culture experiments should be performed in a type II Biological Safety Cabinet under sterile conditions, and cells were incubated in a humidified incubator at 37 °C and 5% CO_2. All experiments were carried out in antibiotic-free conditions, which is best practice.

The following steps describe the volumes required for a T25 flask. The volumes of reagents can be scaled proportionate to the surface area of the desired vessel (*see* Table 1).

Table 1
Standard cell density plating guide for different cell plates/flasks

Culture flask/ plate	Surface area (cm^2)	Medium approx. volume (ml)	Cell suspension approx. volume (μl)				
			20,000 cells/cm^2	30,000 cells/cm^2	40,000 cells/cm^2	50,000 cells/cm^2	60,000 cells/cm^2
T25 flask	25	5	500	750	1000	1250	1500
T75 flask	75	15	1500	2250	3000	3750	4500
6-well	9.6	1–3	200	300	400	500	600
12-well	3.8	1–2	80	120	160	200	240
24-well	1.9	0.5–1	40	60	80	100	120
48-well	1.1	0.2–0.5	20	30	40	50	60
96-well	0.32	0.1–0.2	6.4	9.6	12.8	16	19.2

3.1 hPSC Culture and Maintenance

It is recommended that hiPSCs be cultured in vitro for a maximum of 10 passages to minimize the amount of genetic and epigenetic changes, which others have reported [15].

1. Passage cells by washing the confluent culture with 5 ml D-PBS, followed by incubation with 1 ml accutase for four minutes (*see* **Note 1**).
2. Neutralize accutase by adding 9 ml of warmed E8 medium and spin this cell dissociation mix at 160 × *g* for 4 min.
3. Conduct cell count of the cell dissociation mix by using an automated CEDEX HiRes counter or a hemocytometer for manual counting.
4. Aspirate the supernatant and add an appropriate amount of warmed E8 medium supplemented with 10 μM Y-27632 to generate to create a cell suspension of 1000 cells μl/ml.
5. Transfer 0.5 ml of the previous cell suspension into a new Matrigel-coated T25 flask at approximately 20 × 10^3 cells per cm^2 and add 4.5 ml of E8 so that the total volume is 5 ml in the T25 flask (*see* Table 1).
6. Change 5 ml of fresh E8 medium every 24 h.

3.2 Monolayer hPSC Differentiation into Endothelial Cells

The following protocol outlines hPSC-directed differentiation into cardiac endothelial cells without the requirement of cell sorting or selection. An overview of this protocol is provided in Fig. 2 and appropriate volumes are listed in Table 1.

1. On day 0, passage hPSCs (*see* Subheading 3.1) in warm E8 basal medium supplemented with 10 μM Y-27632 to achieve a seeding density at 20 × 10^3 cells per cm^2. As different plating densities influence differentiation efficiency, we tested four different densities per cell line (*see* **Note 2**).

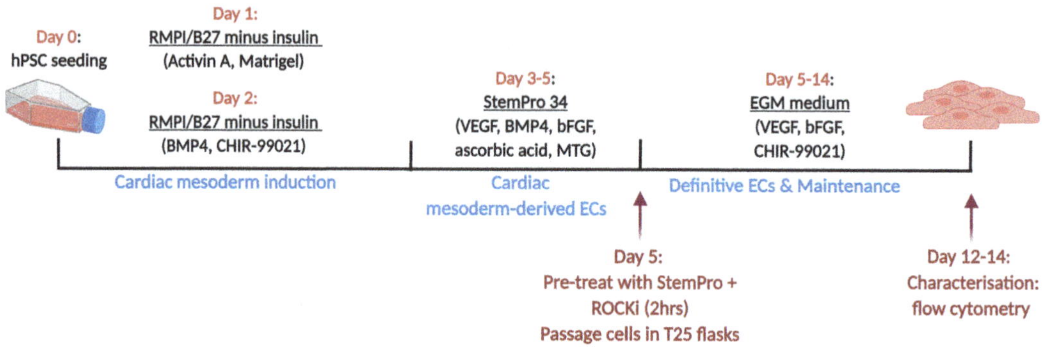

Fig. 2 The differentiation protocol toward cardiac endothelial cell fate. Cardiac mesoderm was induced by Activin A, BMP4, and CHIR-99021 from day 0 to day 3, followed by treatment with VEGF and bFGF. The graphic was created with https:/BioRender.com

2. On day 1, to induce cardiac mesoderm and add 5 ml of the pre-conditioning medium in a T25 flask of hPSCs (*see* Subheading 2.2, **item 4**). Incubate the cells with the pre-conditioning medium for 16 h at 37 °C, 5% CO_2 (*see* Fig. 3a, b).
3. After 16 h incubation, aspirate medium and add 5 ml of a mixture of 37 °C RPMI (supplemented with B27 minus insulin) + 5 ng/ml BMP4 + 1 μM CHIR-99021 to the cells. Place the cells in the incubator for 24 h.
4. The next day, aspirate the spent medium and add 10 ml of 37 °C SP34 medium + 4×10^{-4} M MTG + 2 mM L-glutamine + 50 μg/ml ascorbic acid + 10 ng/ml BMP4 + 5 ng/ml bFGF + 300 ng/ml VEGF to the cells, and incubate for 72 h at 37 °C, 5% CO_2.
5. On day 5, pre-treat cells with 5 ml of StemPro™34-serum free medium supplemented with 10 μM Y-27632 for 2 h. Next, dissociate cells using 37 °C accutase dissociation reagent. Re-plate these cells at a density of 9.0×10^3 cells per cm^2 on matrigel-coated vessels with the appropriate volume of 37 °C EGM + 20 ng/ml VEGF + 20 ng/ml bFGF + 1 μM CHIR-99021. This provides the opportunity for cells to proliferate and expand (Fig. 3c). Refresh the medium every 24 h (*see* **Note 3**).
6. Passage cultures, as above, when they reach 80–90% confluence. Endothelial cells should be maintained in 37 °C EGM + 20 ng/ml VEGF + 20 ng/ml bFGF + 1 μM CHIR-99021. The typical "cobblestone" monolayer morphology should now be observable (Fig. 3d, *see* **Note 4**).
7. After Day 12, medium can be changed every 48 h and cells can be cryopreserved (*see* **Note 5**).

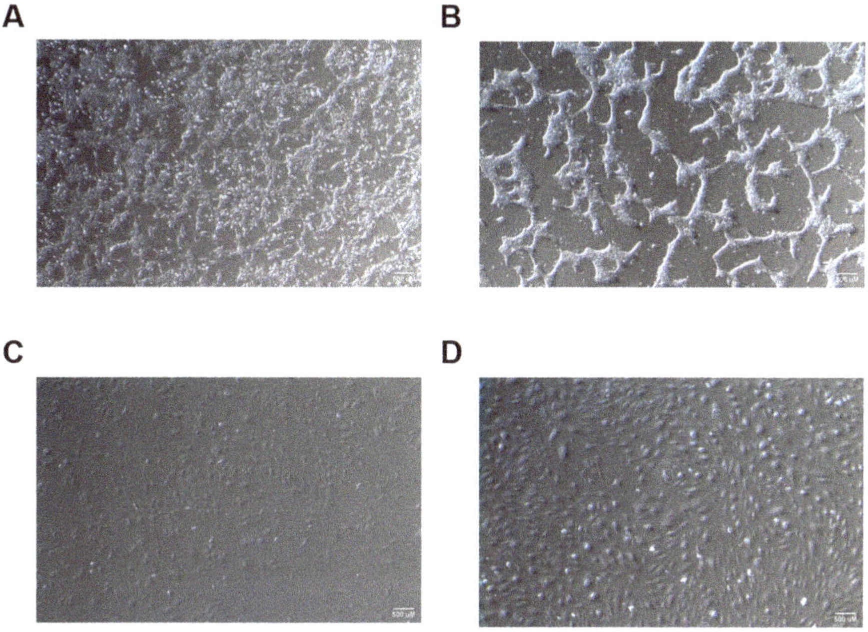

Fig. 3 Brightfield images during hPSC differentiation into cardiac endothelial cells. (**a**) Day 0 hPSCs seeded at 20,000 cells/cm^2. (**b**) Day 1 post-cardiac mesoderm induction. (**c**) Day 6, cells seeded at 9000 cells/cm^2 on matrigel-coated flasks. (**d**) Cardiac hPSC-ECs Day 14, cobblestone-like morphology. Scale bars, 500 μm

8. Perform flow cytometry analysis for the cells between day 12 and day 15 to determine the purity of CD31+/VE-cadherin+/VEGFR2+ cells (*see* Subheading 3.4). Our data show dependency on the plating density at the onset of differentiation to achieve >85% purity (Fig. 4a, b).

3.3 Cryopreservation of Endothelial Cells

1. Dissociate endothelial cells as described in Subheading 3.2 and perform the cell count.
2. Centrifuge the cell suspension to pellet endothelial cells for 5 min at 100 × *g* at room temperature.
3. Aspirate the supernatant and resuspend the endothelial pellet in the appropriate volume of cold CryoStor CS10 medium to achieve the final cell concentration 1 × 10^6 cells/vial.
4. Place the cryo-boxes at −80 °C for 24 h.
5. Transfer cryovial to −150 °C storage tank (*see* **Note 6**).

3.4 Flow Cytometry: Surface Staining of Endothelial Cells (Day 12–14)

1. Detach cells as described in Subheading 3.2 to produce a single cell suspension and avoid clumps.
2. Wash cells with 1 ml PBA.
3. Place 50 μl of cells (10^5 to 10^6 cells/sample) into the bottom of a tube or V-bottom plates.

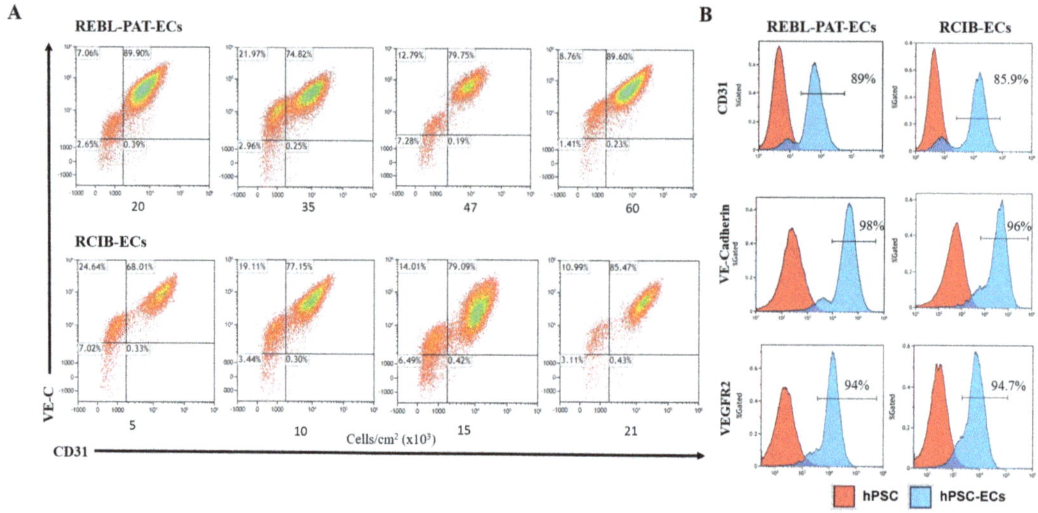

Fig. 4 Endothelial differentiation efficiency on the basis of initial seeding density for two hPSC lines (REBL-PATs and RCiB). (**a**) Representative flow cytometric plots for CD31/VE-Cadherin on day 14 of differentiation. (**b**) Representative flow cytometric histograms showing protein levels of CD31, VE-Cadherin, and VEGFR2 in human hPSC-ECs on day 14 of differentiation (blue) versus undifferentiated hPSCs (red)

4. Add 100 μl of primary antibody to cells and mix by pipetting, refer to Table 2 for antibody dilutions.
5. Incubate samples on ice in the dark for 30 min.
6. Add 100 μl of PBA to each sample, centrifuge samples at 300 × *g* for 5 min in a swing-out rotor (*see* **Note 7**).
7. Remove supernatant and add 100 μl of PBA to each sample, centrifuge samples at 300 × *g* for 5 min in a swing-out rotor.
8. Resuspend in 150 μl PBA or 4% formaldehyde and analyze samples by flow cytometer.

4 Notes

1. When the cell culture reaches approximately 80% confluence, the cells should be passaged and sub-cultured into a new culture vessel.
2. An optimization for seeding density should be conducted for each individual cell line. A suggested seeding density may range from 5–80 × 10^3 cells per cm^2, using Table 2 to generate multiple seeding densities. We recommend the optimization be conducted in 12 well plate format, while the routine endothelial differentiation can be scaled up in T25/T75 or T175 flask.

Table 2
Flow cytometry antibodies, conjugate and dye information for identification of an endothelial phenotype

Endothelial cell markers	*Antibody*	*Vendor (Cat #)*	*Dilution*	*Conjugate*
	CD31	BD Biosciences (55545)	1:5	FITC
	CD144 (VE-Cadherin; 16B1)	eBiosciences (17-1449-42)	1:20	APC
	KDR	BD Biosciences (55545)	1:5	AF647
	Propidium Iodide Ready Flow Reagent	Thermofisher Scientific (R37169)	1:12	N/A
	SYTOX™ Green Ready Flow™ Reagent	Thermofisher Scientific (R37168)	1:12	N/A

3. At day 5 progenitor cells are very sensitive, avoid excessive trituration. If there is high cell death despite pre-treatment with StemPro34 and Y-27632, add 10 μM Y-27632 to the media when seeding cells. As the cells quickly proliferate and expand between day 6 and day 12, the medium should be refreshed in every 24 h. Passage the confluent cultures with accutase and re-plate at 9×10^3 cells per cm^2.
4. bFGF and VEGF aliquots should be stored at −20 °C and thawed for immediate use. Do not refreeze aliquots for future use.
5. Avoid passaging cells more than once every 3 days. As endothelial cell proliferation slows down after day 12, thus medium can be changed every 48 h.
6. To minimize cell death when thawing cells, vials should be immediately transferred to a 37 °C water bath. Slowly dilute cells with EGM medium containing Y-27632 over a 10 min period and change medium after 24 h.
7. Do not use fixed angle Eppendorf centrifuges: these will kill the cells.

References

1. Yau JW, Teoh H, Verma S (2015) Endothelial cell control of thrombosis. BMC Cardiovasc Disord 15:130. https://doi.org/10.1186/s12872-015-0124-z
2. Hamilos M, Petousis S, Parthenakis F (2018) Interaction between platelets and endothelium: from pathophysiology to new therapeutic options. Cardiovasc Diagn Ther 8(5):568–580. https://doi.org/10.21037/cdt.2018.07.01
3. Pober JS, Sessa WC (2014) Inflammation and the blood microvascular system. Cold Spring Harbor Perspect Biol 7(1):a016345. https://doi.org/10.1101/cshperspect.a016345
4. Verweij FJ, Revenu C, Arras G, Dingli F, Loew D, Pegtel DM, Follain G, Allio G, Goetz JG, Zimmermann P, Herbomel P, Del Bene F, Raposo G, van Niel G (2019) Live tracking of inter-organ communication by endogenous exosomes in vivo. Dev Cell 48(4):573–589.e574. https://doi.org/10.1016/j.devcel.2019.01.004
5. Kolka CM, Bergman RN (2012) The barrier within: endothelial transport of hormones. Physiology 27(4):237–247. https://doi.org/10.1152/physiol.00012.2012
6. Collado MS, Cole BK, Figler RA, Lawson M, Manka D, Simmers MB, Hoang S, Serrano F,

Blackman BR, Sinha S, Wamhoff BR (2017) Exposure of induced pluripotent stem cell-derived vascular endothelial and smooth muscle cells in coculture to hemodynamics induces primary vascular cell-like phenotypes. Stem Cells Transl Med 6(8):1673–1683. https://doi.org/10.1002/sctm.17-0004

7. Vanhoutte PM, Shimokawa H, Feletou M, Tang EH (2017) Endothelial dysfunction and vascular disease—a 30th anniversary update. Acta Physiol 219(1):22–96. https://doi.org/10.1111/apha.12646
8. Hsieh PCH, Davis ME, Lisowski LK, Lee RT (2006) Endothelial-cardiomyocyte interactions in cardiac development and repair. Annu Rev Physiol 68:51–66. https://doi.org/10.1146/annurev.physiol.68.040104.124629
9. Gui F, You Z, Fu S, Wu H, Zhang Y (2020) Endothelial dysfunction in diabetic retinopathy. Front Endocrinol 11:591. https://doi.org/10.3389/fendo.2020.00591
10. Tabit CE, Chung WB, Hamburg NM, Vita JA (2010) Endothelial dysfunction in diabetes mellitus: molecular mechanisms and clinical implications. Rev Endocr Metab Disord 11(1):61–74. https://doi.org/10.1007/s11154-010-9134-4
11. Williams IM, Wu JC (2019) Generation of endothelial cells from human pluripotent stem cells. Arterioscler Thromb Vasc Biol 39(7):1317–1329. https://doi.org/10.1161/ATVBAHA.119.312265
12. Lian X, Bao X, Al-Ahmad A, Liu J, Wu Y, Dong W, Dunn KK, Shusta EV, Palecek SP (2014) Efficient differentiation of human pluripotent stem cells to endothelial progenitors via small-molecule activation of WNT signaling. Stem Cell Rep 3(5):804–816. https://doi.org/10.1016/j.stemcr.2014.09.005
13. Giacomelli E, Bellin M, Orlova VV, Mummery CL (2017) Co-differentiation of human pluripotent stem cells-derived cardiomyocytes and endothelial cells from cardiac mesoderm provides a three-dimensional model of cardiac microtissue. Curr Protoc Hum Genet 95(1):21.29.21–21.29.22. https://doi.org/10.1002/cphg.46
14. Palpant NJ, Pabon L, Friedman CE, Roberts M, Hadland B, Zaunbrecher RJ, Bernstein I, Zheng Y, Murry CE (2017) Generating high-purity cardiac and endothelial derivatives from patterned mesoderm using human pluripotent stem cells. Nat Protoc 12(1):15–31. https://doi.org/10.1038/nprot.2016.153
15. Chen KG, Mallon BS, McKay RDG, Robey PG (2014) Human pluripotent stem cell culture: considerations for maintenance, expansion, and therapeutics. Cell Stem Cell 14(1):13–26. https://doi.org/10.1016/j.stem.2013.12.005

Part IV

Making Use of Publically Existing Datasets and Resources

Chapter 28

Finding and Verifying Enhancers for Endothelial-Expressed Genes

Alice Neal, Helena Rodriguez-Caro, and Sarah De Val

Abstract

Identification and analysis of enhancers for endothelial-expressed genes can provide crucial information regarding their upstream transcriptional regulators. However, enhancer identification can be challenging, particularly for people with limited access or experience of bioinformatics, and transgenic analysis of enhancer activity patterns can be prohibitively expensive. Here we describe how to use publicly available datasets displayed on the UCSC Genome Browser to identify putative endothelial enhancers for mammalian genes. Furthermore, we detail how to utilize mosaic Tol2-mediated transgenesis in zebrafish to verify whether a putative enhancer is capable of directing endothelial-specific patterns of gene expression.

Key words Endothelial-expressed genes, Genome, Enhancers, Promoter, ChIP

1 Introduction

Eukaryotic genes typically contain two types of transcriptional regulatory DNA elements: core promoters proximal to the transcriptional start site and more distal regulatory elements including enhancers, insulators, and silencers [1]. Regulation of complex spatio-temporal patterns of gene transcription during mammalian development is often accomplished by enhancers, although those located close to core promoters are sometimes referred to as proximal promoters. A single gene can be regulated by multiple different enhancers, each responsible for a particular aspect of expression [2].

Enhancers are densely clustered aggregations of transcription factor binding motifs able to influence transcription independent of location, distance, or orientation [3]. Identification and analysis of enhancers can prove a direct link between a transcription factor which binds the enhancer and expression of a particular gene. Additionally, detailed analysis of the sequences within a specific enhancer can help identify a cohort of transcription factors required

Andrew V. Benest (ed.), *Angiogenesis: Methods and Protocols*, Methods in Molecular Biology, vol. 2441,
https://doi.org/10.1007/978-1-0716-2059-5_28,

for particular patterns of gene expression (e.g., analysis of an arterial-specific enhancer can reveal factors which collectively regulate arterial differentiation). Analysis of enhancers containing disease associated SNPs (which are most frequently located in non-coding DNA) can also provide crucial insight into both normal development and disease-causal pathways.

Enhancers are typically found relatively close to their target gene promoter, with a recent genome wide analysis of over 600 enhancer-core promoter pairs finding an average distance of 24 kb between enhancer and promoter [4]. However, enhancers can be located upstream, downstream, within introns or even within exons of both target and neighboring genes. Additionally, although apparently less common, enhancers can influence reporter gene expression from much greater distances, and there are many examples of enhancers located around 1 MB from their target gene (e.g., the Shh limb enhancer [5] and the Coup-TFII endothelial enhancer [6]). Consequently, identifying the enhancer (s) regulating your gene of interest in endothelial cells can be a challenging task. However, some hallmarks of enhancers can assist in this endeavor. Active enhancers are usually in more open regions of the chromatin, which can be easily identified by hypersensitivity to DNase digest (DNase-seq or DNase HS), by identifying regions that fail to cross-link to nucleosomes (FAIRE-seq) or by identifying regions with increased accessibility to transposases (ATAC-seq) [7]. Enhancers are also associated with increased levels of the specific histone modifications H3K4Me1 and H3K27Ac, and an absence of the histone modification H3K4Me3 (which instead is associated with promoters). Additionally, enhancers are often bound by EP300 and cell-type specific transcription factors. For example, nearly all known endothelial enhancers are bound by members of the ETS transcription factor family [3, 8]. Direct enhancer-promoter binding can also be detected by analysis of 3D proximity (e.g., chromatin conformation capture techniques) [9], and enhancer-transcribed RNA (eRNAs) can be detected through RNA-seq [10]. However, none of these characteristics either together or alone appears to be able to reliably identify functional enhancers every time, nor can they indicate the expression pattern regulated by the enhancer. Consequently, identification of putative enhancers ideally needs to be combined with some level of in vivo validation to determine both the spatial (e.g., does the enhancer drive ubiquitous or tissue-specific expression) and temporal (e.g., is the enhancer active during a particular developmental window?) characteristics of each putative enhancer.

In this protocol paper, we describe simple ways to identify putative endothelial gene enhancers for those with limited knowledge or access to bioinformatic assistance, using publicly available data displayed on the UCSC Genome Browser. Furthermore, we detail how to verify that these enhancers are able to direct endothelial gene expression in vivo using Tol2-mediated zebrafish transgenesis, which combines cost effectiveness with speed and relative ease of access.

2 Materials

2.1 Non-standard Equipment

1. Programmable Microinjector and micromanipulation system mounted on an inverted microscope (e.g., Eppendorff Femtojet^{R} 4X and Adapter Leica® 1).
2. Capillary glass 1.0 OD × 0.58 ID × 150 L mm (Harvard apparatus e.g., 30-0020/Clark Electromedical Instruments GC100F-15).
3. A micropipette puller (e.g., Flaming Brown P-1000 micropipette puller).
4. 96 well plate for confocal imaging (e.g., Thermo Scientific Nunc MicroWell 96-well Optical-Bottom Plates).
5. Confocal microscope.
6. Access to Sanger sequencing (e.g., Eurofinsgenomics custom DNA sequencing services).
7. 2 Fine tipped forceps (e.g., Jewelers forceps, Dumont No.5).
8. E1b-GFP-Tol2 plasmid (Addgene plasmid).
9. Expression plasmid for production of Tol2 transposase (e.g., pCS2FA CO Tol2 TPase Addgene).
10. pCR™8/GW/TOPO/TA Cloning Kit with One Shot™ TOP10 E. Coli (ThermoFisher Scientific).
11. Gateway™ LR Clonase™ II Enzyme mix (ThermoFisher Scientific).
12. Mini prep kit (e.g., QIAprep Spin Miniprep kit).
13. Midi prep kit (e.g., QIAGEN plasmid plus midi kit).
14. Various restriction enzymes (New England BioLabs).
15. Antibiotics: Spectinomycin and Ampicillin.
16. mMessage mMachine™ SP6 Transcription Kit (ThermoFisher Scientific).
17. RNA purification columns (e.g., Monarch RNA purification Columns NEB T2010).
18. RNAse free H_2O.

NaCl—34.8 g (5 mM).

KCl—1.6 g (0.17 mM).

5.8 g $CaCl_2{\cdot}2H_2O$ (0.33 mM).

9.78 g $MgCl_2{\cdot}6H_2O$ (0.33 mM).

2.2 Non-standard Solutions

2.2.1 60× Embryo Medium (2 L of 60× solution)

Dissolve in H_2O to a final volume of 2 L.

Adjust pH to 7.2 with NaOH. Autoclave and store at room temperature.

2.2.2 1× Embryo Medium (1 L)

Dilute 16.5 mL 60× Embryo medium to 1 L.

To prevent fungal growth add 100 μL of 1% methylene blue.

2.2.3 Tricaine Stock Solution

400 mg Ethyl 3-aminobenzoate methanesulfonate (MS-222/Tricaine) (Sigma-Aldrich).

98 mL ddH_2O.

2 mL 1 M Tris base (pH 9).

Adjust pH to ~7.

Store at −20 °C.

2.2.4 Tricaine Working solution

5 mL tricaine stock solution.

45 mL embryo medium.

3 Methods

3.1 Identifying Putative Endothelial Enhancers In Silico Using Publicly Available Datasets Displayed on UCSC Genome Browser

3.1.1 Accessing Data Readily Available on UCSC Genome Browser to Identify Human Endothelial Enhancers

While there is a myriad of ways of acquiring and interrogating information about the enhancer-associated characteristics for your gene loci of interest, one of the most accessible ways is to use standard tracks and data from the ENCODE project [11] available on the UCSC Genome Browser. Currently, access to required data tracks is best using the Human Feb. 2009 (GRCh37/hg19) assembly. This approach will find human putative enhancer regions, although orthologous regions for other species (e.g., mouse) can easily be inferred from this data where sequence conservation exists. In addition to the protocol here, we have also provided a public session (*see* **Note 1**).

1. At https://genome.ucsc.edu/index.html, click on "Genomes" in the toolbar, select "Human" and build "Feb. 2009 (GRCh37/hg19)" from the drop-down menu. Alternative European (https://genome-euro.ucsc.edu/index.html) and Asian (https://genome-asia.ucsc.edu/index.html) mirror sites are available.
2. Type gene of interest or co-ordinates of the region of interest into the central search bar. Because enhancers can be identified away from the gene of interest, begin your enhancer search by focusing on 20–80 kb at either side of the transcriptional start

site (TSS). This can later be widened if necessary. Our example in Fig. 1 uses a 70 kb region around *EPHB4* (chr7: 100,385,001-100,455,000) (*see* **Note 2**).

3. After selecting "default order" and "hide all" (Fig. 1A) in the display configuration tab, use the track display controls to access required information:
   (a) Gene(s) of interest: Under "Gene and Gene Predictions" go to "UCSC Genes" and switch from "hide" to "pack" (Fig. 1B). You will now see all protein-coding and non-coding RNA genes, including splice variants and predictions. For a simpler display, click on "UCSC Genes" to access the configuration page and de-select "non-coding genes" and "splice variants", although this will remove potential e-RNAs from view.
   (b) Enhancer-associated histone modifications: Under "Regulation" go to "ENCODE Regulation" (Fig. 1C), switch to "show" and click through to access configuration page. From there, select "Layered H3K4Me1," "Layered H3K4Me3," and "Layered H3K27Ac" subtracks and switch to "full" display mode. Click through "Layered H3K4Me1" to access settings page, where you can specifically select HUVEC and untick all others. Repeat for H3K4Me3 and H3K27Ac subtracks. Light blue enrichment peaks indicate enhancer-associated H3K4Me1 and H3K27Ac and promoter-associated H3K4Me3 histone marks in HUVECS for your region of interest (Fig. 1D) (*see* **Note 3**).
   (c) Enhancer-associated open chromatin regions: Under "Regulation" go to "ENC DNase/FAIRE" (Fig. 1E), switch to "show" then click the link to access configuration page. Select "UW DNAseI DGF" subtrack (which provides DNase I Digital Genomic Footprinting from ENCODE/University of Washington and includes a selection of endothelial cell lines) and switch to "dense." Leave all other subtrack options as "hide," click the link to UW DNaseI DGF settings page and select cell lines of interest (*see* **Notes 4** and **5**). Ensure maximum display mode is set to "dense" and return to the main genome browser. The DNase HS footprint will be displayed as a greyscale heatmap (*see* Fig. 1F).
   (d) Sequence conservation: Under "Comparative Genomics," set "Conservation" to "pack," and click through to track settings (Fig. 1G). It is up to you to decide which species are important for your research. We find that mouse, tenrec, opossum, chicken, *X. tropicalis*, and zebrafish are good choices for comparative genomics. Select the

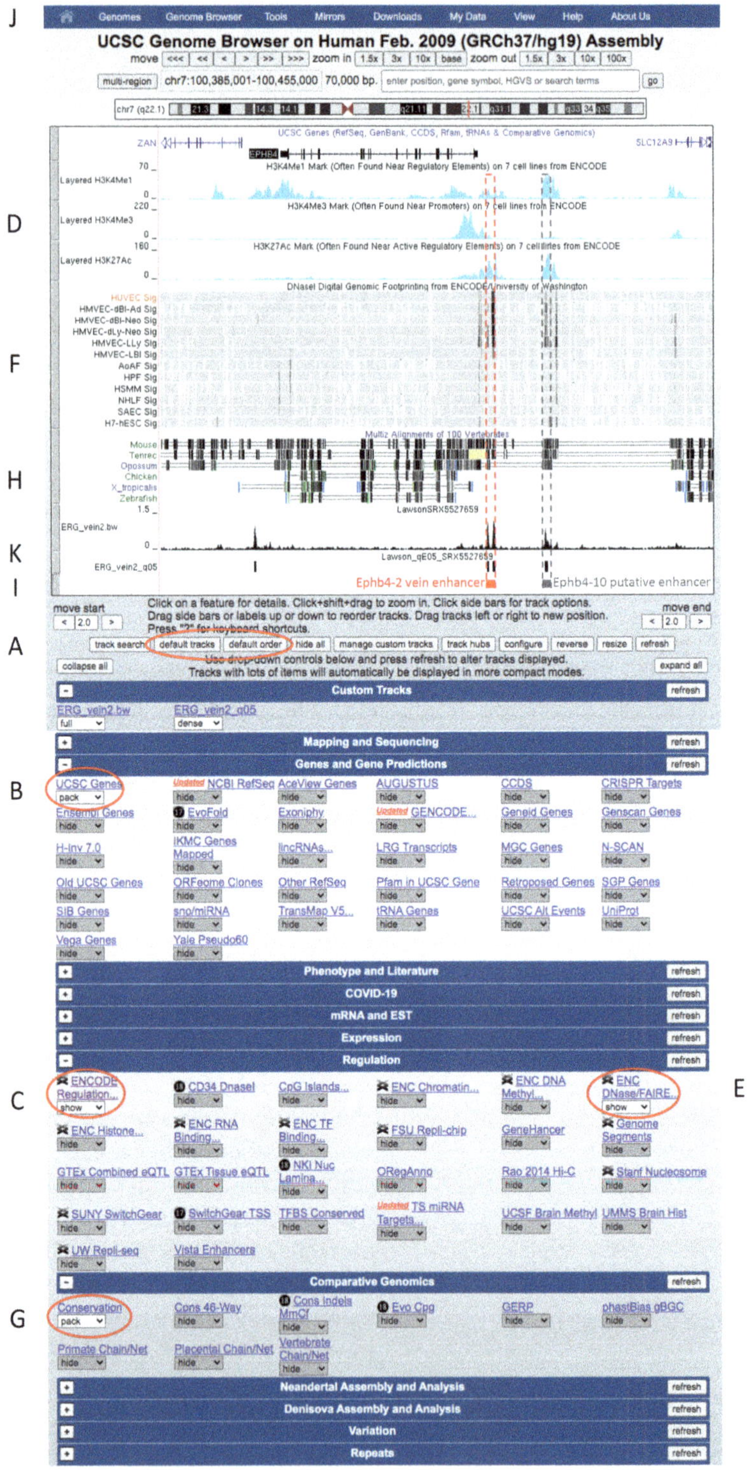

Fig. 1 Screenshot of the USCS browser showing relevant portions discussed in text. Ephb4-2 and Ephb4-10 enhancers are described in detail in [6]. ERG ChIP-seq data comes from [13]

species of interest, set Multiz Alignments to "pack" and leave the other options as "hide." Conservation will be displayed on browser by relative intensity or scarcity of black vertical lines, each indicating a conserved nucleotide (Fig. 1H).

(e) ChIP-seq data for transcription factor binding patterns: While the UCSC Genome Browser main page includes publicly available ChIP-seq data tracks, there are currently few directly relevant to endothelial enhancer identification. However, you can browse ChIP-seq data via "ENCODE Regulation" through the "TF ChIP" subtrack or by searching for specific transcription factors or cell lines using the "track search" feature in the display configuration buttons in the main browser page.

3.1.2 Using Datasets Accessed via ChIP-Atlas, the Sequence Read Archive (SRA) or via the Gene Expression Omnibus (GEO)

While the standardized tracks on UCSC can be very informative, there are increasing amounts of public ChIP-seq data available that can also assist in the identification of enhancers. In particular, all investigated endothelial enhancers contain binding motifs for the ETS family of transcription factors, and in virtually all cases these are essential for the activity of the enhancer [8]. Consequently, publicly available ChIP-seq datasets detailing peaks of different ETS factors can be used alongside histone marks and open chromatin patterns to aid the identification of functional enhancers. Examples of ETS factor ChIP-seq datasets in the public domain include ETS1 [12], ERG [13, 14], FL1 [15], and ETV2 [16]. Additionally, there are a number of ChIP-seq datasets for EP300 (a transcriptional coactivator bound to most tissue specific enhancers) binding in endothelial cells (e.g., [14] which compares HUVEC with HUAEC [17], which uses human pulmonary arterial ECs, and [18] in Tie2 Cre+ mouse endothelial and blood cells). Additional endothelial H3K4Me1 and H3K27Ac ChIP-seq datasets are also available for a variety of human and mouse cell lines/tissues cultured under different conditions.

These datasets can be accessed via the accession numbers provided within the published papers (usually through GEO https://www.ncbi.nlm.nih.gov/geo/). However, we find the simplest way to find the required ChIP-seq is often via ChIP Atlas https://chip-atlas.org/. It is important to note that the quality of the ChIP data available on both ChIP-Atlas and GEO is highly variable, therefore datasets must be selected and interpreted with caution after considering the methods sections of the relevant papers.

1. To find ChIP-seq datasets using ChIP-Atlas, select "Dataset Search" from the top toolbar and enter transcription factor of interest and/or cell type of interest into the search window. Click on the SRX ID of each entry to access details about each

dataset (e.g., culture conditions, cell line). Once the appropriate ChIP-seq sample is selected, you can visualize the data directly on IGV (Integrative Genomics Viewer) as a BigWig file (to visualize a histogram) or a peak-called Bed file (to see the location of statistically significant binding sites). Alternatively, you can note the SRXID number and genome build for use as a custom track on UCSC browser or note the GEO ID number to access and download the data directly from GEO.

2. To access ChIP-seq datasets using GEO, enter the accession number provided within the paper into the search window. BigWig, BED, and other files can then be downloaded directly from the results page. Alternatively, click through the SRA number (often provided in the "Relations" section on the GEO results page) to access the SRXID number for each run. Note that data downloaded from GEO is often of higher quality than raw reads accessed via SRX.

3. To visualize ChIP-seq data on IGV, first open an IGV browser window (this can be downloaded from https://software.broadinstitute.org/software/igv/). You can now directly look at data previously downloaded from GEO or ChIP-Atlas by going to "file" and selecting "download file." Alternatively, return to your page of interest in ChIP-Atlas, select "Visualize" and chose either BigWig or Peak Call at the required statistical significance. The relevant track will automatically open on your IGV browser.

4. To visualize ChIP-seq data on UCSC Browser alongside your histone and DNase HS data, return to your UCSC browser and click on "My Data" in the top toolbar (Fig. 1I), and select "Custom Tracks." Ideally, you want to visualize both the BigWig data (to see the pattern of binding as a histogram) and the BED files (which will show the statistically significant peak calls). You can upload a file previously downloaded from GEO or elsewhere or paste a URL from ChIP-Atlas (*see* **Notes 6–8**).

5. Once custom tracks are loaded on UCSC Browser, you can alter the way the ChIP-seq information is displayed by right-clicking within the track and selecting "configure." "Dense" displays data as a heatmap and "full" displays data as a histogram (*see* **Note 9** and Fig. 1J–K). Always check that the data looks similar to that shown in the original paper, as mislabelling of GEO deposited data can easily occur accidently. It is important to remember that local peak heights are not comparable across different ChIP-seq data tracks, as they depend heavily on experimental design and analysis pipelines. ENCODE data is the exception to this rule as it includes rigorous controls and has been processed using standardized pipelines that ensure data quality and replicability [11].

3.1.3 Selecting Your Putative Human Endothelial Enhancers

1. Use the information now displayed on the Genome Browser (and/or IGV) to identify putative endothelial enhancers for your gene of interest. Putative endothelial enhancers should be defined as regions enriched in endothelial-specific H3K4Me1 and H3K27Ac marks alongside EC-specific patterns of DNase hypersensitivity and will likely be bound by ETS factors and EP300 in endothelial cells. In Fig. 1, the validated Ephb4-2 venous enhancer is marked in red, while a second putative enhancer is marked in grey (Fig. 1I).
2. Enhancers can be located either within a gene (usually in introns but occasionally in exons) or either 5′ upstream or 3′ downstream. Additionally, endothelial genes are often regulated by more than one endothelial enhancer. Consequently, it is advised to test all putative enhancers within the gene loci. In cases where the neighboring genes are not expressed in endothelial cells (this can be detected by lack of endothelial H3K4Me3 and/or absence of DNAse HS in endothelial cells lines), all putative enhancer regions within at least 100 kb either side of the TSS should be tested, and further if obvious putative enhancers are observed. We have found that enhancers are often more distal from the TSS when the gene has no neighboring genes or contains large introns.
3. Once identified, you need to obtain the full DNA sequence of your human endothelial enhancer. While enhancers have no well-defined size limits, validation in zebrafish works optimally with putative enhancers between 100 and 900 bp. To obtain the sequence of your putative enhancer in UCSC browser, center your region of interest using DNase HS region, zoom in until you are displaying the desired size of DNA, select "View" from the toolbar at the top of the page (Fig. 1I), "DNA" in the dropdown menu, and click "get DNA" button at the bottom of the window.

3.1.4 Finding Endothelial Enhancers for Other Species

A considerable number of publicly available ChIP-seq datasets are now available for enhancer-associated traits in many species, particularly mouse and zebrafish.

1. The UCSC Browser Mouse July 2007 (NCBI37/mm9) Assembly includes a "LICR Histone" track (in the "Expression and Regulation" section) with H3K4Me1, H3K4Me3, and H4K27Ac marks for a number of mouse tissues including embryonic and adult heart.
2. Other public ChIP-seq datasets that can be accessed via GEO or ChIP-Atlas, including ETS factor and EP300 binding in mouse endothelial cells (e.g., [16, 18]), TF binding and histone marks in embryonic mouse cardiac cells [19], and ATAC-seq in endothelial vs non endothelial cells in zebrafish [20].

3. If your human enhancer region is conserved between species, the orthologous enhancer sequence from an alternative species can be obtained from the Multiz Alignment track in the human UCSC genome browser. Begin at the most conserved 100–200 bp region of your human enhancer and click directly on the alignment display (Fig. 1H) to go to the "Multiz Alignments of 100 Vertebrate" track page. Select a region of high conservation, find the required species and select the "B" at the far left-hand side to go to the orthologous region on the UCSC genome browser. Alternatively, clicking "D" on the Multiz Alignment will take you directly to the "Get DNA" page, but you will have to adjust the genome coordinates to get the entire putative enhancer sequence.

3.2 Validating Enhancer Activity in Zebrafish

While the exact definition of a functional enhancer can vary, enhancers are generally considered validated in vivo if they are shown capable of affecting gene expression levels. This is usually done either by cloning the putative enhancer upstream of a reporter gene and generating transgenic animals, or through germline deletion or mutation of the enhancer followed by analysis of expression levels of the assumed target gene [21]. In the case of endothelial enhancers, an in vivo validated endothelial-expressed enhancer would need to either drive reporter gene expression (e.g., GFP) in endothelial cells or show reduced endothelial gene expression after germline deletion/mutation. Here, we detail how to use Tol2-driven mosaic transgenic zebrafish embryos [22, 23] to validate enhancers. Both we and others have used this to successfully validate many endothelial enhancers, both using mammalian and teleost enhancer sequences [20, 24].

3.2.1 Gateway Cloning to Create an Enhancer:E1b: Reporter Tol2 Plasmid

First, you need to use pCR™8/GW/TOPO to create an entry vector plasmid containing your putative enhancer region:

1. Perform PCR with forward and reverse primers designed to amplify the putative enhancer region.
2. Using 10 μL of PCR product, check the amplified region is the expected size using agarose gel electrophoresis and a standard DNA 1 kb ladder.
3. Using the pCR™8/GW/TOPO TA Cloning Kit, combine

   2 μL of PCR product (your putative enhancer).

   1 μL of pCR™8/GW/TOPO plasmid.

   1 μL of salt solution (from kit).

   2.5 μL of ddH_2O.

   Incubate at room temperature for 10 min.

4. Transfer 2 μL of pCR™8-enhancer/GW/TOPO mixture to a thawed vial of TOP10-competent *E. coli*. Mix well by gently flicking the bottom of the tube. Incubate on ice for 10 min.
5. For transfection by heat shock, place *E. coli* in a 37.5 °C water bath for 2 min, immediately on ice for 1 min, then add 200 μL of L.B broth to *E. coli* and incubate in a 37.5 °C shaking incubator for 30 min.
6. Plate transfected *E. coli* by spreading the entire 250 μL volume onto a pre-warmed agar plate containing 50 μg/mL spectinomycin.
7. Incubate plates overnight at 37 °C or until there are clearly visible but distinct colonies.
8. Using a pipette tip, pick 6 colonies and transfer each to an individual 15 mL plastic canonical tube containing 2.5 mL LB broth with 50 μg/mL spectinomycin, 1 colony per tube. Grow overnight in a 37 °C shaking incubator.
9. Perform DNA miniprep on bacteria from all overnight cultures to isolate plasmid DNA.
10. Assess putative enhancer integration into pCR™8/GW/TOPO plasmid by restriction enzyme digestion (*see* **Note 10**).
11. Select 2 pCR™8-enhancer/GW/TOPO plasmids containing the correct enhancer insert and Sanger sequence to check for any errors that may be introduced by PCR amplification.

    Now that your putative enhancer sequence is in the pCR™8/GW/TOPO Gateway compatible entry vector, you can create an Tol2-enhancer:E1b:GFP-Tol2 expression plasmid. E1b is a minimal promoter commonly used to validate enhancer activity in zebrafish transgenesis, as it is does not drive notable expression of GFP alone. An E1b:GFP-Tol2 plasmid can be obtained from Addgene (*see* Subheading 2). However, it should be noted that some researchers have reported better enhancer activity when using the endogenous promoter, particularly when looking at teleost enhancers [13].

    The Tol2-enhancer:E1b:GFP-Tol2 expression plasmid is created using LR clonase:

12. Combine

    1 μL of 37 ng/μL pCRTM8:enhancer/GW/TOPO mid.

    1 μL of 37.5 ng/μL pE1B:GFP-Tol2 expression plasmid.

    0.5 μL GatewayTM LR ClonaseTM II Enzyme mix.

    Spin briefly.

    Incubate at room temperature for 1 h.

13. To stop LR Clonase™ activity, add 0.5 μL of proteinase K (from Invitrogen LR clonase kit) and incubate at 55 °C for 15 min.

14. Transfect plasmid into TOP10 competent *E. coli* by repeating above **steps 4–8** replacing spectinomycin with ampicillin throughout.
15. Perform a DNA miniprep from bacteria from all overnight cultures.
16. Assess integration of putative enhancer region into E1b:GFP-Tol2 expression plasmid by restriction digest. The LR reaction is directional, so insert will be in the same orientation as in the pCR™8-enhancer/GW/TOPO plasmid.
17. Select an overnight culture that contains the desired plasmid and midiprep for a greater concentration of plasmid.
18. Sequence midiprep of your putative enhancer:E1b:GFP-Tol2 plasmid.
19. Store at −20 °C.

3.2.2 Making Tol2 mRNA

1. Linearize your Tol2 expression plasmid by restriction enzyme digestion and column purify linearized plasmid.
2. Using the mMessage mMachine™ SP6 Transcription Kit combine:

   10 μL 2× NTP/CAP.

   2 μL 10× reaction buffer.

   2 μL enzyme mix.

   1 μg linear expression plasmid.

   RNase-free H_2O to 20 μL.

   Mix, spin and incubate for 2 h at 37 °C.
3. Add 1 μL of TURBO DNase. Incubate for 15 min at 37 °C.
4. Column purify RNA and measure concentration on spectrophotometer.
5. Check mRNA is not denatured by running ~100 ng mRNA on a 1% agarose gel in 0.5× TBE buffer (*see* **Note 11**).
6. Dilute to 1 μg/μL in RNAse free H_2O. 1 μg/μL aliquots are enough for the injection of 2 constructs.
7. Store aliquots at −80 °C.

3.2.3 Creating Enhancer: E1b:GFP-Tol2 Transgenic Zebrafish Embryos

You are now ready to inject fertilized zebrafish eggs with a combination of your putative enhancer:E1b:GFP-Tol2 plasmid and Tol2 mRNA. These embryos will become F0 transgenic zebrafish (sometimes referred to as transient transgenic zebrafish). Remember that these embryos will be mosaic, as Tol2-induced transgenesis is not uniform.

This protocol requires long taper microinjection pipettes, normally made in-house by pulling capillary glass in a micropipette puller. It is important that this step is optimized for the specific capillaries/capillary puller/microinjector set up being used, as a good needle is crucial for injection success. Needles with a very thin shaft will break easily and be difficult to use, while needles with a shaft that is too thick will damage the chorion as they enter (*see* **Note 12**).

1. Thaw 1 μg/μL aliquot of Tol2 mRNA and add 9 μL of RNAse free H_2O to create a 100 ng/μL solution.
2. Combine Enhancer:E1b:GFP-Tol2 plasmid with Tol2 mRNA for a 10 μL solution of 60 ng/μL plasmid 50 ng/μL Tol2mRNA (e.g., 5 μL of 120 ng/μL plasmid and 5 μL of 100 ng/μL mRNA).
3. Load the mixture into a fine tipped microinjection needle, break the tip of the needle with fine-tipped forceps, and mount on micromanipulation system.
4. Using a calibration micrometre slide, adjust injection pressure and time on microinjector controls to deliver an injection volume of 0.5 nL.
5. Collect zebrafish eggs immediately after fertilization and arrange eggs in a single line along the long edge of a microscope slide in a petri dish orientated such that the single cell of the embryo is pointing away from the slide.
6. Inject plasmid mRNA mixture directly into the single cell. The needle should pierce the chorion and directly enter the cell but should not go through the yolk.
7. Once the line has been injected collect embryos with a plastic Pasteur pipette and transfer to a petri dish. Keep the embryos in embryo medium at 27 °C until analysis.

3.2.4 Analyzing Zebrafish Embryos

You can now assess putative enhancer activity by looking for GFP expression in the injected zebrafish embryos as they develop. We usually restrict our analysis to 22–96 h post fertilization (hpf), although later time points will be required if you are interested in the lymphatic vasculature. In our experience, validated mammalian enhancers (e.g., ones that also drive expression in mouse transgenic embryos) usually drive GFP expression in between 10% and 70% of injected zebrafish. Remember these F0 *tg(enhancer:E1b:GFP)* zebrafish will have only mosaic integration of the transgene, so GFP activity will not be in every endothelial cell. Figure 2 shows an example of an F0 *tg(Ephb4-2:E1b:GFP)* zebrafish, but note that sometimes GFP expression may be restricted to a small number of endothelial cells.

1. If imaging prior to 48 hpf, embryos require manually dechorionating. Keeping zebrafish embryos in E3 medium, use fine tipped forceps and a suitable microscope to tear a small hole in

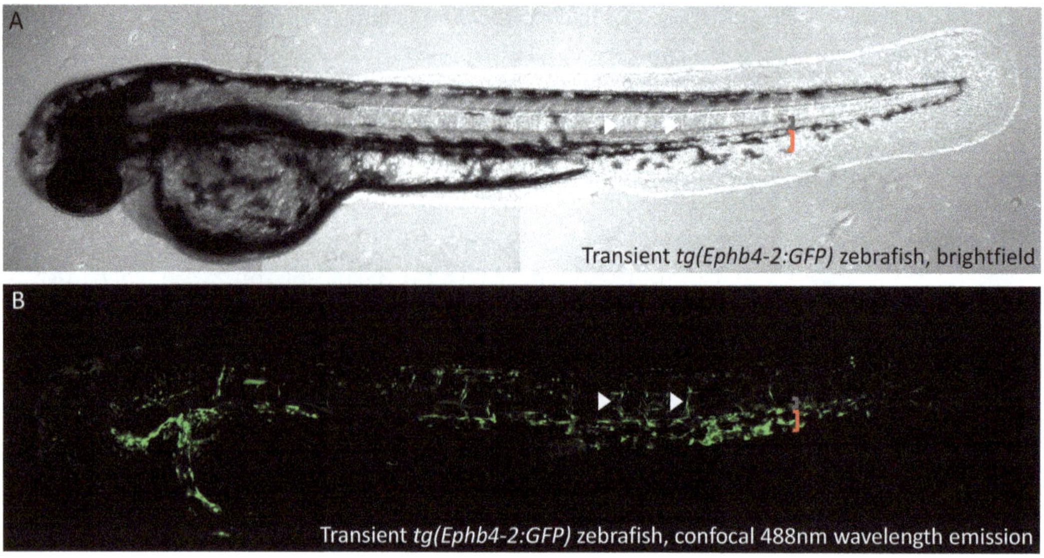

Fig. 2 (**a**) Brightfield and (**b**) GFP (488 nm wavelength emission) images of an anesthetized transgenic zebrafish embryo at 48 h post-fertilization (hpf). This embryo was injected at the 1 cell stage with an E1b-GFP-Tol2 expression plasmid containing the Ephb4-2 enhancer [6] along with Tol2 mRNA, making an F0 *tg(Ephb4-2:E1b:GFP)* transgenic zebrafish. The GFP signal demonstrates that the Ephb4-2 enhancer can drive gene expression in the venous endothelium. Red brackets indicate the caudal vein expressing GFP, grey brackets indicate the aorta which is GFP negative, white arrow heads indicate GFP expression in venous intersegmental vessels

the chorion and release the embryo. Care must be taken not to apply excessive force as this can easily damage the embryo.

2. Anesthetize fish by immersing in tricaine working solution and use a plastic Pasteur pipette place 1 fish in each well of a 96 well plate suitable for confocal microscopy. Ensure fish remain submerged in tricaine throughout.
3. Image fish on a confocal microscope. Multiple Z-stacks will be required to assess GFP reporter gene expression in the zebrafish vasculature.
4. When imaging is complete, remove fish from 96 well plates back into E3 medium and keep at 27 °C. Fish will recover quickly and can be re-anesthetized and re-imaged as often as required.

4 Notes

1. We have created a UCSC public session with the basic tracks for endothelial enhancer identification. This session contains the UCSC genes, histone modifications, DNA accessibility, and comparative genomics tracks described here as well as relevant

Chip-seq custom tracks of HUVEC cells extracted from Chip-Atlas. To access this public session, click on "My Data" in the toolbar, then click "Public Sessions," and search for it writing the session name (AngiogenesisProtocol_2021_DeVal) or the author (HRC) in the search space on the right. This public session is available at the main UCSC site and the European mirror site.

2. The configuration of the USCS browser can be altered (e.g., to increase font size, width of image or label location) by selecting "view" from the top tool bar, then selecting "configure browser."
3. It is often very helpful to compare the histone marks in endothelial cells with those of other cells. To do this, go into the track settings page for each histone mark (e.g., Layered H3K4Me1 Track Settings, accessed from "ENCODE regulation") and select any other cell line. The marks will be displayed in the color indicated next to the cell line name.
4. The endothelial cell lines on offer include HUVEC (human umbilical vein endothelial cells), HMVEC-dBl-Ad (human dermal derived adult blood microvascular endothelial cells), HMVEC-dBL-Neo (human dermal derived neonatal blood microvascular endothelial cells), HMVEC-dLy-Neo (human dermal-derived neonatal lymphatic microvascular endothelial cells), HMVEC-LBl (human lung-derived blood microvascular endothelial cells), and HMVEC-LLy (human lung-derived lymphatic microvascular endothelial cells).
5. It is advisable to also select some non-endothelial cell lines for comparative purposes.
6. To add a BigWig file from ChIP-Atlas into UCSC as a custom track:

   track type=bigWig name=track name.bw description=track

   description

   bigDataUrl=http://dbarchive.biosciencedbc.jp/kyushu-

   u/hg19 or hg38 /eachData/bw/SRXID.bw

7. To add a BigBed file from ChIP-Atlas into UCSC as a custom track (red text is for you to change as needed, green text can be

adjusted for required q):

track type=bigBed name=track name_q05

description=track description

bigDataUrl=http://dbarchive.biosciencedbc.jp/kyushu-

u/hg19 or hg38/eachData/bb05/SRXID.05.bb

* q refers to the MACS2 *Q*-value threshold for statistical significance and can displayed as 05, 10, or 20. As a reference: if the MACS2 threshold of significance is set to 50, Chip-Atlas shows a q value $<1E-05$.

8. An example URL to use in UCSC browser as a custom track, using part of a dataset from [13], showing binding of the ETS factor ERG in HUVECs:

track type=bigWig name=ERG_vein2.bw

description=LawsonSRX5527659

bigDataUrl=http://dbarchive.biosciencedbc.jp/kyushu-

u/hg19/eachData/bw/SRX5527659.bw

track type=bigBed name=ERG_vein2_q05

description=Lawson_qE05_SRX5527659

bigDataUrl=http://dbarchive.biosciencedbc.jp/kyushu-

u/hg19/eachData/bb05/SRX5527659.05.bb

9. Within the "configure" window for each custom track there are also options for scaling of the *y*-axis. We often find it useful to autoscale the axis in order to view locally enriched binding, but it is important to note that the apparent size of any local peak is relevant only to what is currently within the browser window and therefore these "sizes" are not comparable across regions.
10. As pCRTM8/GW/TOPO/TA cloning is not directional it is helpful to choose enzymes that will give distinct band sizes for each integration option, that is; no insert, enhancer inserted in 5′–3′ orientation, enhancer inserted in 3′–5′ orientation. Typically, we select the 5′–3′ orientation for enhancers located in

the 5′ upstream region, and the 3′–5′ orientation for enhancers located 3′ of the TSS of the relevant gene.

11. Specific RNA gel protocols are available, but in our experience standard agarose gels are sufficient to visualize Tol2 mRNA at a high concentration. DNA/RNA dye must be added directly to mRNA.
12. As a **guide only**, Flaming/Brown micropipette puller model P-1000 with capillary glass 1.0 OD × 0.58 ID × 150 L (mm) uses the following program:
    (a) Heat 500.
    (b) Pull 110.
    (c) Velocity 75.
    (d) Time 90.
    (e) Pressure 500.
    (f) This will pull a needle of ~0.5 μm that is ~150–200 μm long.

References

1. Maston GA, Evans SK, Green MR (2006) Transcriptional regulatory elements in the human genome. Annu Rev Genomics Hum Genet 7:29–59. https://doi.org/10.1146/annurev.genom.7.080505.115623
2. Visel A, Rubin EM, Pennacchio LA (2009) Genomic views of distant-acting enhancers. Nature 461(7261):199–205. https://doi.org/10.1038/nature08451
3. Shlyueva D, Stampfel G, Stark A (2014) Transcriptional enhancers: from properties to genome-wide predictions. Nat Rev Genet 15(4):272–286. https://doi.org/10.1038/nrg3682
4. Gasperini M, Hill AJ, McFaline-Figueroa JL, Martin B, Kim S, Zhang MD et al (2019) A genome-wide framework for mapping gene regulation via cellular genetic screens. Cell 176(1–2):377–90.e19. https://doi.org/10.1016/j.cell.2018.11.029
5. Lettice LA, Heaney SJ, Purdie LA, Li L, de Beer P, Oostra BA et al (2003) A long-range Shh enhancer regulates expression in the developing limb and fin and is associated with preaxial polydactyly. Hum Mol Genet 12(14):1725–1735. https://doi.org/10.1093/hmg/ddg180
6. Neal A, Nornes S, Payne S, Wallace MD, Fritzsche M, Louphrasitthiphol P et al (2019) Venous identity requires BMP signalling through ALK3. Nat Commun 10(1):453. https://doi.org/10.1038/s41467-019-08315-w
7. Wu X, Zheng Y, Liu M, Li Y, Ma S, Tang W et al (2021) BNIP3L/NIX degradation leads to mitophagy deficiency in ischemic brains. Autophagy 17(8):1934–1946. https://doi.org/10.1080/15548627.2020.1802089
8. De Val S, Black BL (2009) Transcriptional control of endothelial cell development. Dev Cell 16(2):180–195. https://doi.org/10.1016/j.devcel.2009.01.014
9. Sewitz SA, Fahmi Z, Lipkow K (2017) Higher order assembly: folding the chromosome. Curr Opin Struct Biol 42:162–168. https://doi.org/10.1016/j.sbi.2017.02.004
10. Sartorelli V, Lauberth SM (2020) Enhancer RNAs are an important regulatory layer of the epigenome. Nat Struct Mol Biol 27(6):521–528. https://doi.org/10.1038/s41594-020-0446-0
11. Consortium EP (2012) An integrated encyclopedia of DNA elements in the human genome. Nature 489(7414):57–74. https://doi.org/10.1038/nature11247
12. Chen J, Fu Y, Day DS, Sun Y, Wang S, Liang X et al (2017) VEGF amplifies transcription through ETS1 acetylation to enable angiogenesis. Nat Commun 8(1):383. https://doi.org/10.1038/s41467-017-00405-x
13. Sissaoui S, Yu J, Yan A, Li R, Yukselen O, Kucukural A et al (2020) Genomic

characterization of endothelial enhancers reveals a multifunctional role for NR2F2 in regulation of arteriovenous gene expression. Circ Res 126(7):875–888. https://doi.org/10.1161/CIRCRESAHA.119.316075

14. Kalna V, Yang Y, Peghaire CR, Frudd K, Hannah R, Shah AV et al (2019) The transcription factor ERG regulates super-enhancers associated with an endothelial-specific gene expression program. Circ Res 124(9): 1337–1349. https://doi.org/10.1161/CIRCRESAHA.118.313788
15. Wang S, Chen J, Garcia SP, Liang X, Zhang F, Yan P et al (2019) A dynamic and integrated epigenetic program at distal regions orchestrates transcriptional responses to VEGFA. Genome Res 29(2):193–207. https://doi.org/10.1101/gr.239053.118
16. Liu F, Li D, Yu YY, Kang I, Cha MJ, Kim JY et al (2015) Induction of hematopoietic and endothelial cell program orchestrated by ETS transcription factor ER71/ETV2. EMBO Rep 16(5):654–669. https://doi.org/10.15252/embr.201439939
17. Miyagawa K, Shi M, Chen PI, Hennigs JK, Zhao Z, Wang M et al (2019) Smooth muscle contact drives endothelial regeneration by BMPR2-Notch1-mediated metabolic and epigenetic changes. Circ Res 124(2):211–224. https://doi.org/10.1161/CIRCRESAHA.118.313374
18. Zhou P, Gu F, Zhang L, Akerberg BN, Ma Q, Li K et al (2017) Mapping cell type-specific transcriptional enhancers using high affinity, lineage-specific Ep300 bioChIP-seq. elife 6: e22039. https://doi.org/10.7554/eLife.22039
19. Akerberg BN, Gu F, VanDusen NJ, Zhang X, Dong R, Li K et al (2019) A reference map of murine cardiac transcription factor chromatin occupancy identifies dynamic and conserved enhancers. Nat Commun 10(1):4907. https://doi.org/10.1038/s41467-019-12812-3
20. Quillien A, Abdalla M, Yu J, Ou J, Zhu LJ, Lawson ND (2017) Robust identification of developmentally active endothelial enhancers in zebrafish using FANS-assisted ATAC-Seq. Cell Rep 20(3):709–720. https://doi.org/10.1016/j.celrep.2017.06.070
21. Li K, Liu Y, Cao H, Zhang Y, Gu Z, Liu X et al (2020) Interrogation of enhancer function by enhancer-targeting CRISPR epigenetic editing. Nat Commun 11(1):485. https://doi.org/10.1038/s41467-020-14362-5
22. Kawakami K (2007) Tol2: a versatile gene transfer vector in vertebrates. Genome Biol 8 (Suppl 1):S7. https://doi.org/10.1186/gb-2007-8-s1-s7
23. Suster ML, Kikuta H, Urasaki A, Asakawa K, Kawakami K (2009) Transgenesis in zebrafish with the tol2 transposon system. Methods Mol Biol 561:41–63. https://doi.org/10.1007/978-1-60327-019-9_3
24. Sacilotto N, Monteiro R, Fritzsche M, Becker PW, Sanchez-Del-Campo L, Liu K et al (2013) Analysis of Dll4 regulation reveals a combinatorial role for Sox and Notch in arterial development. Proc Natl Acad Sci U S A 110(29): 11893–11898. https://doi.org/10.1073/pnas.1300805110

Chapter 29

Endothelial Cell RNA-Seq Data: Differential Expression and Functional Enrichment Analyses to Study Phenotypic Switching

Guillermo Díez Pinel, Joseph L. Horder, John R. King, Alan McIntyre, Nigel P. Mongan, Gonzalo Gómez López, and Andrew V. Benest

Abstract

RNA-seq is a common approach used to explore gene expression data between experimental conditions or cell types and ultimately leads to information that can shed light on the biological processes involved and inform further hypotheses. While the protocols required to generate samples for sequencing can be performed in most research facilities, the resulting computational analysis is often an area in which researchers have little experience. Here we present a user-friendly bioinformatics workflow which describes the methods required to take raw data produced by RNA sequencing to interpretable results. Widely used and well documented tools are applied. Data quality assessment and read trimming were performed by FastQC and Cutadapt, respectively. Following this, STAR was utilized to map the trimmed reads to a reference genome and the alignment was analyzed by Qualimap. The subsequent mapped reads were quantified by featureCounts. DESeq2 was used to normalize and perform differential expression analysis on the quantified reads, identifying differentially expressed genes and preparing the data for functional enrichment analysis. Gene set enrichment analysis identified enriched gene sets from the normalized count data and clusterProfiler was used to perform functional enrichment against the GO, KEGG, and Reactome databases. Example figures of the functional enrichment analysis results were also generated. The example data used in the workflow are derived from HUVECs, an in vitro model used in the study of endothelial cells, published and publicly available for download from the European Nucleotide Archive.

Key words Endothelial transcriptomics, RNASeq, Differential gene expression analysis, DESEQ2

1 Introduction

High-throughput RNA sequencing (RNA-seq) has become a widely used approach in the study of genome-wide transcriptomes, with common application in differential expression (DE) analyses [1]. The key steps in the preparation of RNA-seq datasets include extraction of purified, high-quality RNA, conversion to cDNA, libraries construction from cDNA fragments, and subsequent

Andrew V. Benest (ed.), *Angiogenesis: Methods and Protocols*, Methods in Molecular Biology, vol. 2441,
https://doi.org/10.1007/978-1-0716-2059-5_29,

sequencing to generate short-reads collections and raw data output files [2, 3]. The raw data files are then processed through computational methods in order to quantify the short-read counts to evaluate transcripts abundance. First, the raw data files undergo quality control to provide information on the quality of the short-reads obtained. Following this, the short-reads are mapped to a reference genome and the counts for each feature of the genome are quantified, normalized, and subjected to DE analysis [1, 4]. Differentially expressed genes (DEGs) resulting from DE analysis are those genes whose normalized read counts significantly differ between the experimental conditions, reflecting a statistically relevant change in gene expression levels between the conditions compared [5, 6]. Statistical significance is measured by a *p*-value generally adjusted with the Benjamini–Hochberg or similar methods [7] to address the randomness effects derived from multiple testing [7]. Size and direction of DE are defined as a log2 fold change, with positive fold changes signifying upregulated or more abundantly expressed DEGs and negative downregulated DEGs or those genes with lower expression. While DEGs provide useful information on the transcriptomic changes between experimental conditions as individual genes of interest, it is often the pathways, groups of genes or genesets that are enriched in the list of DEGs that lead to further biological insight. This is carried out via functional enrichment analysis that typically takes two forms: Over-Representation Analysis (ORA), which identifies gene pathways that are enriched in the list of DEGs compared to all genes in the genome and Gene Set Enrichment Analysis (GSEA) in which the normalized counts of all genes are ranked and tested to identify enriched pathways and gene sets [8].

We used multiple bioinformatics tools to analyze a publicly available RNA-seq dataset [9] containing raw reads from six human umbilical vein endothelial cells (HUVECs) samples, three per group, as an example of an RNA-seq analysis workflow applied to a recent endothelial cell (EC) study. The control samples were transduced with lentiviruses encoding FLAG-CAS9 nuclease, while cMYC KO samples were generated by the lentiviruses also encoding gRNAs targeting the *MYC* gene [9]. HUVECs are commonly used as an EC in vitro model to study blood vessels and angiogenesis due to their ease of culture and manipulation compared with other (particularly) microvessel EC populations [10]. The analysis process carried out by the researchers behind the dataset can be found in the methods section of the original publication [9]. Here, FastQC [11, 12] was used for raw data quality control and Cutadapt [13] for low-quality calls and adapter trimming. Once the quality was checked before and after trimming, the STAR [14] aligner was used for read mapping to the human reference genome (GRCh38.p13 assembly) and Qualimap [15] for mapped reads was used to assess alignment quality. Next, a gene expression matrix was generated with featureCounts [16] and the differential expression

(DE) analysis was carried out using DESeq2 [17]. Finally, functional enrichment analyses were performed in R using the clusterProfiler [18] and ReactomePA [19] packages, as well as using Gene Set Enrichment Analysis (GSEA) software [20, 21].

This work aims to provide the reader with little to no bioinformatics experience a user-friendly workflow (Fig. 1) that can be followed in order to carry out RNA-seq analysis of both publicly available datasets and their own data. Each stage of the analysis is broken down into the commands and lines of code that are used with an explanation alongside, all provided in the chronological order of the analysis process. The output of the commands/code is shown where relevant to aid with the following of the example analysis. Once the workflow has been followed, the reader should be equipped with the tools and a greater understanding of the steps required to achieve robust, interpretable results from a dataset of raw reads.

2 Materials

This example workflow runs on Ubuntu 20.04.2 LTS. Ubuntu and Bio-Linux (built on an Ubuntu base) are among the most common operating systems used in bioinformatics and they constitute beginner-friendly environments for users without any prior Linux experience (*see* **Note 1**).

2.1 Software Summary

A list with the main programs used with a brief description is shown in the following. Furthermore, Fig. 1 illustrates the general workflow followed by this protocol. Table 1 provides the URLs to the software and packages used in this analysis where the user/reference manuals can be found and referred to throughout this workflow.

2.1.1 FastQC [11, 12]

This tool allows to perform quality control of raw reads in FASTQ format resulting from sequencing by high-throughput sequencing (HTS) platforms. The output is a report with several modules assessing a variety of parameters such as sequencing quality, sequence content or the presence of adapters that should be checked before continuing with the rest of the analysis to ensure that the quality of reads is sufficient.

2.1.2 Cutadapt [13]

This tool is used to trim low-quality bases and adapters from read ends before proceeding with read mapping.

2.1.3 STAR [14]

This software is used to map reads to a reference genome. First, a genome index is created once using the reference genome sequence in FASTA file format as well as a genome annotation file in GTF or GFF format. The index can then be used to identify the location of the reads contained in the FASTQ files in the reference genome.

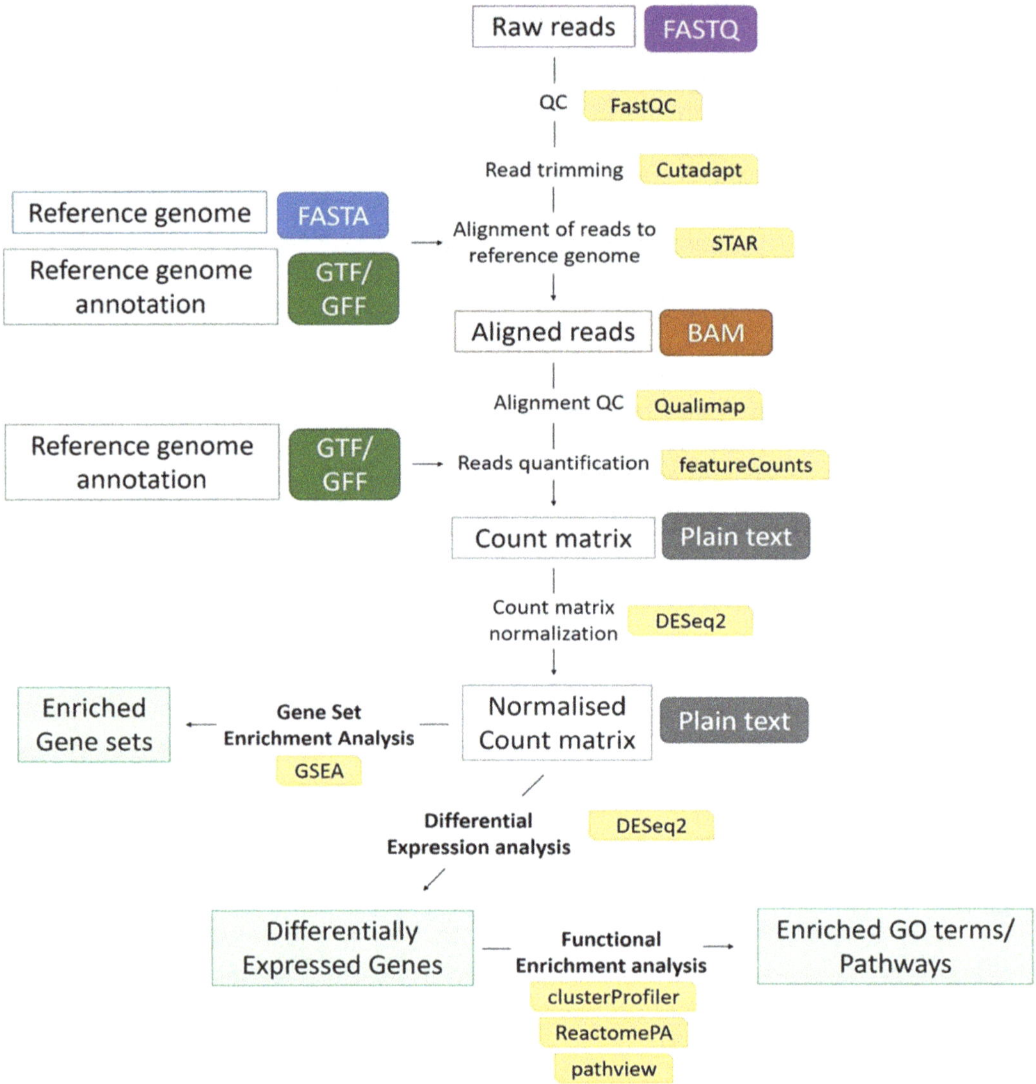

Fig. 1 Analysis workflow. The steps of workflow are shown with the tools used in yellow boxes alongside. Data required/produced in the analysis are contained in white boxes with the file format in purple, blue, dark green, orange, and grey boxes. Results obtained are shown in light green boxes

2.1.4 Qualimap [15]

A quality control tool used to assess the quality of mapped reads in SAM or BAM format. It provides the percentage of mapped reads and the transcript coverage profile.

2.1.5 featureCounts [16]

Using the GTF or GFF genome annotation file and the BAM mapped reads, featureCounts [16] generates an expression matrix containing the expression values of each sample for each gene. This software is included in the Subread package.

Table 1
Software, R [29] packages and databases used in the workflow

Software	URL
Ubuntu	https://ubuntu.com/
FastQC [11, 12]	https://www.bioinformatics.babraham.ac.uk/projects/fastqc/
Cutadapt [13]	https://github.com/marcelm/cutadapt
STAR [14]	https://github.com/alexdobin/STAR
Qualimap [15]	http://qualimap.conesalab.org/
Subread (featureCounts [16])	http://subread.sourceforge.net/
R [29]	https://www.r-project.org/
Rstudio [30]	https://www.rstudio.com/
DESeq2 [17]	https://bioconductor.org/packages/release/bioc/html/DESeq2.html
clusterProfiler [18]	https://bioconductor.org/packages/release/bioc/html/clusterProfiler.html
pathview [31]	http://www.bioconductor.org/packages/release/bioc/html/pathview.html
ReactomePA [19]	https://bioconductor.org/packages/release/bioc/html/ReactomePA.html
enrichplot [42]	https://bioconductor.org/packages/release/bioc/html/enrichplot.html
biomaRt [33, 34]	https://bioconductor.org/packages/release/bioc/html/biomaRt.html
ggplot2 [35]	https://ggplot2.tidyverse.org/
GO [22, 23]	http://geneontology.org/
KEGG [24–26]	https://www.genome.jp/kegg/
Reactome [27]	https://reactome.org/
GSEA [20, 21]	https://www.gsea-msigdb.org/gsea/index.jsp

2.1.6 DESeq2 [17]

An R-Bioconductor package which normalizes the raw counts via the median of ratios method and performs differential expression analysis based on the negative binomial distribution of the data.

2.1.7 ClusterProfiler [18] and ReactomePA [19]

These tools are used to perform functional enrichment analysis using the most upregulated and downregulated genes identified in the differential expression analysis. ClusterProfiler [18] is used in this protocol to identify overrepresented gene sets from the Gene Ontology (GO) [22, 23] and Kyoto Encyclopedia of Genes and Genomes pathway (KEGG) [24–26] databases, while ReactomePA [19] those included in the Reactome [27] pathway database. Only the gene IDs are used, not their expression values.

2.1.8 GSEA [20, 21]

GSEA [20, 21] allows the identification of differentially expressed gene sets between two biological states. These gene sets can be those present in the MsigDB [20, 28] database or custom gene sets. GSEA [20, 21] uses the expression values of all genes as an input.

3 Methods

3.1 Workspace Preparation

In order to generate the directories tree used in this workflow, we first created the directory rnaseq_exp under /home/user. This will be our main working directory. When a path containing "user" is shown (e.g., /home/user/rnaseq_exp), "user" represents the user's name and it should be replaced by it. When executable commands are displayed, they will be preceded by different symbols depending on the program they are executed in or the type of code line they are:

1. Linux terminal commands are preceded by the "$" symbol.
2. Inline comments are preceded by the "#" symbol. These are not executed but are used to describe subsequent Linux and R commands.
3. R commands are preceded by the ">" symbol.
4. R output is preceded by "##".

Open a Linux terminal and execute the commands below to build the workspace. It is key to ensure that the workspace is suitably organized, as this will allow the downstream processes and outputs to be easily run and accessible.

```
#Preparing the workspace
$ cd /home/user
$ mkdir rnaseq_exp
$ cd rnaseq_exp
$ mkdir output raw_data resources programs
```

- raw_data directory will contain the FASTQ files (*see* Subheading 2.2.10) consisting of the read raw data resulting from RNA sequencing.
- output directory will contain subsequent directories that will be created during the analysis, each storing the results of a particular piece of software or tool.
- resources directory will store other files that will be used as input that are not the raw sequencing data.

- programs directory includes programs that are not installed from Ubuntu repositories (an Ubuntu repository is a server in which program packages are stored).

3.2 Software Installation and Minimum Requirements

All software used is open source and available under the GNU General Public License. R packages were installed via the `install.packages()` function or R-Bioconductor (`BiocManager::install()` function). A 64-bit machine with at least 32 Gb of RAM is recommended for the majority of steps in the process. Some of the tools and analyses used, such as DESeq2 [17], require less. However, the RAM usage by STAR [14] for the reference genome indexing and read mapping is considerably higher, as mammal genomes require at least 16 Gb of RAM, with 32 Gb or more being the recommended amount of available memory (*see* **Note 2**). In order to install the required software and tools, follow the following steps. We have included the sudo (SuperUser domain) prefix in front of commands that require SuperUser privileges, as well as the software versions used in the protocol. If any program installation asks for permission to continue, follow the instructions printed in the terminal to accept. This often requires typing "Y" followed by enter. A message prompting the user to enter their sudo password may also appear in the terminal for some installations. Simply type the user password and press enter (the password characters will not show up in the terminal while typing).

First of all, run the below commands in order to update the package databases and therefore be able to obtain the latest versions of the packages installed using the Linux terminal:

```
#Updating package databases:
$ sudo apt-get update
```

3.2.1 FastQC [11, 12]

FastQC [11, 12] is a java application and requires Java Runtime Environment (JRE) to be installed. JRE is also required by other programs used in the analysis workflow. You can check if it is installed in your machine with the following commands:

```
#Checking JRE installation
$ java -version
```

If the output is similar to the one shown below, you are ready to install FastQC [11, 12].

```
openjdk version "11.0.10" 2021-01-19
 OpenJDK Runtime Environment (build 11.0.10+9-Ubuntu-0ubun-
```

```
tu1.20.04)
 OpenJDK 64-Bit Server VM (build 11.0.10+9-Ubuntu-0ubun-
tu1.20.04, mixed mode, sharing)
```

If an error message appears, you need to install JRE. To do so, run the installation command and check the version again:

```
#JRE installation
$ sudo apt-get install default-jdk
$ java -version
```

Once JRE is installed, install FastQC [11, 12] with the command shown below:

```
#FastQC installation
$ sudo apt-get install fastqc
```

3.2.2 Cutadapt [13]

Cutadapt [13] (package cutadapt) can be installed from Ubuntu repositories using the Linux terminal:

```
#Cutadapt v2.8 installation
$ sudo apt-get install cutadapt
```

3.2.3 STAR [14]

Open a Linux terminal and install the latest version of STAR [14] (package rna-star) with the following command:

```
#STAR v2.7.3a installation command:
$ sudo apt-get install rna-star
```

3.2.4 Qualimap [15]

Like FastQC [11, 12], Qualimap [15] has Java Runtime Environment (JRE version 6 or above) as a requirement. In addition, it also requires R [29] (version 3.1 or above). The installation of JRE and R [30] are described in the FastQC [11, 12] (Subheading 3.2.1) and R [17] (Subheading 3.2.6) installation instructions, respectively.

After those dependencies are installed, Qualimap [15] can be downloaded from http://qualimap.conesalab.org/. Download the zip file described as "Latest package for GNU Linux, MacOS and MS Windows" and place it in the /home/user/rnaseq_exp/

programs directory. Then, use a terminal to unzip and install Qualimap [15] as shown in the following.

```
#Qualimap v2.2.1 installation
$ cd /home/user/rnaseq_exp/programs
$ unzip qualimap_v2.2.1.zip
```

3.2.5 Subread (featureCounts [16])

The R subread package includes the function *featureCounts* and can be installed with the following command in a Linux terminal:

```
#Subread v2.0.0 installation commands:
sudo apt-get install subread
```

3.2.6 R [29] and Rstudio [30]

Several of the tools used are part of R packages. In order to install R [29] and Rstudio [30], follow the steps below:

1. Install R (https://cran.r-project.org/).
   (a) Select "Download R for Linux".
   (b) In the following page, select the "ubuntu/" directory.
   (c) Run the commands indicated in the section "Install R" in your terminal, prefixing sudo.
   (d) Check R installation. The R version and a welcome message should be shown. Exit R and do not save the workspace:

```
#Start R in the terminal to check the installation
$ R
# Quit R and do not save the workspace when asked
$ q()
$ n
```

   (e) Install external libraries, needed by some R packages using the following commands in a Linux terminal:

```
#R packages dependencies installation
$ sudo apt-get install libxml2-dev
$ sudo apt-get install libcurl4-openssl-dev
$ sudo apt-get install r-base-dev
$ sudo apt-get install libssl-dev
$ sudo apt-get install libopenblas-base
$ sudo apt-get install libopenblas-dev
$ sudo apt-get install libpq5
```

```
$ sudo apt-get install libclang-dev
```

(f) If the above commands results in an error ending in "E: Unmet dependencies. Try 'apt --fix-broken install' with no packages (or specify a solution)," type the following command, then redo **step 1e**:

```
$ sudo apt --fix-broken install
```

2. Download and install Rstudio Desktop (https://www.rstudio.com/products/rstudio/download/).
   (a) Click on the "DOWNLOAD" button under "Rstudio Desktop."
   (b) Download the .deb file located in the "Download" column next to "Ubuntu18/Debian 10" in your Downloads folder.
   (c) Move the .deb file to /home/user/rnaseq_exp/programs.
   (d) Open a Linux terminal and run the commands below to install Rstudio:

```
#Rstudio installation commands
$ cd /home/user/rnaseq_exp/programs
$ sudo dpkg -i rstudio-1.4.1106-amd64.deb
```

3.2.7 Bioconductor and Bioconductor Packages: DESeq2 [17], c4lusterProfiler [18], pathview [31], ReactomePA [19], apeglm [32], biomaRt [33, 34], ggplot2 [35]

Most R packages used in this protocol can be obtained from the R-Bioconductor repositories (*see* **Note 3**).

First, open Rstudio [30]:

```
#Open RStudio
$ cd ~
$ rstudio
```

The following command will be executed in the Rstudio [30] window console command line. To obtain the latest version of Bioconductor run the installation commands indicated below. Check the bioconductor webpage (https://www.bioconductor.org/install/) for the latest version and change the command accordingly.

```
#Installation of Bioconductor packages (whole section). See
the above URL for the latest version.
>if (!requireNamespace("BiocManager", quietly = TRUE))
install.packages("BiocManager")
>BiocManager::install(version = "3.13")
```

Once Bioconductor has been installed, run the commands below to install the required R packages.

```
#Installation of DESeq2 v1.30.1. DESeq2 requires many depen-
dencies and may take a while to install.
>BiocManager::install("DESeq2")

#Installation of clusterProfiler. clusterProfiler requires
many dependencies and may take a while to install.
>BiocManager::install("clusterProfiler")

#Installation of pathview
>BiocManager::install("pathview")

#Installation of ReactomePA
>BiocManager::install("ReactomePA")

#Installation of apeglm
>BiocManager::install("apeglm")

#Installation of biomaRt
>BiocManager::install("biomaRt")
```

ggplot2 [35] is in the regular R CRAN repository, not in Bioconductor, and can be downloaded with the following command:

```
#Installation of ggplot2 v3.3.2
>install.packages("ggplot2")
```

Close the RStudio window.

3.2.8 GSEA [20, 21]

To download GSEA [20, 21] software, access http://www.gsea-msigdb.org/gsea/index.jsp and follow the steps below:

1. Select the "Downloads" tab located in the upper region of the page.
2. In the log-in screen select "Click here" to register as a user.

3. Enter your e-mail address and organization and select "Register".
4. Select "Continue" after receiving a registration confirmation.
5. Go to the "Downloads" tab again.
6. In the "Software" section, download the .zip file containing the program for Linux.
7. Place the downloaded file in /home/user/rnaseq_exp/programs, open a terminal, set that directory as the working directory and decompress the file with the following commands:

```
#GSEA for Linux v4.1.0 downloaded file decompression. Replace
the version number of #the zip file if needed.
$ cd /home/user/rnaseq_exp/programs
$ unzip GSEA_Linux_4.1.0.zip
```

8. You can now safely delete the .zip file. Set the decompressed GSEA folder as the working directory in the terminal and launch GSEA with these commands whenever you want to use it:

```
#Launching GSEA from the terminal. Example using the 4.1.0
version of GSEA for Linux:
$ cd /home/user/rnaseq_exp/programs/GSEA_Linux_4.1.0
$ bash gsea.sh
```

3.2.9 RNA-Seq Data

We used a publicly available dataset from a published study [9] on HUVECs as the input for our analysis. The raw RNA-seq data in FASTQ format (*see* **Note 4**) was obtained from the European Nucleotide Archive (ENA) [36] (https://www.ebi.ac.uk/ena/browser/view/PRJNA679567). The dataset consists of six FASTQ files, each containing the raw reads of the transcriptome of each of the six samples analyzed in the experiment. The accession number for each run from the PRJNA679567 BioProject (see link above) and the group they belong to are shown in Table 2.

The data can be easily downloaded from the ENA [36] (https://www.ebi.ac.uk/ena/browser/view/PRJNA679567). To do so, access the PRJNA679567 BioProject page and click on "Download All" above the "FASTQ FTP" column (Fig. 2). Once downloaded, place the files in the raw_data folder, open a terminal and execute the following commands to decompress the files:

```
#Decompressing raw data files:
$ cd /home/user/rnaseq_exp/raw_data
```

Table 2
Sample attributes

SRA run accession number	Group
SRR13086510	RNAseq gCTL transduced cells (replicate 1)
SRR13086511	RNAseq gCTL transduced cells (replicate 2)
SRR13086512	RNAseq gCTL transduced cells (replicate 3)
SRR13086513	RNAseq gMYC transduced cells (replicate 1)
SRR13086514	RNAseq gMYC transduced cells (replicate 2)
SRR13086515	RNAseq gMYC transduced cells (replicate 3)

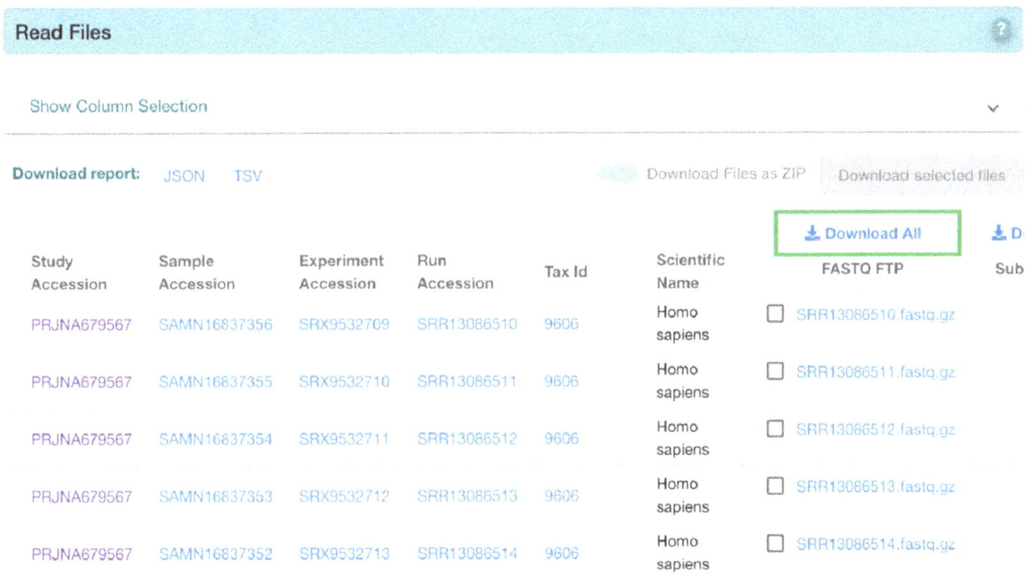

Fig. 2 The European Nucleotide Archive (ENA) [36] PRJNA679567 BioProject page (https:/www.ebi.ac.uk/ena/browser/view/PRJNA679567). The "Download All" button is highlighted by the green box

```
$ unzip ena_files.zip
$ mv ./*/*.gz .
$ gunzip *.gz
$ rmdir *
```

3.2.10 Reference Genome

The reference genome sequence in FASTA format is needed to map the raw RNA-seq reads. In order to measure gene expression, the annotation of the reference genome in GTF or GFF format is required, as it contains the location of genes in the reference genome. Both can be readily obtained from Ensembl [37] FTP

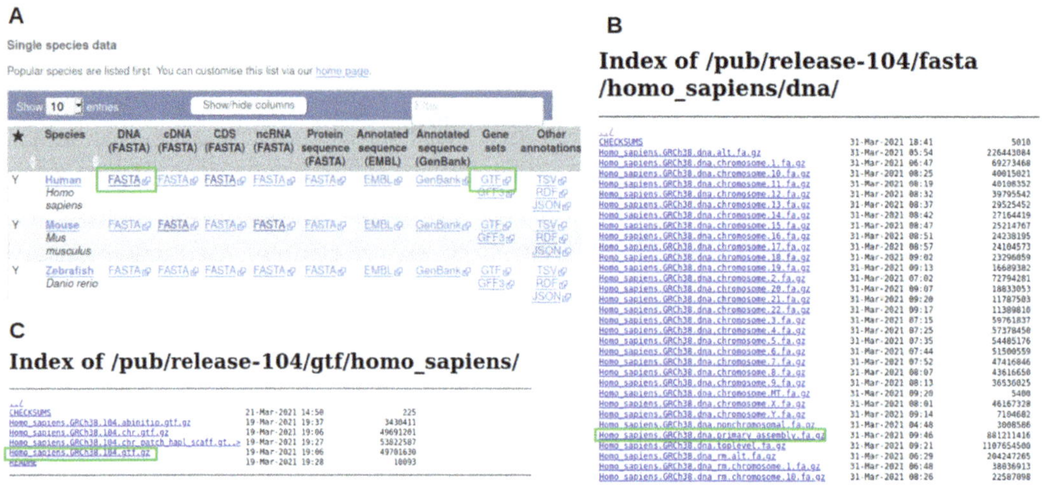

Fig. 3 Navigation process in Ensembl [37] FTP Download site (http:/www.ensembl.org/info/data/ftp/index.html) to download FASTA and GTF files. The links to access the human DNA index (contains the human reference genome, left green rectangle) and the gene sets index (contains the genome annotation, right green rectangle) in the "Single species data" section are shown in (**a**). (**b**) Shows the reference human genome file download link in the DNA index page. (**c**) Shows the genome annotation file download link in the Gene sets index

Download (http://www.ensembl.org/info/data/ftp/index.html) via an FTP client (*see* **Note 5**, Fig. 3).

Once downloaded, save both files in the resources directory. To decompress them, run the following commands in a Linux terminal:

```
#Reference Genome sequence and annotation files decompres-
sion:
$ cd /home/user/rnaseq_exp/resources
$ gunzip Homo_sapiens.GRCh38.104.gtf.gz
$ gunzip Homo_sapiens.GRCh38.dna.primary_assembly.fa.gz
```

3.3 Raw Data Quality Control

3.3.1 FastQC [11, 12]

We run the FastQC [11, 12] tool on the FASTQ files located in the raw_data directory to generate a quality control statistics report of the raw data in HTML format. An additional .zip file containing the report files is also generated. We can indicate the desired output directory using -o option. To illustrate the QC report interpretation, we will use the *SRR13086510.fastq* file.

```
#FastQC quality control of raw data
#First, we create the output directory
$ cd /home/user/rnaseq_exp
$ mkdir output/fastqc
```

```
#Now we run fastqc on every file ending in '.fastq'
$ fastqc raw_data/*.fastq -o output/fastqc

#Decompressing the report files (optional)
$ cd /home/user/rnaseq_exp/output/fastqc
$ unzip SRR13086510_fastqc.zip
```

3.3.2 FastQC [11, 12] Report Interpretation

FastQC [11, 12] provides a detailed explanation of the analysis modules used to assess the quality of the raw data. Below, we have condensed the information described by FastQC [11, 12] and supplemented with plots resulting from the quality control of the data used in this workflow. If further information on a particular module is required, see https://www.bioinformatics.babraham.ac.uk/projects/fastqc/Help/.

3.3.3 Summary

FastQC [11, 12] summary includes an HTML report with different modules. Each section is accompanied by a valid results (green tick), warnings (orange exclamation mark) or QC failures (red cross) symbol. Warnings and failures should be interpreted, as they do not necessarily translate into real problems. For example, the per-base sequence content will always raise a failure when analyzing RNA-seq reads. The reason why this happens is described in its corresponding module below.

3.3.4 Basic Statistics

This module contains some descriptive information about the file that was analyzed (Fig. 4). It will never raise a warning or failure [11, 12].

3.3.5 Per Base Sequence Quality

The Per Base Sequence Quality module shows the distribution of quality values for each position of the reads (Fig. 5). There is a box plot for each position, with a central red line indicating the median value, a yellow box containing the interquartile range (25–75%), upper and lower whiskers containing the 10% and 90% points, and a blue line representing the mean quality. If there are many quality values in a certain position that are very high, the box might be missing and the red line might appear in the top part of the plot [11, 12].

The quality scores are plotted on the *y*-axis. This axis is divided in three zones: the green area contains the calls (nucleotides inferred from the signals generated during sequencing) with a very good quality, the orange represents calls of reasonable quality, and the red zone contains calls of poor quality. Note that due to how most sequencing platforms function, a gradual decrease in base call quality is expected along the *x*-axis. However, large sudden drops in quality or large percentages of poor quality reads could suggest there was an issue during the sequencing process [11, 12].

Basic Statistics

Measure	Value
Filename	SRR13086510.fastq
File type	Conventional base calls
Encoding	Sanger / Illumina 1.9
Total Sequences	31375428
Sequences flagged as poor quality	0
Sequence length	60-75
%GC	47

Fig. 4 FastQC basic statistics module [11, 12]

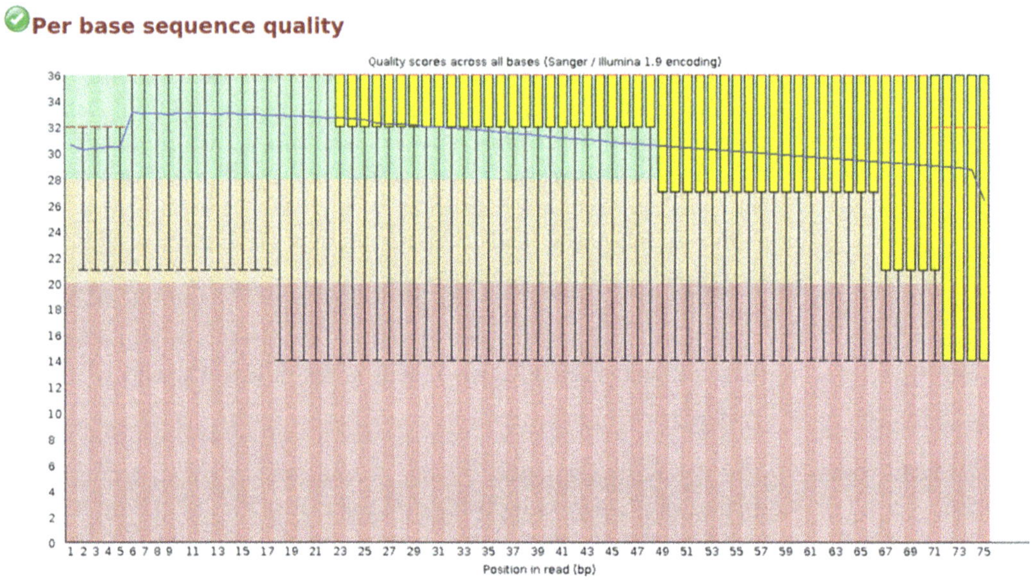

Fig. 5 FastQC Per base sequence quality plot [11, 12]

3.3.6 Per Sequence Quality Scores

This plot indicates the mean quality score values (*x*-axis) and the number of reads scoring those values (*y*-axis) (Fig. 6). A warning will be raised if the most frequent average quality is under 27 (sequencing error rate of 0.2%) and a failure if it is under 20 (1% error rate). Furthermore, if there is a subset of reads with low mean quality scores, they will appear as a bump in the plot. This can arise from technical issues during sequencing but does not represent a problem if they constitute a small percentage of sequences [11, 12].

Fig. 6 FastQC Per sequence quality scores plot [11, 12]

3.3.7 Per Base Sequence content

In this plot, the proportion of each base is given for every position of the reads (Fig. 7). In an ideal library, the sequenced transcripts would be random and the difference between bases would be none or minimal, resulting in all lines running parallel. However, this module will always raise a "Failure" in RNA-seq, due to alterations in the content in the 5′ end (usually up to 12 first bases) of reads. This is due to some of the random primers used in the cDNA synthesis binding messenger RNA molecules more efficiently than others, introducing an unavoidable technical bias enriching transcripts that contain their complementary sequences [11, 12]. This cannot be solved by trimming the 5′ ends, as these sequences come from real well-sequenced transcripts that are predominantly primed [38].

If the difference between any of the bases is greater than 10% in any position, a warning will be raised. If it is greater than 20%, a failure will appear instead. If the difference is observed other than in the 5′ end, it could be due to the presence of overrepresented sequences, biased composition libraries or aggressive adapter trimming. In our example analysis, we observed a 0% proportion of A in position 75 (Fig. 7). This could be due to the data we

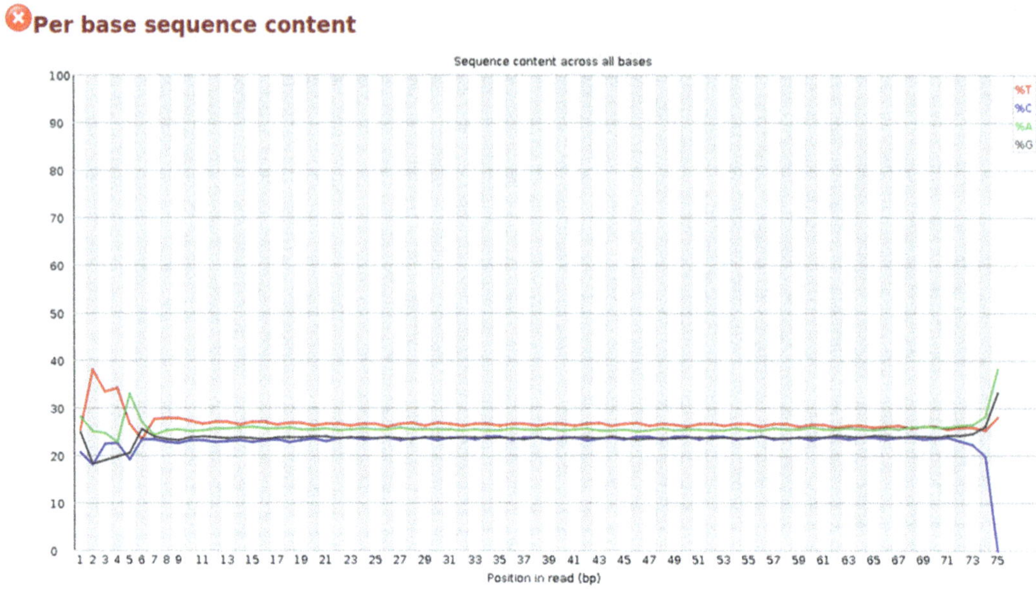

Fig. 7 FastQC Per Base Sequence Content plot [11, 12]

downloaded having already undergone adapter trimming, and all 3′ terminal adenines having been removed as the Illumina universal adapter begins with A [11, 12].

3.3.8 Per Sequence GC Content

This plot compares the distribution of the mean GC content in our reads to an expected theoretical distribution (Fig. 8). A normal distribution is always expected. Sharp peaks can appear due to the presence of overrepresented sequences and wide peaks can result from contamination with another organism. The theoretical distribution is automatically generated based on the input data, but the normal GC content for the organism being analyzed is unknown to FastQC [11, 12]. Therefore, it is advisable to check if the central peak of the plot matches the normal GC content of the genome of the species being studied [11, 12].

3.3.9 Per Base N Content

Sequencers often introduce an "N" instead of one of the DNA bases when unable to make a base call confidently. The percentage of N for each position is represented in this plot (Fig. 9). Ns are more common at the end of reads, but if their proportion is over 5% (warning) or 10% (failure) a base-calling quality issue could be happening and other quality modules should be analyzed together with this one. In addition, a high proportion of N at a small number of early positions could result from high bias in the library sequence composition [11, 12].

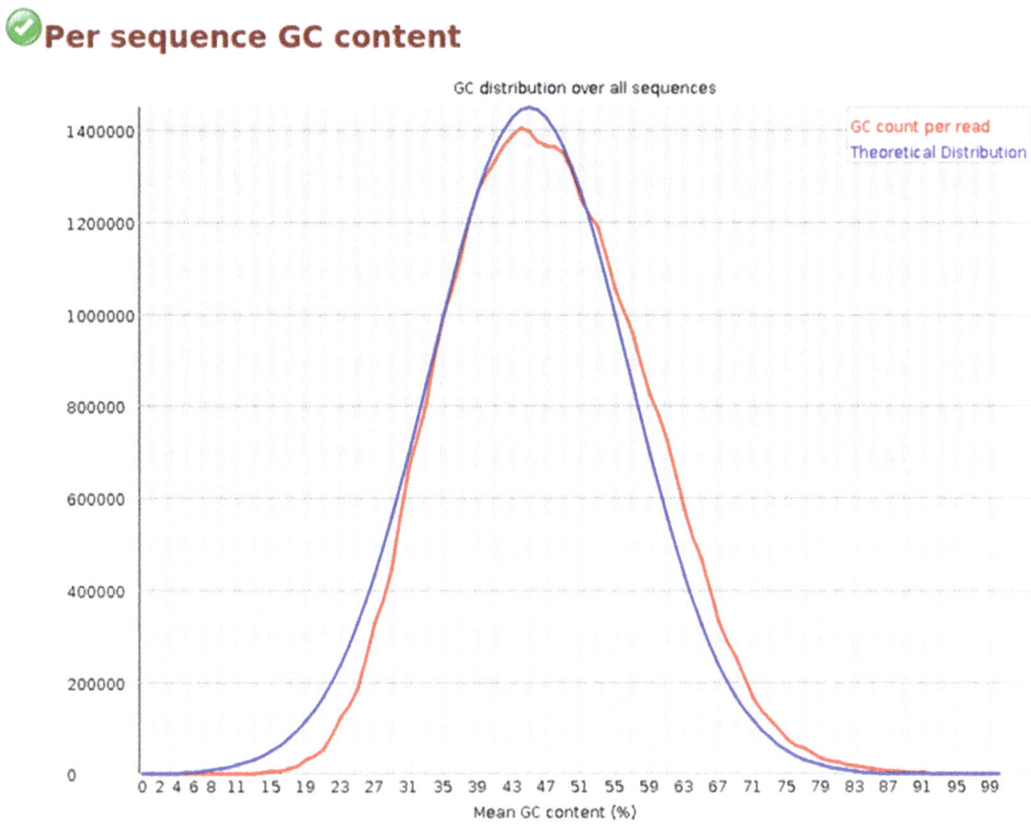

Fig. 8 FastQC Per Sequence GC content plot [11, 12]

3.3.10 Sequence Length Distribution

Depending on the sequencing platform used, fragments can either be all of the same length or generate reads of varying lengths. Even when fragments are equally long, some pipelines trim low-quality base calls located at the end of reads. In this module's plot, the distribution of read lengths is shown (Fig. 10). A warning will be raised if more than one read length is detected, but this is not necessarily of concern unless the sequencing platform used is not supposed to generate reads of different sizes. A failure will be raised if the length of any sequences is 0 [11, 12].

3.3.11 Sequence Duplication Levels

In this module plot, the proportion of duplicated sequences is shown in two ways. The blue line shows the percentage of sequences (*y*-axis) that belong to unique sequences, to duplicates, triplicates (*x*-axis) and so on (Fig. 11). In contrast, when those sequences that appear multiple times are deduplicated but remain in the same position on the *x*-axis, the red line values can be calculated as the percentage for each position relative to the total amount of deduplicated sequences. A broader explanation with an example was provided by Biostars main maintainer Istvan Albert (https://www.biostars.org/p/107402/). This module assumes

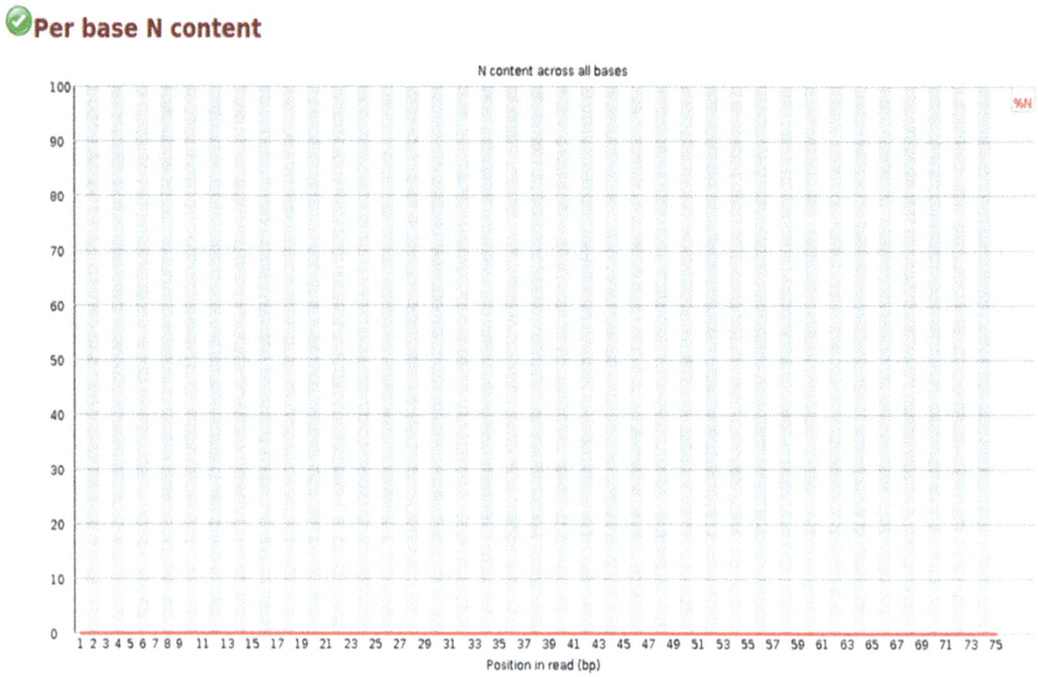

Fig. 9 FastQC per base N content plot [11, 12]

that the library used is unenriched. Even in those cases, two reads can be exactly the same by chance. However, RNA-seq libraries are commonly enriched by PCR amplification in order to allow detection of lowly expressed genes, creating technical duplicates and raising a warning or a failure in FastQC [11, 12] when they represent over 20% and 50% of sequences respectively. This plot can be used in preliminary experiments to help adjust the optimal number of PCR cycles [11, 12].

3.3.12 Overrepresented Sequences

If any sequences appear in this module, it could suggest that they play a highly relevant biological role. Alternatively, it can also be due to the library being contaminated or having a poor diversity of transcripts. A sequence is considered overrepresented if it constitutes over 0.1% of the total. However, only the 100,000 first sequences of each file are analyzed for this module and overrepresented sequences might be missed. FastQC [11, 12] uses a database of common contaminants and will determine the best hit for each overrepresented sequence found [11, 12]. In our FASTQ file, no overrepresented sequences were found.

3.3.13 Adapter Content

This module reports those K-mers in the library that are not evenly distributed throughout the length of reads (Fig. 12). This includes the presence of adapters. If a sequence appears in more than 5% of reads, a warning will be raised. If this percentage is over 10%, a

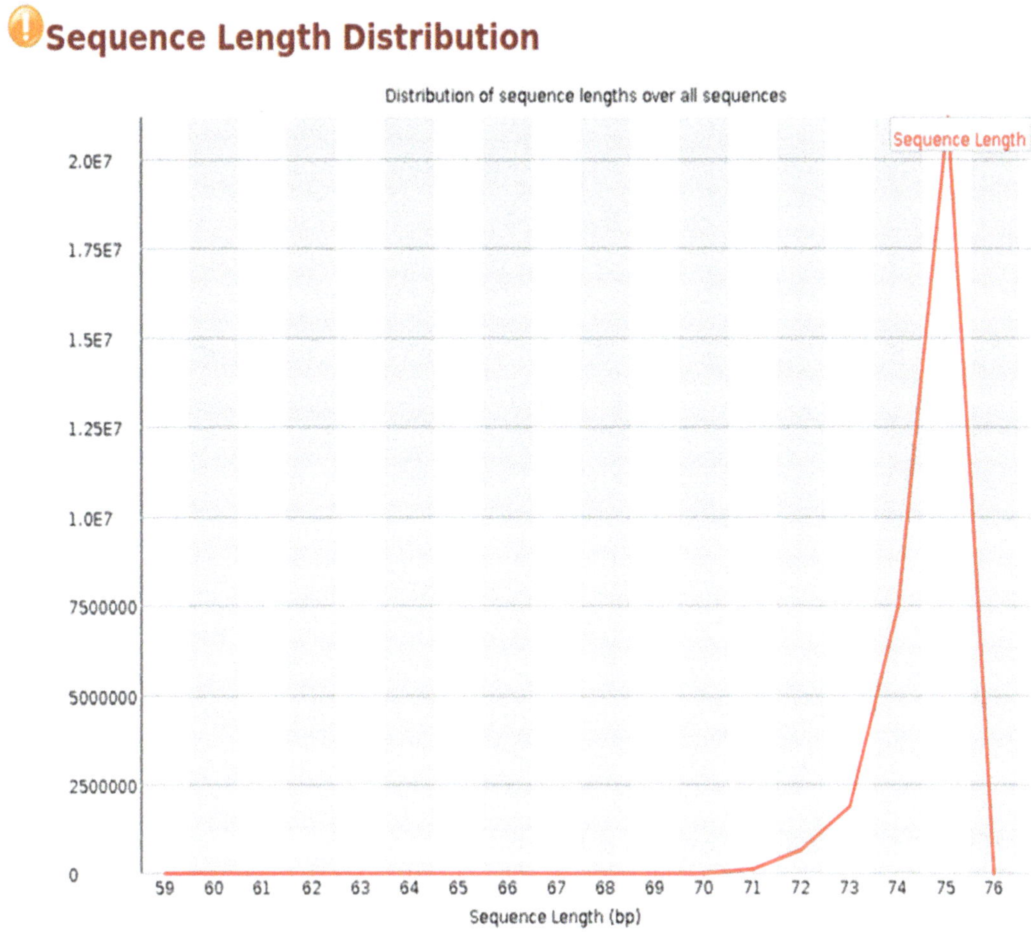

Fig. 10 FastQC Sequence Length Distribution plot [11, 12]

failure will be raised instead. It is preferable to remove contaminating adapter sequences as these can interfere with alignment accuracy [11, 12]. The adapter sequences that FastQC [11, 12] checks can be found in the *adapter_list.txt* file located in */etc/fastqc/Configuration* (Table 3).

3.4 Read Trimming: Cutadapt [13]

Once FastQC [11, 12] statistics have been analyzed, Cutadapt [13] can be used to remove adapters detected in the *Adapter Content* module or low-quality bases. The -a option should be followed by the 3′ adapter sequence. If using other types of adapter, the appropriate option can be found in the Cutadapt [13] documentation page (https://cutadapt.readthedocs.io/en/stable/guide.html#adapter-types). The -q option is followed by a number specifying the minimum quality value a base needs to have to avoid trimming. Trimming by quality is performed before adaptor trimming. --minimum-length 1 directs Cutadapt [13] to remove all reads that are shorter than the specified length, one in this case.

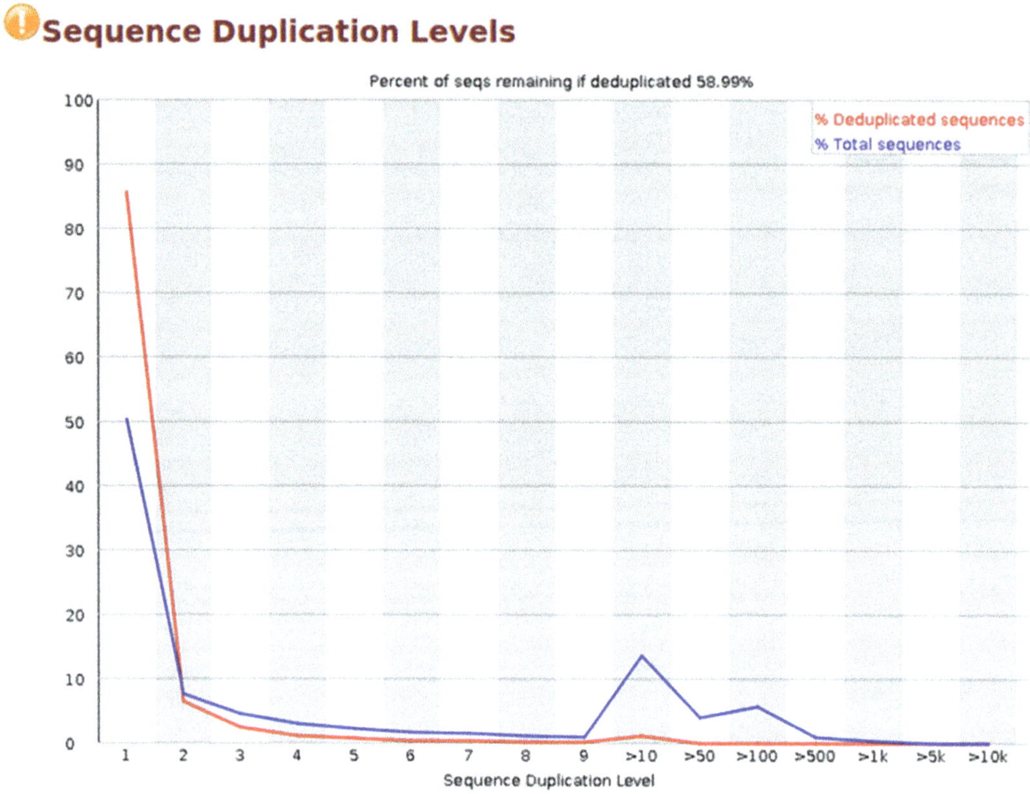

Fig. 11 FastQC Sequence Duplication Levels plot [11, 12]

This is essential, as the presence of reads with a length of zero will cause errors to arise in the use of downstream programs. The -o option should be followed by the desired output file name, a space and the input file. If you only want to remove adapters, simply do not include the "*-q value*" option.

```
#Trimming single-end reads with Cutadapt
$ cd /home/user/rnaseq_exp/
$ mkdir output/cutadapt
$ cutadapt -a AGATCGGAAGAG -q 20 --minimum-length 1 -o output/
cutadapt/SRR13086510_trimmed.fastq raw_data/SRR13086510.fastq
```

For paired-end reads, the quality option is applied to both reads, but two separate adapters have to be indicated for the forward and reverse read. The command should have the following structure:

```
#Trimming paired-end reads with Cutadapt
$ cd /home/user/rnaseq_exp/raw_data
$ cutadapt -a adapter_forward -A adapter_reverse --minimum-
```

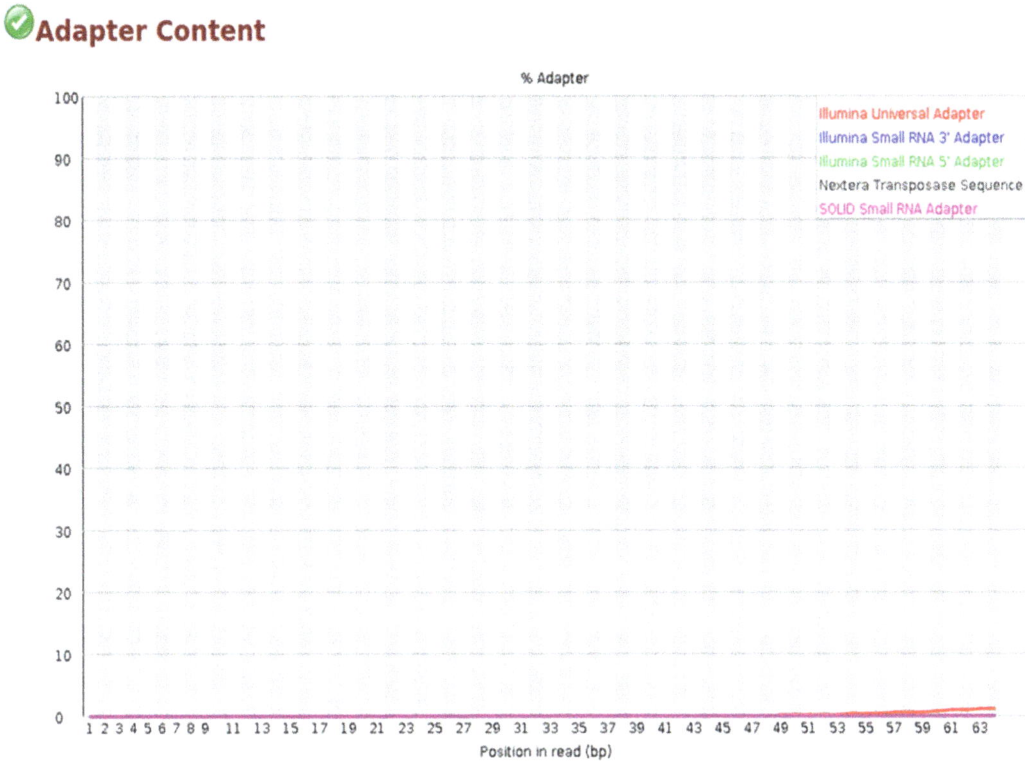

Fig. 12 FastQC Adapter Content plot [11, 12]

Table 3
Adapter sequences analyzed by FastQC [11, 12] in the "Adapter Content" module

Adapter	Sequence
Illumina universal adapter	AGATCGGAAGAG
Illumina small RNA 3′ adapter	TGGAATTCTCGG
Illumina small RNA 5′ adapter	GATCGTCGGACT
Nextera transposase sequence	CTGTCTCTTATA
SOLID small RNA adapter	CGCCTTGGCCGT

```
length 1 -o output1.fastq -p output2.fastq input1.fastq in-
put2.fastq
```

When the trimming process is complete, check your results again using FastQC [11, 12]:

```
#FastQC on the trimmed data
$ cd/home/user/rnaseq_exp
$ fastqc output/cutadapt/*.fastq -o output/fastqc/
```

3.5 Read Mapping: STAR [14]

Successful quality control checks are followed by the mapping of reads to the reference genome. This process involves two steps in STAR [14]. First, an index of the reference genome is generated from the reference genome sequence in FASTA format and its annotation in GTF or GFF format. This index allows much faster mapping of the reads to the reference genome. These two steps are the longest and most memory demanding in the workflow.

3.5.1 Genome Index Files Generation [14]

The command for the genome index generation can be found below. It is only necessary to build a reference genome once. In the following commands the absolute paths (i.e., starting with /home/user) are used rather than the relative paths to files and directories to avoid any issues that may arise when loading the reference genome. The option --runThreadN specifies the number of threads, set this option in accordance with your CPU. --runMode should be set to "genomeGenerate." The output directory where the index files will be stored is indicated with the --genomeDir option. The input FASTA files location is indicated after --genomeFastaFiles. --sjdbGTFfile should be followed by the GTF file location. If using a GFF3 annotation file instead of a GTF file, the option "--sjdbGTFtagExonParentTranscript Parent" should also be included. Lastly, the ideal value for the option --sjdbOverhang is the read length −1 (or maximum read length −1 if the read length is variable). However, the default value of 100 will work well in most cases. As the read length of our dataset is 75, we will use 74. Setting --genomeSAsparseD to 2 instead of the default 1 can greatly decrease the amount of memory needed at the cost of a slight reduction in mapping speed in the next step. If an error message appears that includes "EXITING because of INPUT ERROR: the file format of the genomeFastaFile: /home/user/rnaseq_exp/resources/Homo_sapiens.GRCh38.dna.primary_assembly.fa is not fasta" (*see* **Note 6**).

```
#Reference Genome Index Generation with STAR
$ cd /home/user/rnaseq_exp
$ mkdir -p output/star/genome_index
$ STAR --runThreadN 12 --runMode genomeGenerate --genomeDir /
home/user/rnaseq_exp/output/star/genome_index --genomeFasta-
Files /home/user/rnaseq_exp/resources/Homo_sapiens.GRCh38.
dna.primary_assembly.fa --sjdbGTFfile /home/user/rnaseq_exp/
resources/Homo_sapiens.GRCh38.104.gtf --sjdbOverhang 74 --
genomeSAsparseD 2
```

3.5.2 Mapping Reads to the Reference Genome [14]

The reads in FASTQ format and the genome index files generated in the previous step are provided to map the reads to the reference genome. Only one genome index is needed to map the reads of all samples. The command for read mapping is shown in the following. The --runThreadN option allows the user to set the number of threads used. --genomeDir should be followed by the directory containing the genome index files. The location of the FASTQ file to map is indicated with the --readFilesIn option. The alignment format is chosen with the --outSAMtype option. We chose the BAM format (see https://samtools.github.io/hts-specs/SAMv1.pdf for a full description of this file type) to avoid the need to convert the default SAM uncompressed format into a compressed BAM file (most tools are able to handle BAM files and they occupy much less disk space). Several tools require BAM files to be sorted by genomic location. To obtain a sorted BAM alignment file, use "--outSAMtype BAM SortedByCoordinate." Finally, indicate the location and prefix of the output alignment file after --outFileNamePrefix. For paired-end data, simply add the FASTQ files of both reads separated by a space after the --readFilesIn option.

```
#Read Mapping with STAR
$ cd /home/user/rnaseq_exp/
$ mkdir -p output/star/alignments
$ STAR --runThreadN 14 --genomeDir output/star/genome_index/
--readFilesIn output/cutadapt/SRR13086510_trimmed.fastq --
outSAMtype BAM SortedByCoordinate --outFileNamePrefix output/
star/alignments/SRR13086510
```

Repeat the last command for the rest of FASTQ files.

3.6 Alignment Files Quality Control: Qualimap [15]

The quality of the BAM files containing mapped reads can be assessed using Qualimap [15]. To utilize Qualimap [15], run the following commands. The option --java-mem-size sets the amount of memory to be used. Set this option to 70–80% of your memory. BAM files as well as the GTF file used for mapping are required as an input. If your data were paired-end, add the -pe option to count pair fragments instead of single reads. Qualimap [15] results can be easily explored by opening the report in HTML format in the selected output directory. Sample SRR13086510 is used as an example to interpret the report.

```
#Alignment files QC with Qualimap
$ cd /home/user/rnaseq_exp
$ mkdir output/qualimap
$ ./programs/qualimap_v2.2.1/qualimap rnaseq --java-mem-si-
ze=24G -bam output/star/alignments/SRR13086510Aligned.sorted-
```

A

Transcript coverage profile

5' bias:	0.68
3' bias:	0.62
5'-3' bias:	1.15

B

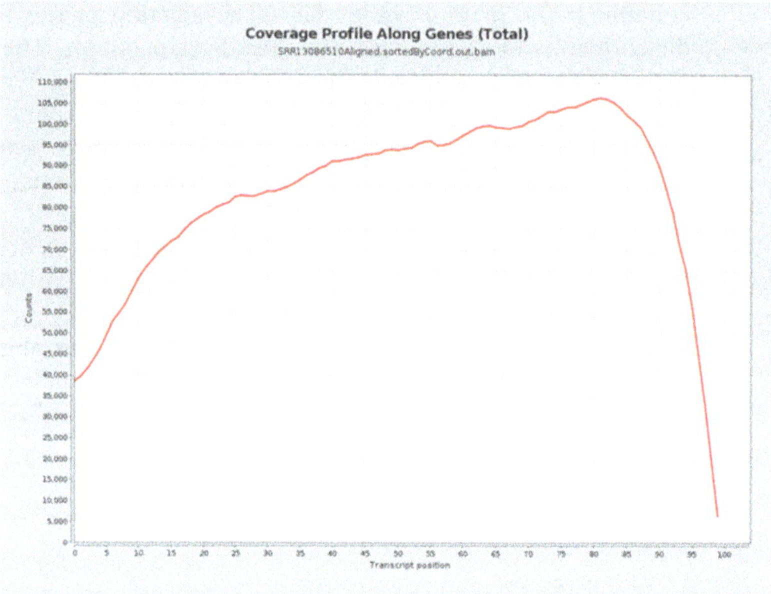

C

Strand specificity estimation (fwd/rev):	0.04 / 0.96

Fig. 13 (**a**) Qualimap [15] Transcript coverage profile table. (**b**) Coverage Profile Along Genes plot. A slight 3′ bias can be observed. (**c**) Strand specificity information, which in this case indicates the data is reverse stranded

```
ByCoord.out.bam -gtf resources/Homo_sapiens.GRCh38.104.gtf
-outdir output/qualimap/
```

Repeat the last command for all other BAM files.

One of the most important parameters in the report is the percentage of mapped reads, calculated as (Number of mapped reads/Total number of alignments) × 100, which is usually 70–90% for human genomes. In addition, a significant amount of reads should be multi-mapping reads that can be aligned to more than one region of the genome (Number of secondary alignments). Furthermore, the presence of marked 3′ coverage bias from poly-A-selected transcripts is an indicator of low RNA quality [4]. This can be accessed via the "Transcript coverage profile" (Fig. 13a) and "Coverage Profile Along Genes (Total)" plot (Fig. 13b). Qualimap [15] also provides useful information regarding the strand specificity of the data (Fig. 13c), knowledge of which is required when running featureCounts to quantify the reads. Non-strand specificity would be indicated by ~0.5 fwd/~0.5 rev, forward strand specificity ~0.9 fwd/~0.1 rev and reverse strand specificity ~0.1 fwd/~0.9 rev [15]. See http://qualimap.conesalab.org/doc_html/analysis.html for further information.

3.7 Quantification of Gene Expression Level: featureCounts [16]

Following the mapping of reads to the reference genome, the reads for each gene need to be quantified in order to create an expression matrix that is required for count normalization and subsequent DE analysis. This protocol employs featureCounts [16] to summarize the reads. featureCounts [16] requires the GTF file used in the read mapping step (Homo_sapiens.GRCh38.104.gtf) as well as the output BAM files (found in /home/user/rnaseq_exp/output/star/alignments). In this case, as we have multiple BAM files, one from each of the datasets downloaded from ENA [36], we can run one command to encompass all of them.

In the command below, the -s option specifies whether the library is not stranded (-s 0), stranded (-s 1) or reverse stranded (-s 2) and should be adjusted accordingly to the library generated in the experiment (*see* Subheading 3.6). Each read has a corresponding feature type, indicated in the GTF file. "-t exon" is used to only count those reads whose feature type is exon. To indicate that exons (features) have to be grouped into genes (meta-features), the option "-g gene_id" is used. The location path of the GTF file is pointed to using -a. -o is followed by the output file. The BAM files are listed at the end of the command.

```
#Quantification of gene expression level with featureCounts
for single-read data.
$ mkdir /home/user/rnaseq_exp/output/featurecounts
$ cd /home/user/rnaseq_exp/output/star/alignments
$ featureCounts -s 2 -t exon -g gene_id -a /home/user/
rnaseq_exp/resources/ Homo_sapiens.GRCh38.104.gtf -o /home/
user/rnaseq_exp/output/featurecounts/counts.txt SRR13086510A-
ligned.sortedByCoord.out.bam SRR13086511Aligned.sortedBy-
Coord.out.bam SRR13086512Aligned.sortedByCoord.out.bam
SRR13086513Aligned.sortedByCoord.out.bam SRR13086514Aligned.
sortedByCoord.out.bam SRR13086515Aligned.sortedByCoord.out.
bam
```

In this instance we are working with single-read data. In the case of paired-end data, one must apply the -p argument to the command as seen below. In this situation, fragments are counted as opposed to reads. For paired-end data, paired reads need to be positioned together in the BAM file. If this is not already the case, featureCounts [16] will automatically sort the reads file by name for the user.

```
#Quantification of gene expression level with featureCounts
for paired-end data.
$ mkdir /home/user/rnaseq_exp/output/featurecounts
$ cd /home/user/rnaseq_exp/output/star/alignments
$ featureCounts -p -s 2 -t exon -g gene_id -a /home/user/
rnaseq_exp/resources/ Homo_sapiens.GRCh38.104.gtf -o /home/
user/rnaseq_exp/output/featurecounts/counts.txt SRR13086510A-
ligned.sortedByCoord.out.bam SRR13086511Aligned.sortedBy-
Coord.out.bam SRR13086512Aligned.sortedByCoord.out.bam
SRR13086513Aligned.sortedByCoord.out.bam SRR13086514Aligned.
sortedByCoord.out.bam SRR13086515Aligned.sortedByCoord.out.
bam
```

Two files are the output of featureCounts [16] in /home/user/rnaseq_exp/output/star/alignments, the read count table (*counts.txt*) and a summary of the alignment results (*counts.txt.summary*). The read count table will be taken forward for differential expression analysis.

3.8 Differential Expression Analysis: DESeq2 [17]

Once the *counts.txt* file has been generated by featureCounts [16], DE analysis can be performed with the R package DESeq2 [17].

3.8.1 Workspace Preparation

First, open a terminal window and create a new directory to save the outputs from DESeq2 [17] and functional enrichment analysis and a directory for the pathview [31] results.

```
#New directory for DESeq2 outputs.
$ mkdir /home/user/rnaseq_exp/output/deseq2_funcenrich
$ mkdir /home/user/rnaseq_exp/output/pathview
```

Open RStudio [30]. In the RStudio [30] window select the console tab and make sure the environment is clear by pressing the broom icon in the environment window (Fig. 14).

In the RStudio [30] console, the command line is indicated by a ">" symbol. Once a command has been typed out the "enter" key is pressed and the command is executed. When the command has finished running a new command line is preceded by ">," indicating that a new command can be entered. As is the case with the Ubuntu terminal, RStudio [30] inline comments are marked by "#." The lines of the command output are denoted with "##" in text.

Fig. 14 Rstudio [30] window. The Console tab (red box) and the environment tab (green box) are selected. The environment is cleared by clicking the broom icon (blue box)

The working directory needs to be set in order to specify the save location of the DE analysis results.

```
#Setting the working directory
>setwd("/home/user/rnaseq_exp/output/deseq2_funcenrich/")
```

The packages that were previously installed require loading into the current R [29] session with the following commands.

```
#Loading all required packages.
>library(DESeq2)
>library(ggplot2)
>library(clusterProfiler)
>library(pathview)
>library(ReactomePA)
>library(biomaRt)
```

3.8.2 Read and Prepare the Data

The data from the *counts.txt* file generated by featureCounts [16] needs to be read into R [29] as a variable, here named `read_-counts`. We also require the row names of the resulting data frame to be set as the gene ids and extract only the count information by selecting certain columns in the *counts.txt* data file, which vary depending on the number of samples. The sample names in the counts.txt file are cumbersome (e.g., `SRR13086510Aligned.sortedByCoord.out.bam`). These columns were renamed to something more appropriate. The ENA [36] entries for the samples give the individual sample information, for example "SRR13086510" is "gCTL transduced cells (replicate 1)" (Table 2), therefore a suitable name for this sample is `gCTL1`. It is important to check the order of the samples in the counts.txt file to avoid mislabeling when renaming. The top lines of the `counts` variable were printed before and after renaming for clarity.

```
#Reading the featureCounts data.
>read_counts_file <-file.path("/home/user/rnaseq_exp/output/
featurecounts/counts.txt")
>read_counts <- read.table(read_counts_file, header = TRUE,
stringsAsFactors = FALSE)

#Setting the row names of the counts file to be the gene IDs.
>rownames(read_counts)<-read_counts$Geneid

#Removing all unnecessary data.
>counts <- read_counts[-c(1:6)]

#Printing the top lines of the counts.
>head(counts)

## SRR13086510Aligned.sortedByCoord.out.bam
##ENSG00000284662 0
##ENSG00000186827 0
##ENSG00000186891 1
##ENSG00000160072 1145
##ENSG00000041988 374
##ENSG00000260179 2
## SRR13086511Aligned.sortedByCoord.out.bam
##ENSG00000284662 0
##ENSG00000186827 0
##ENSG00000186891 0
##ENSG00000160072 1246
##ENSG00000041988 363
##ENSG00000260179 6
## SRR13086512Aligned.sortedByCoord.out.bam
##ENSG00000284662 0
##ENSG00000186827 2
##ENSG00000186891 0
##ENSG00000160072 924
##ENSG00000041988 288
##ENSG00000260179 3
## SRR13086513Aligned.sortedByCoord.out.bam
##ENSG00000284662 0
##ENSG00000186827 1
##ENSG00000186891 0
##ENSG00000160072 680
##ENSG00000041988 359
##ENSG00000260179 2
## SRR13086514Aligned.sortedByCoord.out.bam
```

```
##ENSG00000284662 0
##ENSG00000186827 0
##ENSG00000186891 0
##ENSG00000160072 486
##ENSG00000041988 313
##ENSG00000260179 1
## SRR13086515Aligned.sortedByCoord.out.bam
##ENSG00000284662 0
##ENSG00000186827 0
##ENSG00000186891 0
##ENSG00000160072 695
##ENSG00000041988 392
##ENSG00000260179 6

#Renaming the columns based on the group
>colnames(counts)<-c("gCTL1", "gCTL2", "gCTL3", "gMYC1",
"gMYC2", "gMYC3")
#Printing the top lines of the data frame after renaming
columns
>head(counts)

## gCTL1 gCTL2 gCTL3 gMYC1 gMYC2 gMYC3
##ENSG00000284662 0 0 0 0 0 0
##ENSG00000186827 0 0 2 1 0 0
##ENSG00000186891 1 0 0 0 0 0
##ENSG00000160072 1145 1246 924 680 486 695
##ENSG00000041988 374 363 288 359 313 392
##ENSG00000260179 2 6 3 2 1 6
```

Along with the counts matrix, DESeq2 [17] requires a design matrix in order to carry out the analysis. Design matrices contain information about the independent variables. In this case, each row represents one of the samples, and the "status" column states their genotype. The following commands build the design matrix and store it in a variable called `groups`. A column named `status` is created in the `groups` variable in which the sample group is specified, which here is `control` or `cMYC_KO`. It is imperative that the order of columns in the `counts` variable seen above and the row names in the `groups` variable below match, otherwise downstream analysis will not be correct.

```
#Create an experimental design matrix.
>groups <- data.frame(row.names=colnames(counts))
>groups$status <- c("control", "control", "control",
"cMYC_KO", "cMYC_KO", "cMYC_KO")
>groups$status <- as.factor(groups$status)
```

```
#Print the groups variable to display the sample information
matrix.
>groups

## status
##gCTL1 control
##gCTL2 control
##gCTL3 control
##gMYC1 cMYC_KO
##gMYC2 cMYC_KO
##gMYC3 cMYC_KO

#Check the columns and row names match. The output should be
TRUE.
>counts<-counts[, rownames(groups)]
>all(rownames(groups) == colnames(counts))

##[1] TRUE
```

3.8.3 Normalized Counts and Running Differential Expression Analysis

The following command builds a DESeq2 [17] object using the `DESeqDataSetFromMatrix()` function, which is appropriate in this case as featureCounts [16] returns counts data in a matrix (*counts.txt*). Three arguments are given: `countData`, which is the count data matrix (`counts`), `colData` refers to the sample information matrix (`groups`), and `design` is the column in the sample matrix which will be used to design the analysis (`status`). Red text may appear in the RStudio [30] console window discussing the reference level. This will be set later.

```
#Build the dds object.
>dds <- DESeqDataSetFromMatrix(countData = counts, colData =
groups, design = ~ status)
```

Normalization is a crucial step when analyzing RNA-seq data with several methods currently in existence [39]. DESeq2 [17] normalizes the count data via the median-of-ratios method, accounting for library size and allowing the counts to be compared across samples [16]. The library size factors are calculated using the `estimateSizeFactors()` function, requiring the DESeq2 [17] object as the input. The `counts()` function returns the sample counts in a matrix using the DESeq2 [17] object, `dds`, and the `normalized` argument, which here is `TRUE` to specify the normalized counts. This normalized count matrix is stored in the `normalized_counts` variable and saved as a CSV file. The normalized counts are taken forward for DE analysis and GSEA [20, 21].

```
#Calculating the library size factor.
>dds <- estimateSizeFactors(dds)
#Printing the size factors for each sample.
>sizeFactors(dds)

##gCTL1 gCTL2 gCTL3 gMYC1 gMYC2 gMYC3
##0.8905162 1.0159641 0.9287804 1.1572972 0.8864420 1.2155742

#Extracting the normalized counts and saving it as a CSV file.
>normalized_counts <- counts(dds, normalized=TRUE)
>normalized_counts <- as.data.frame(normalized_counts)
>write.csv(as.data.frame(normalized_counts), file=file.path
("normalized_counts.csv"))

#Printing the top lines of the normalized counts.
>head(normalized_counts)

## gCTL1 gCTL2 gCTL3 gMYC1 gMYC2
##ENSG00000284662 0.000000 0.00000 0.000000 0.0000000
0.000000
##ENSG00000186827 0.000000 0.00000 2.153362 0.8640823
0.000000
##ENSG00000186891 1.122944 0.00000 0.000000 0.0000000
0.000000
##ENSG00000160072 1285.771122 1226.42124 994.853033
587.5759738 548.259213
##ENSG00000041988 419.981135 357.29608 310.084062 310.2055509
353.096983
##ENSG00000260179 2.245888 5.90572 3.230042 1.7281646
1.128105
## gMYC3
##ENSG00000284662 0.000000
##ENSG00000186827 0.000000
##ENSG00000186891 0.000000
##ENSG00000160072 571.746256
##ENSG00000041988 322.481342
##ENSG00000260179 4.935939
```

The analysis is performed using the commands below using the `DESeq()` function. Prior to running the analysis, the reference level is set to `control` as the base for which DESeq2 [17] will use to compare against in the DE analysis. This is performed with the `relevel()` function, which requires the user to specify in which column of the `groups` variable the reference level is found, which is `status` in this case.

```
#Running the analysis.
>dds$status <- relevel(dds$status, ref = "control")
>dds <- DESeq(dds)
```

Independent filtering of the DE results based on the mean of normalized counts is carried out via the `results()` function. This includes filtering genes with a total read count of zero and a low mean count that is determined by DESeq2 [17]. To specify the DE results comparing `cMYC_KO` and `control` samples the `contrast` argument is used. The argument `alpha` refers to the adjusted *p*-value cut off that will be used to determine the significance of the results. This is set to 0.1 as default.

```
#Calculate the results for cMYC_KO against control.
>contrast <- c("status", "cMYC_KO", "control")
>res <- results(dds, contrast=contrast, alpha = 0.05)
```

The `lfcShrink()` function of DESeq2 [17] shrinks the log2-FoldChange (log2FC) values to give a more accurate estimate for genes with limited information, i.e., low counts [17]. The shrunken values are useful when ranking differentially expressed genes (DEGs), particularly by log2FC [17]. The function takes the DESeq2 [17] object previously generated, `dds`, the results that were calculated in the previous commands, `res`, the coefficient which can be identified with the `resultsNames()` function, `coef`, and the type of shrinkage estimator to be used, `type`, which here is `apeglm` [32]. The results include the mean of normalized counts across all samples (`baseMean`), the `log2Fold-Change` and the Benjamini-Hochberg [7] adjusted *p*-value (`padj`) which is used to determine significance. The shrunken log2FC results will be taken forward for further analysis. The `summary()` function below gives information of the results and the independent filtering described above.

```
#Identify the coefficient for log2FC shrinkage.
>resultsNames(dds)

##[1] "Intercept" "status_cMYC_KO_vs_control"

#Calculate the shrunk log2FC results.
>resLFC <- lfcShrink(dds, res = res, coef = "status_cMYC_K-
O_vs_control", type = "apeglm")

#Print the top lines of the shrunken log2FC results.
```

```
>head(resLFC)

##log2 fold change (MAP): status cMYC_KO vs control
##Wald test p-value: status cMYC KO vs control
##DataFrame with 6 rows and 5 columns
## baseMean log2FoldChange lfcSE pvalue padj
## <numeric> <numeric> <numeric> <numeric> <numeric>
##ENSG00000284662 0.000000 NA NA NA NA
##ENSG00000186827 0.502907 -0.0240911 0.410430 6.98901e-01 NA
##ENSG00000186891 0.187157 -0.0234397 0.413298 7.77767e-01 NA
##ENSG00000160072 869.104474 -1.0195211 0.110127 3.07827e-21
5.82268e-20
##ENSG00000041988 345.524192 -0.1297103 0.128335 2.86242e-01
4.07071e-01
##ENSG00000260179 3.195643 -0.0725264 0.391272 6.16626e-01
7.24399e-01

#Summary of shrunken log2FC results.
>summary(resLFC)

##out of 31389 with nonzero total read count
##adjusted p-value < 0.05
##LFC > 0 (up) : 4533, 14%
##LFC < 0 (down) : 4509, 14%
##outliers [1] : 0, 0%
##low counts [2] : 12152, 39%
##(mean count < 3)
##[1] see 'cooksCutoff' argument of ?results
##[2] see 'independentFiltering' argument of ?results
```

3.8.4 Extracting and Visualizing Differential Expression Analysis Results

In order for the DE analysis results easily interpretable, the Ensembl [37] gene IDs can be converted to gene symbols using biomaRt [33, 34]. This is performed only for visualization in the output CSV files of DE analysis results. The Ensembl [37] IDs will continue to be used for downstream analysis (*see* **Note 7**). biomaRt [33, 34] requires first connecting to a BioMart database, which in this case is hosted by Ensembl [37] through applying the `useEnsembl()` function. The first argument signifies the Ensembl [37] BioMart service, which is `"genes"` in this case. The next parameter `dataset` specifies the dataset to be used. The `getBM()` function retrieves the desired information. The `attributes` argument are the attributes the user wishes to retrieve. `values` are the gene IDs the user requires converting, `filters` are the type of gene ID to be mapped to the desired gene ID type and `mart` refers to the Mart object built using the `useEnsembl()` function. The resulting variable needs to have any duplicate values resulting from multiple

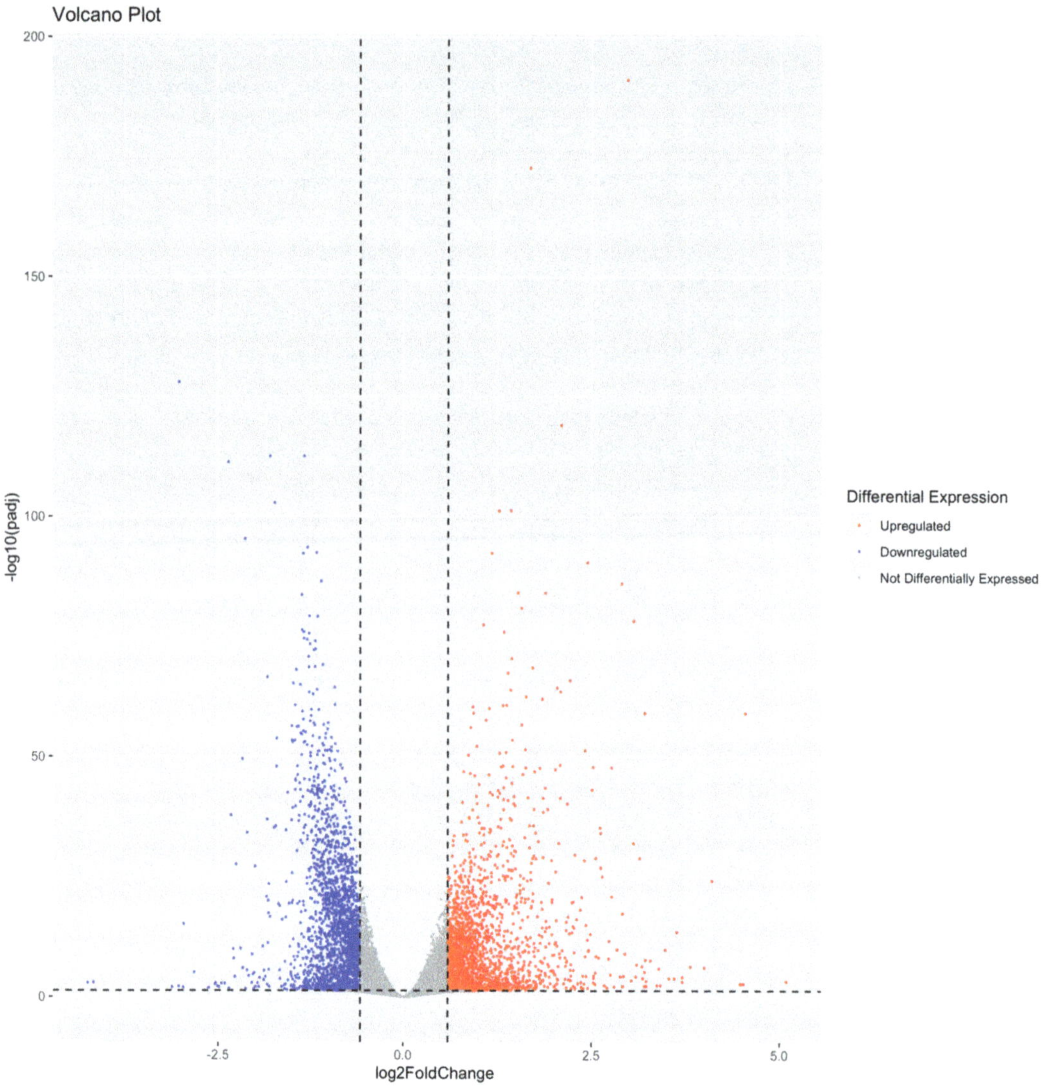

Fig. 15 Volcano plot of the log2FoldChange and −log10(padj) values of the differential expression analysis created using ggplot2 [35]. Each dot represents a transcript. Grey dots are not differentially expressed, blue are downregulated DEGs and red dots are upregulated DEGs

mapping of Ensembl [37] IDs to gene symbols removed. The gene symbols corresponding to the Ensembl [37] IDs present in the data frame of shrunken DE results can be added to the data frame as a new column.

```
#Building a data frame of the shrunken DE results.
>de_res <- as.data.frame(resLFC)

#Building a Mart object.
>mart_data <- useEnsembl("genes", dataset = "hsapiens_gen-
```

```
e_ensembl")

#Converting ENSEMBL IDs to gene symbols.
>ens_to_symbol <- getBM(attributes = c("ensembl_gene_id",
"hgnc_symbol"), values = rownames(de_res), filters = "en-
sembl_gene_id", mart = mart_data)

#Removing duplicate results resulting from multiple mapping
of ENSEMBL IDs.
>ens_to_symbol <- ens_to_symbol[!duplicated(ens_to_symbol
$ensembl_gene_id), ]

#Adding the gene symbols to the shrunken DE results data
frame.
> de_res$symbol <- ens_to_symbol$hgnc_symbol[match(rownames
(de_res), ens_to_symbol$ensembl_gene_id)]
```

The results of the DE analysis can be ordered by `padj` value and saved to a CSV file. The results can also be filtered to identify significant up and down regulated genes based on a `log2FoldChange` and `padj` threshold decided by the user and ordered by `log2FoldChange`. The commands below specify a `padj` cut-off of 0.05 and a Fold Change (FC) threshold of log2(1.5). Upregulated DEGs are those with a `log2FoldChange` > 1.5 and `padj` < 0.05. Downregulated DEGs are identified with a `log2FoldChange` < −1.5 and `padj` < 0.05. Both up and downregulated DEGs data are written to CSV files.

```
#Order the results by the adjusted p-value and write to a CSV
file.
>de_res <- de_res[order(de_res$padj), ]
>write.csv(as.data.frame(de_res), file=file.path("de_re-
sults.csv"))

#Set the user thresholds.
>fc_threshold <- log2(1.5)
>padj_threshold <- 0.05

#CSV file for all significant DEGs ordered by adjusted
p-value.
>de_res_sig <- subset(de_res, padj < padj_threshold & abs
(log2FoldChange) > fc_threshold)
>write.csv(as.data.frame(de_res_sig), file=file.path("sig-
nificant_de_results.csv"))

#Print the top lines of all significant DEG results.
>head(de_res_sig)
```

```
## baseMean log2FoldChange lfcSE pvalue
##ENSG00000149591 1770.7451 2.968410 0.09970407 4.571929e-196
##ENSG00000100292 10472.7909 1.679230 0.05956038 2.044698e-177
##ENSG00000099937 734.1335 -3.017358 0.12357106 1.348795e-132
##ENSG00000117595 1028.9119 2.092364 0.08894815 1.633119e-123
##ENSG00000116133 2005.4699 -1.800585 0.07869048 6.478862e-117
##ENSG00000128050 8503.8162 -1.334500 0.05830185 6.400575e-117
## padj symbol
##ENSG00000149591 8.795019e-192 TAGLN
##ENSG00000100292 1.966693e-173 HMOX1
##ENSG00000099937 8.648922e-129 SERPIND1
##ENSG00000117595 7.854080e-120 IRF6
##ENSG00000116133 2.077231e-113 DHCR24
##ENSG00000128050 2.077231e-113 PAICS

#CSV file for all significant upregulated DEGs ordered by log2FC
>up_res_sig <- subset(de_res, padj < padj_threshold & log2FoldChange > fc_threshold)
>up_res_sig <- up_res_sig[order(-up_res_sig$log2FoldChange),]
>write.csv(as.data.frame(up_res_sig), file=file.path("significant_up_de_results.csv"))

#Print the top lines of the upregulated DEG results.
>head(up_res_sig)

## baseMean log2FoldChange lfcSE pvalue
##ENSG00000135298 4.105870 5.093049 2.5187182 1.681736e-04
##ENSG00000178573 169.695983 4.540831 0.2736062 2.594023e-62
##ENSG00000239389 3.226172 4.520573 2.5321469 5.534736e-04
##ENSG00000123612 3.202881 4.489799 2.5453183 6.111462e-04
##ENSG00000018236 3.175264 4.476468 2.5430648 6.206696e-04
##ENSG00000145242 67.334876 4.100599 0.3918255 2.214349e-26
## padj symbol
##ENSG00000135298 5.462944e-04 ADGRB3
##ENSG00000178573 7.232060e-60 MAF
##ENSG00000239389 1.656374e-03 PCDHA13
##ENSG00000123612 1.814853e-03 ACVR1C
##ENSG00000018236 1.840577e-03 CNTN1
##ENSG00000145242 6.245959e-25 EPHA5

#CSV file for all significant downregulated DEGs ordered by log2FC.
>down_res_sig <- subset(de_res, padj < padj_threshold &
```

```
log2FoldChange < -fc_threshold)
 >down_res_sig <- down_res_sig[order(down_res_sig$log2Fold-
Change),]
 >write.csv(as.data.frame(down_res_sig), file=file.path("sig-
nificant_down_de_results.csv"))

 #Print the top lines of the downregulated DEG results.
 >head(down_res_sig)

 ## baseMean log2FoldChange lfcSE pvalue
 ##ENSG00000196126 6.386386 -4.257406 1.6066570 3.718895e-04
 ##ENSG00000197249 5.901932 -4.181709 1.5418466 3.344456e-04
 ##ENSG00000172137 3.828608 -3.122920 1.8053004 3.116593e-03
 ##ENSG00000153292 5.258807 -3.030824 1.4130284 2.385704e-03
 ##ENSG00000099937 734.133488 -3.017358 0.1235711 1.348795e-
132
 ##ENSG00000275216 3.539021 -2.967018 1.8049588 3.688191e-03
 ## padj symbol
 ##ENSG00000196126 1.144646e-03 HLA-DRB1
 ##ENSG00000197249 1.037196e-03 SERPINA1
 ##ENSG00000172137 8.097524e-03 CALB2
 ##ENSG00000153292 6.334547e-03 ADGRF1
 ##ENSG00000099937 8.648922e-129 SERPIND1
 ##ENSG00000275216 9.432296e-03
```

The `log2FoldChange` values can be plotted against −log10 (`padj`) for each DEG and visualized in a volcano plot. In order to achieve the desired color coding, each gene is categorized as `Not Differentially Expressed`, `Upregulated`, or `Downregulated`. Dashed lines can be drawn on the plot to show the boundaries of the user defined thresholds. The title is set by modifying the contents of the `ggtitle("")` argument. The volcano plot is saved as a PNG file (Fig. 15).

```
 #Add a column to the data that specifies if a gene is
Upregulated, Downregulated, or Not differentially expressed.
 >de_res_plot <- de_res
 >de_res_plot$direction <- "Not Differentially Expressed"

 >de_res_plot$direction[de_res_plot$log2FoldChange >
fc_threshold & >de_res_plot$padj < padj_threshold] <- "Upre-
gulated"

 >de_res_plot$direction[de_res_plot$log2FoldChange <
-fc_threshold & >de_res_plot$padj < padj_threshold] <- "Down-
regulated"
```

```
#Volcano plot of results.
&gt;volcano &lt;- ggplot(data=de_res_plot, aes(log2Fold-
Change, -log10(padj))) +
geom_point(aes(col=direction), size=0.2) +
scale_colour_manual(name='Differential Expression', values =
setNames(c('red', 'blue', 'grey'), c('Upregulated', 'Downre-
gulated', 'Not Differentially Expressed'))) +
geom_vline(xintercept = c(fc_threshold, fc_threshold*-1),
linetype = "dashed") +
geom_hline(yintercept = -log10(padj_threshold), linetype =
"dashed") +
ggtitle("Volcano Plot")
>ggsave("volcano.png",volcano, dpi = 300, width = 10, height
= 10)
```

3.9 Functional Enrichment Analysis

Following DE analysis by DESeq2 [17], functional enrichment analysis can be carried out to annotate putative biological relevance to the data by identifying groups of genes from various databases that are enriched within the list of DEGs, which may play a role in the differences observed between experimental conditions. Enrichment analysis will be carried out against the GO [22, 23], KEGG [24–26], and Reactome [27] databases. GO [22, 23] consists of groups of gene functions that are organized into classes and placed in a hierarchical structure linking related groups (terms) of genes [40] with enrichment analysis revealing significant GO terms related to the dataset. KEGG [24–26] is a database of genes pathways and provides schematics of these pathways [41], with enriched pathways being those that are over-represented in the list of DEGs. The Reactome [27] is also a pathway database, with a focus on the relationship of signalling and metabolic molecules grouped into processes [27] and provides further functional information alongside GO [22, 23] and KEGG [24–26]. Functional enrichment analysis of GO [22, 23] terms and KEGG [24–26] pathways will be performed via clusterProfiler [18] R package and the ReactomePA [19] R package will be employed to perform the functional analysis of the Reactome [27] database. Both clusterProfiler [18] and ReactomePA [19] import another R package called enrichplot [42], which provides several functions to display the results from functional enrichment analysis. Below we employ the `dotplot()` and `barplot()` functions to create dot plots and bar plots respectively. However, the enrichplot [42] package provides other functions to visualize results that can be found in the package documentation (Table 1). pathview [31] R package will be employed to further visualize the enrichment results obtained from KEGG [24–26].

3.9.1 Declaring Variables

Variables of gene ids need to be declared as an input for clusterProfiler [18]. A reference set of gene ids is also required which contains all the gene ids that were present in the DE analysis result file prior to filtering based on the user defined thresholds.

```
#All significant differentially expressed genes.
>all_DEGs <- rownames(de_res_sig)

#Upregulated differentially expressed genes.
>up_DEGs <- rownames(up_res_sig)

#Downregulated differentially expressed genes.
>down_DEGs <- rownames(down_res_sig)

#Reference gene list.
>reference <- rownames(de_res)
```

3.9.2 GO [22, 23] Functional Enrichment Analysis

The `enrichGO()` function of clusterProfiler is used to analyze the GO [22, 23] database with the DEGs identified from DE analysis. The key arguments are the variable of gene IDs (`all_DEGs` in this case), `OrgDb` refers to the organism database, which is human in this case as we are analyzing HUVEC data, `keyType` signifies the type of gene ID that is being used, the `ont` argument allows the user to select the GO [22, 23] ontology (Biological Process (`BP`) as seen below, Cellular Component (`CC`), Molecular Function (`MF`) and all three ontologies (`ALL`)) and `universe` is the `reference` gene list that was created above. The results from the functional analysis can be converted into a data.frame object and written to a CSV file. The same format is used for functional analysis of all significant DEGs (`all_DEGs`), upregulated (`up_DEGs`), and downregulated DEGs (`down_DEGs`) separately.

```
#Functional enrichment analysis against GO for all significant DEGs.
>all_enrichGO<-enrichGO(
all_DEGs,
OrgDb = "org.Hs.eg.db",
keyType = "ENSEMBL",
ont = "BP",
universe = reference
)

#Write the functional analysis results to a CSV file.
>all_enrichGO_data <- data.frame(all_enrichGO)
```

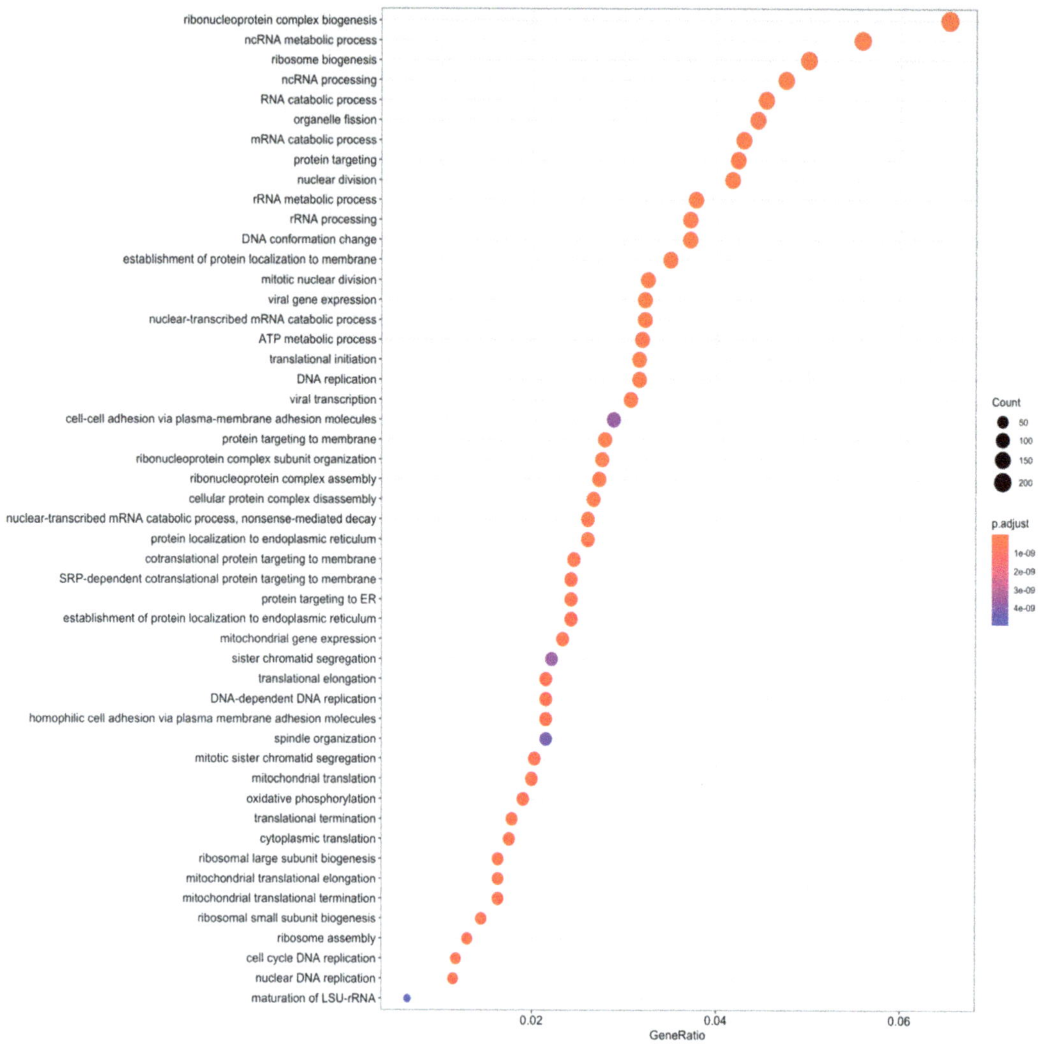

Fig. 16 Dot plot of the top 50 GO [22, 23] terms from the Biological Process ontology ranked by the ratio of DEGs that are related to a particular GO [22, 23] term to the total number of DEGs, termed the GeneRatio. The adjusted *p*-value (p.adjust) is represented by a color scale and the size of the dot refers to the number of DEGs assigned to the GO [22, 23] term (Count). Created using the dotplot()function of clusterProfiler [18] R package

```
>write.csv(all_enrichGO_data, file=file.path("all_enrich-
GOBP.csv"))

#Functional enrichment analysis against GO for significant
upregulated DEGs.
>up_enrichGO <- enrichGO(
up_DEGs,
OrgDb = "org.Hs.eg.db",
keyType = "ENSEMBL",
ont = "BP",
```

```
universe = reference
)

#Write the functional analysis results to a CSV file.
>up_enrichGO_data <- data.frame(up_enrichGO)
>write.csv(up_enrichGO_data, file=file.path("up_enrichGOBP.
csv"))

#Functional enrichment analysis against GO for significant
downregulated DEGs.
>down_enrichGO <- enrichGO(
down_DEGs,
OrgDb = "org.Hs.eg.db",
keyType = "ENSEMBL",
ont = "BP",
universe = reference
)

#Write the functional analysis results to a CSV file.
>down_enrichGO_data <- data.frame(down_enrichGO)
>write.csv(down_enrichGO_data, file=file.path("down_enrich-
GOBP.csv"))
```

The clusterProfiler [18] package includes useful functions that can visualize the enrichment results in a dot plot for example. The `dotplot()` function takes the gene list used in the `enrichGO()` function above and the number of GO terms to plot is specified by the `showCategory` argument, which is `50` in the following commands. The dot plot is saved as a PNG file (Fig. 16). Note that if the `ont = "ALL"` argument was given in the `enrichGO()` function above, then the dot plot may result in a mixture of GO [22, 23] terms from different ontologies.

```
#Dot plot of the top 50 enriched GO Biological Process terms
ordered by gene ratio for all significant DEGs.
>GOBP_dotplot <- dotplot(all_enrichGO, showCategory=50)
>ggsave("all_enrichGO_dotplot.png", GOBP_dotplot, dpi=300,
width=15, height=14)
```

3.9.3 KEGG [24–26] Functional Enrichment Analysis

Functional enrichment against the KEGG [24–26] database is carried out via the `enrichKEGG()` function that is similar to the `enrichGO()` function and takes the gene list, `organism` and `reference` gene set as input. `enrichKEGG()` requires the gene IDs to be converted to Entrez [43] IDs. clusterProfiler [18] has a

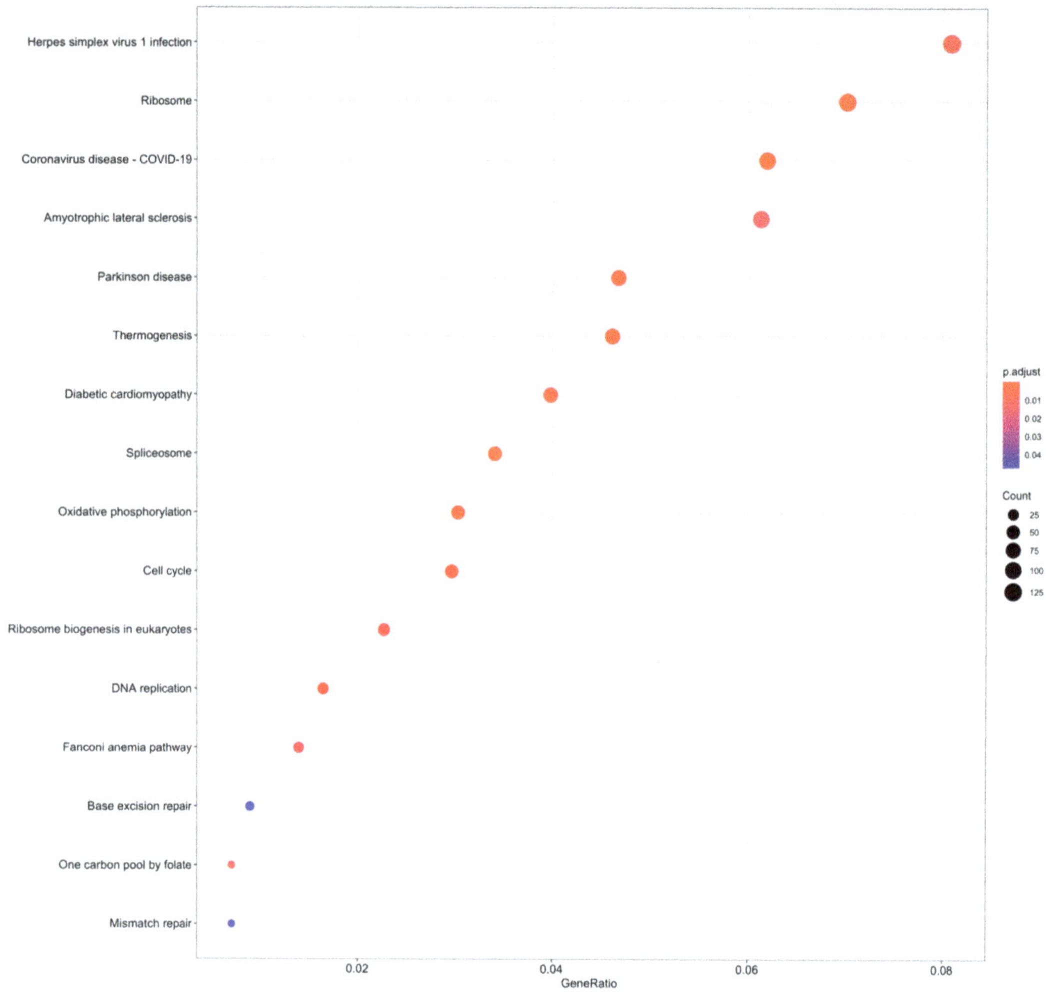

Fig. 17 Dot plot of the statistically significant KEGG [24–26] pathways ranked by the ratio of DEGs that are related to a particular KEGG [24–26] pathway to the total number of DEGs, termed the GeneRatio. The adjusted *p*-value (p.adjust) is represented by a color scale and the size of the dot refers to the number of DEGs assigned to each KEGG [24–26] pathway (Count). Created using the dotplot() function of clusterProfiler [18] R package

function called `bitr()` to perform this conversion which requires the gene list, current ID type (`fromType`), desired ID type (`toType`), and the organism database (`OrgDb`) as arguments. A dot plot containing the significant pathways ordered by gene ratio can be generated using the `dotplot()` function (Fig. 17).

```
#Convert gene ID to entrez to be able to run KEGG and Reactome
enrichment.
>all_symbol_entrez <- bitr(all_DEGs, fromType = "ENSEMBL",
toType = "ENTREZID", OrgDb = "org.Hs.eg.db")
>all_entrez <- all_symbol_entrez$ENTREZID
```

```
>up_symbol_entrez <- bitr(up_DEGs, fromType = "ENSEMBL",
toType = "ENTREZID", OrgDb = "org.Hs.eg.db")
>up_entrez <- up_symbol_entrez$ENTREZID

>down_symbol_entrez <- bitr(down_DEGs, fromType = "ENSEMBL",
toType = "ENTREZID", OrgDb = "org.Hs.eg.db")
>down_entrez <- down_symbol_entrez$ENTREZID

>reference_symbol_entrez <- bitr(reference, fromType = "EN-
SEMBL", toType = "ENTREZID", OrgDb = "org.Hs.eg.db")
>reference_entrez <- reference_symbol_entrez$ENTREZID

#Functional enrichment analysis against KEGG for all signifi-
cant DEGs.
>all_enrichKEGG <- enrichKEGG(
all_entrez,
organism = "hsa",
universe = reference_entrez
)

#Writing the functional analysis results to a CSV file.
>all_enrichKEGG_data <- data.frame(all_enrichKEGG)
>write.csv(all_enrichKEGG_data, file=file.path("all_enrich-
KEGG.csv"))

#Functional enrichment analysis against KEGG for significant
upregulated DEGs.
>up_enrichKEGG <- enrichKEGG(
up_entrez,
organism = "hsa",
universe = reference_entrez
)

#Write the functional analysis results to a CSV file.
>up_enrichKEGG_data <- data.frame(up_enrichKEGG)
>write.csv(up_enrichKEGG_data, file=file.path("up_enrich-
KEGG.csv"))

#Functional enrichment analysis against KEGG for significant
downregulated DEGs.
>down_enrichKEGG <- enrichKEGG(
down_entrez,
organism = "hsa",
universe = reference_entrez
)

#Write the functional analysis results to a CSV file.
>down_enrichKEGG_data <- data.frame(down_enrichKEGG)
```

```
>write.csv(down_enrichKEGG_data, file=file.path("down_en-
richKEGG.csv"))

#Dot plot of the KEGG reults ordered by gene ratio for all
significant DEGs
>KEGG_dotplot <- dotplot(all_enrichKEGG , showCategory=50)
>ggsave("all_enrichKEGG_dotplot.png", KEGG_dotplot, dpi =
300, width = 15, height = 14)
```

Along with the clusterProfiler [18] and enrichplot [42] R packages providing options to plot the functional enrichment results as seen above, the pathview [31] package can be implemented to create figures of enriched KEGG [24–26] pathways with DEGs colored by log2FC. In order to do this, the Entrez [43] IDs of the DEGs and the corresponding log2FC are required as an input. The `pathview()` function can then be executed for an enriched KEGG [24–26] pathway identified by the `enrichKEGG()` function. KEGG [24–26] pathway hsa04110 (cell cycle) was significantly enriched in the functional analysis and is visualized as an example by specifying the `pathway.id` argument of the `pathview()` function. All significant pathways (adjusted *p*-value < 0.05) can be displayed and visualized. The results are automatically saved in the working directory. As pathview [31] outputs several results files, the working directory is set to /home/user/rnaseq_exp/output/pathview. Following the use of pathview [31], the working directory is set back to /home/user/rnaseq_exp/output/deseq2_funcenrich to continue with functional enrichment analysis. Permission from KEGG [24–26] is required to use the figures resulting from the pathview [31] R package in a publication (https://www.kegg.jp/feedback/copyright.html).

```
#Extracting the Entrez ids and corresponding log2FC values.
>all_symbol_entrez$log2FC <- >de_res_sig$log2FoldChange
[match(all_symbol_entrez$ENSEMBL, rownames(de_res_sig))]
entrez_log2FC <- all_symbol_entrez$log2FC
>names(entrez_log2FC) <- all_symbol_entrez$ENTREZID

#Printing the top lines of the extracted Entrez ids and log2FC
values.
>head(entrez_log2FC)

#Setting the working directory to the pathview output
directory.
>setwd("/home/user/rnaseq_exp/output/pathview")

#Creating a KEGG pathway figure for the enriched pathway
```

```
hsa04110.
 >hsa04110 <- pathview(gene.data = entrez_log2FC, pathway.id =
"hsa04110", species = "hsa")

 #Using a function to create figures for all significantly
enriched KEGG pathways.
 >KEGG_pathview <- function(x){
 pathview(gene.data = entrez_log2FC, pathway.id = all_enrich-
KEGG$ID[x], species = "hsa")
 }
 >KEGG_pathview()

 #Returning the working directory to the DESeq2 output file.
 setwd("/home/user/rnaseq_exp/output/deseq2_funcenrich/")
```

3.9.4 Reactome Functional Enrichment Analysis

The functional analysis of the Reactome [27] database is performed using ReactomePA [19] package. The `enrichPathway()` function operates in a similar way to `enrichGO()` and `enrichKEGG()`. The gene ID input is also required to be in the Entrez [43] ID format. Another useful visualization of enrichment analysis is a bar plot, an example of which is shown below with the top-30 Reactome [27] results ranked by adjusted *p*-value (p.adjust) (Fig. 18).

```
 #Functional enrichment analysis against Reactome for all
significant DEGs.
 all_enrichReactome <- enrichPathway(
 all_entrez,
 organism = "human",
 universe = reference_entrez
 )

 #Write the functional analysis results to a CSV file.
 all_enrichReactome_data <- data.frame(all_enrichReactome)
 write.csv(all_enrichReactome_data, file=file.path("all_enri-
chReactome.csv"))

 #Functional enrichment analysis against Reactome for signifi-
cant upregulated DEGs.
 up_enrichReactome <- enrichPathway(
 up_entrez,
 organism = "human",
 universe = reference_entrez
 )

 #Write the functional analysis results to a CSV file.
```

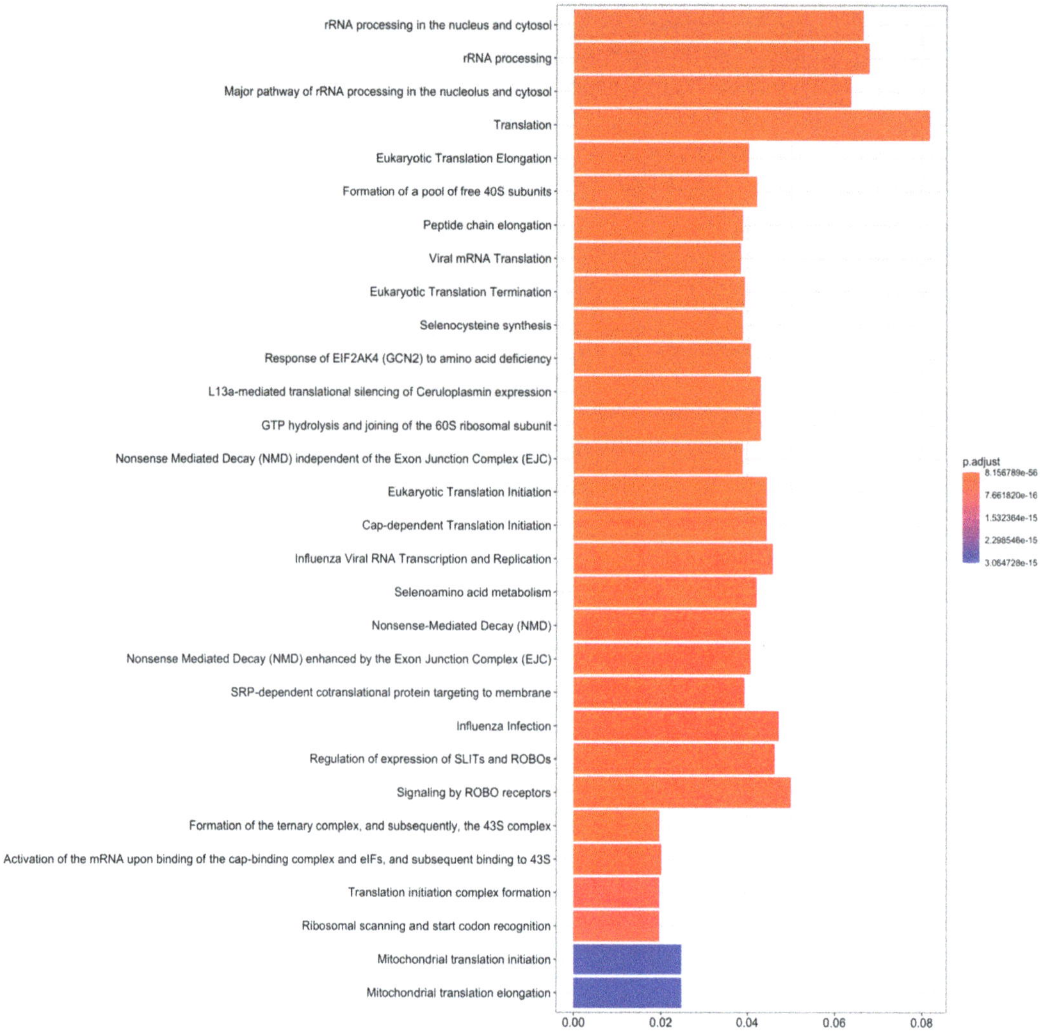

Fig. 18 Bar plot of the top 30 Reactome [27] results ranked by adjusted *p*-value. The *x*-axis represents the GeneRatio as previously described. The color scale bar reflects the adjusted *p*-value of each result. Plot created using the barplot() function of clusterProfiler [18] R package

```
up_enrichReactome_data <- data.frame(up_enrichReactome)
write.csv(up_enrichReactome_data, file=file.path("up_enri-
chReactome.csv"))

#Functional enrichment analysis against Reactome for signifi-
cant downregulated DEGs.
down_enrichReactome <- enrichPathway(
down_entrez,
organism = "human",
universe = reference_entrez
)
```

```
#Write the functional analysis results to a CSV file.
down_enrichReactome_data <- data.frame(down_enrichReactome)
write.csv(down_enrichReactome_data, file=file.path("down_en-
richReactome.csv"))

#Bar plot of the top 30 Reactome results ranked by p.adjust.
Reactome_barplot <- barplot(all_enrichReactome, showCategory
= 30, x = "GeneRatio")
ggsave("all_enrichReactome_barplot.png", Reactome_barplot,
dpi=300, width=15, height=14)
```

3.9.5 R [29] Package Versions

It is good practice when carrying out analyses in R [29] to take note of the versions of the packages used. Not only is this information required when forming the methods section of a manuscript, knowing which version of a package is being implemented can also aid with troubleshooting any issues that may arise. This information is easily gathered with the `sessionInfo()` function.

```
#Information on the package versions used in the analysis.
>sessionInfo()
```

3.10 GSEA [20, 21]

A Gene Set Enrichment Analysis can be performed once the normalized expression matrix has been generated by DESeq2 [17]. Two files following a specific format need to be generated. The first is a file in GCT format, which contains the normalized expression matrix output from DESeq2 [17] and the second is a CLS file, which contains sample information. These files can be created manually following the description provided by the GSEA [20, 21] site: https://software.broadinstitute.org/cancer/software/gsea/wiki/index.php/Data_formats. The CLS file and the top ten lines of the GCT file prepared for this analysis are shown in Fig. 19. These files were saved in the /home/user/rnaseq_exp/resources directory with the filenames rnaseq.gct and rnaseq.cls.

1. If GCT and CLS files are properly formatted, GSEA [20, 21] analysis can be performed. First of all, launch GSEA [20, 21] GUI from a Linux terminal with the following command:

```
#Launching GSEA from the terminal. Example using the 4.1.0
version of GSEA for Linux:
$ cd /home/user/rnaseq_exp/GSEA_Linux_4.1.0
$ bash gsea.sh
```

A

	1	2	3	4	5	6	7	8
1	#1.2							
2	60664	6						
3	NAME	Description	gCTL1	gCTL2	gCTL3	gMYC1	gMYC2	gMYC3
4	ENSG00000284662	na	0	0	0	0	0	0
5	ENSG00000186827	na	0	0	2.15336154	0.86408231	0	0
6	ENSG00000186891	na	1.12294421	0	0	0	0	0
7	ENSG00000160072	na	1285.77112	1226.42124	994.853033	587.575974	548.259213	571.746256
8	ENSG00000041988	na	419.981135	357.296077	310.084062	310.205551	353.096983	322.481342
9	ENSG00000260179	na	2.24588842	5.90572027	3.23004231	1.72816463	1.12810538	4.9359389
10	ENSG00000234396	na	0	0	0	0	0	0

B

```
6 2 1
# control cMYC_KO
control control control cMYC_KO cMYC_KO cMYC_KO
```

Fig. 19 The top ten lines of the GCT file (**a**) and the CLS file (**b**) used in this analysis as the input for GSEA [20, 21]. For the full description of how to manually create these files, see https:/software.broadinstitute.org/cancer/software/gsea/wiki/index.php/Data_formats

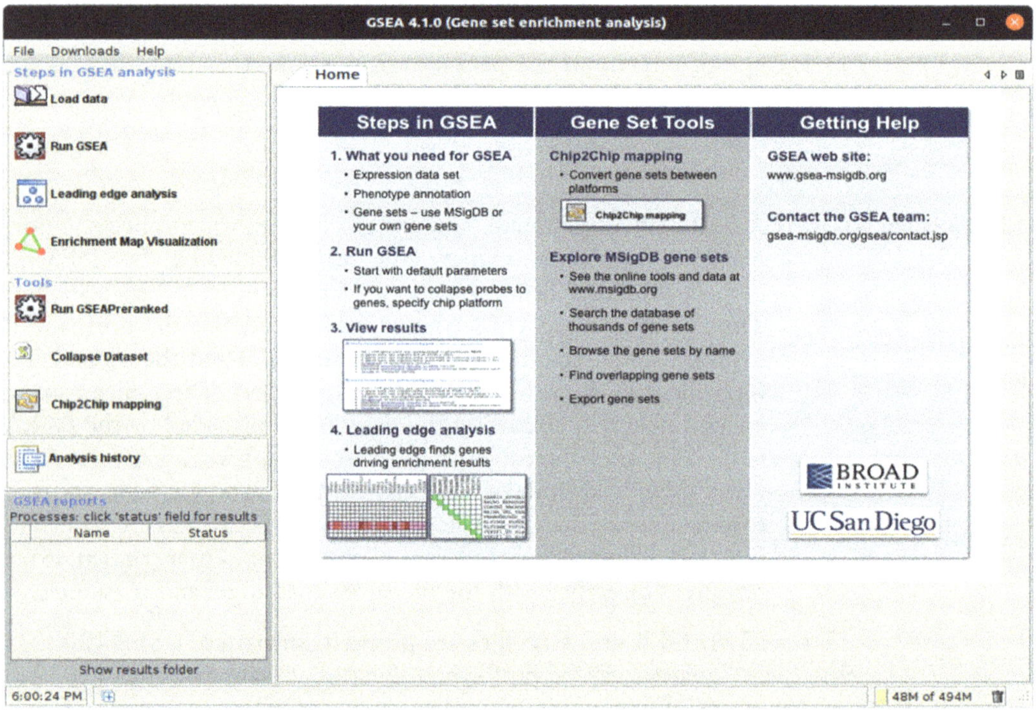

Fig. 20 GSEA [20, 21] Home screen

You should see the following window (Fig. 20):

2. Select "Load data" from the left panel to indicate the files used as input.

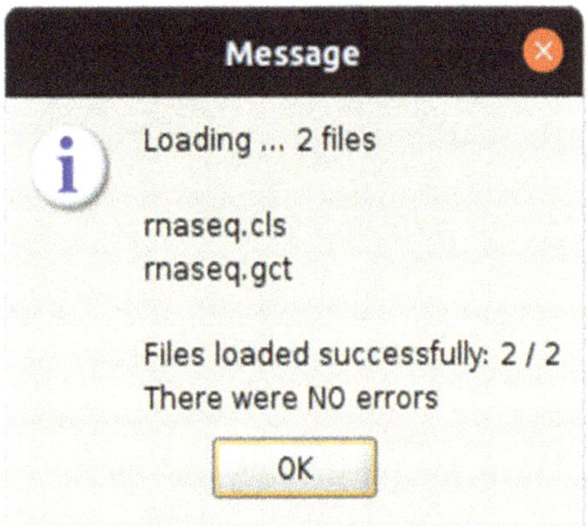

Fig. 21 GSEA [20, 21] Loading files success message

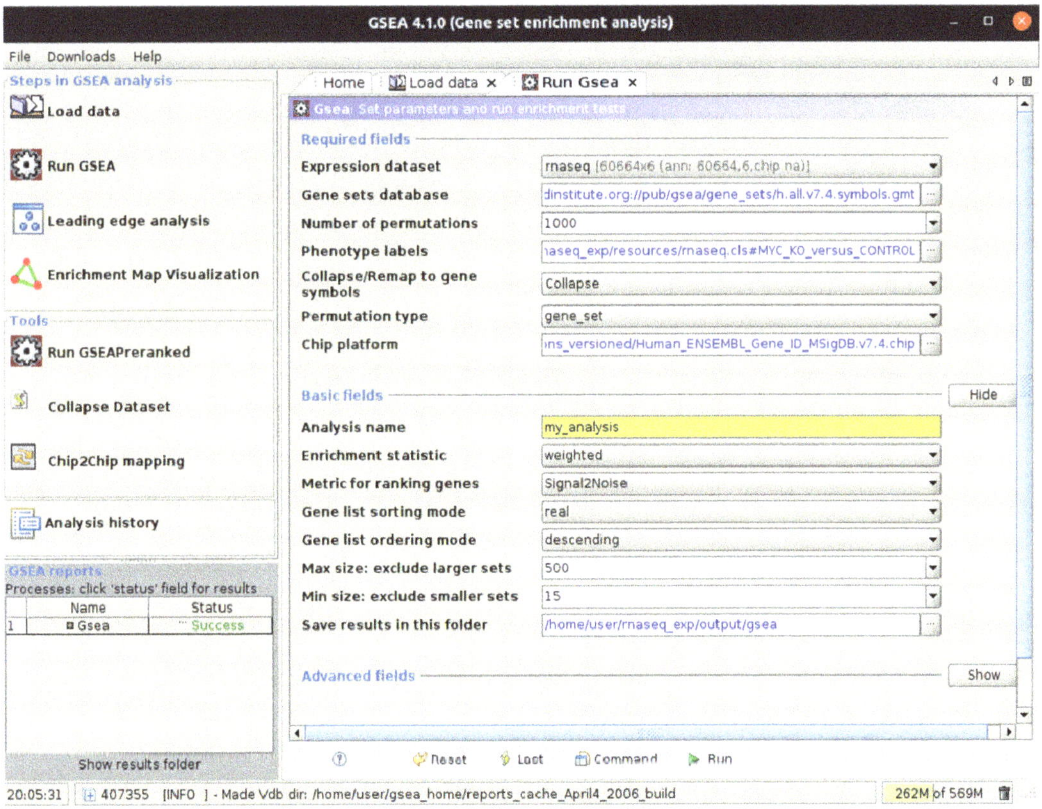

Fig. 22 Example GSEA [20, 21] window showing the parameters used in this analysis as well as the analysis run success message ("GSEA reports" section in bottom left corner)

(a) Select "Browse for files" and open *rnaseq.cls* and *rnaseq.gct* located in the *resources* directory. The success message shown below should appear (Fig. 21). If an error message

GSEA Report for Dataset rnaseq

Enrichment in phenotype: MYC_KO (3 samples)

- 29 / 50 gene sets are upregulated in phenotype **MYC_KO**
- 26 gene sets are significant at FDR < 25%
- 12 gene sets are significantly enriched at nominal pvalue < 1%
- 16 gene sets are significantly enriched at nominal pvalue < 5%
- Snapshot of enrichment results
- Detailed enrichment results in html format
- Detailed enrichment results in TSV format (tab delimited text)
- Guide to interpret results

Enrichment in phenotype: CONTROL (3 samples)

- 21 / 50 gene sets are upregulated in phenotype **CONTROL**
- 13 gene sets are significantly enriched at FDR < 25%
- 11 gene sets are significantly enriched at nominal pvalue < 1%
- 12 gene sets are significantly enriched at nominal pvalue < 5%
- Snapshot of enrichment results
- Detailed enrichment results in html format
- Detailed enrichment results in TSV format (tab delimited text)
- Guide to interpret results

Dataset details

- The dataset has 60664 native features
- After collapsing features into gene symbols, there are: 40265 genes

Gene set details

- Gene set size filters (min=15, max=500) resulted in filtering out 0 / 50 gene sets
- The remaining 50 gene sets were used in the analysis
- List of gene sets used and their sizes (restricted to features in the specified dataset)

Fig. 23 GSEA [20, 21] html report main page

pops-up, check that the format of the GCT and CLS files is correct.

3. Select "Run GSEA" from the left panel.
   (a) Select the expression dataset, which has the same name as the GCT file.
   (b) Select the desired gene sets database. We used h.all.v7.4.symbols.gmt for this example. You can also use your own custom gene sets in GMX or GMT format (see https://software.broadinstitute.org/cancer/software/gsea/wiki/index.php/Data_formats.)

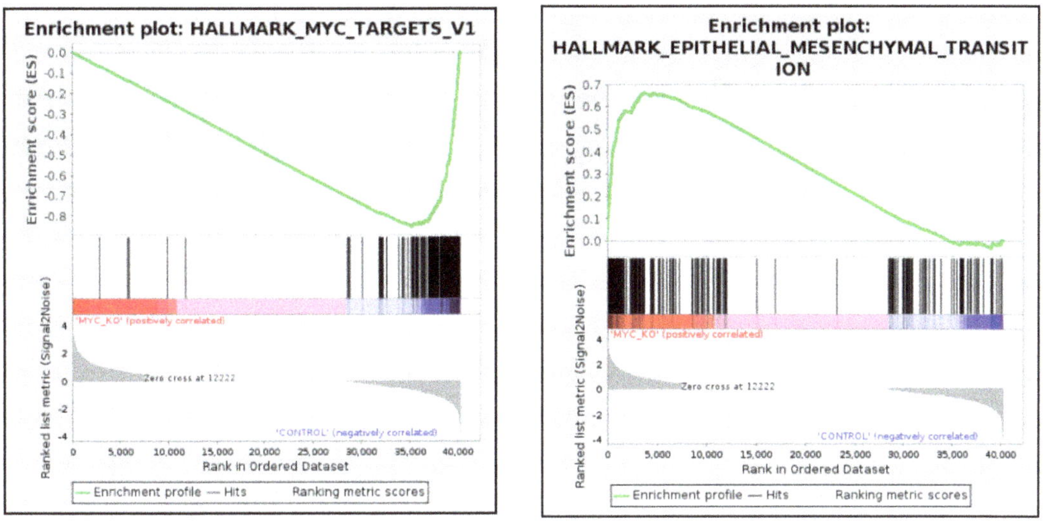

Fig. 24 Enrichment plots of the most significantly downregulated (left) and upregulated (right) gene sets from the Hallmark gene sets [28]

Table: Gene sets enriched in phenotype MYC_KO (3 samples) [plain text format]

	GS follow link to MSigDB	GS DETAILS	SIZE	ES	NES	NOM p-val	FDR q-val	FWER p-val	RANK AT MAX	LEADING EDGE
	HALLMARK_EPITHELIAL_MESENCHYMAL_TRANSITION	Details ...	200	0.67	2.05	0.000	0.000	0.000	3752	tags=45%, list=9%, signal=49%
	HALLMARK_UV_RESPONSE_DN	Details ...	144	0.61	1.83	0.000	0.000	0.001	5416	tags=51%, list=13%, signal=58%
	HALLMARK_KRAS_SIGNALING_DN	Details ...	200	0.55	1.72	0.000	0.001	0.004	7542	tags=40%, list=19%, signal=48%
	HALLMARK_MYOGENESIS	Details ...	200	0.55	1.72	0.000	0.001	0.004	5017	tags=37%, list=12%, signal=42%
	HALLMARK_TNFA_SIGNALING_VIA_NFKB	Details ...	200	0.53	1.63	0.000	0.004	0.021	5453	tags=39%, list=14%, signal=44%
	HALLMARK_TGF_BETA_SIGNALING	Details ...	54	0.61	1.59	0.002	0.006	0.038	4747	tags=43%, list=12%, signal=48%
	HALLMARK_KRAS_SIGNALING_UP	Details ...	200	0.50	1.56	0.000	0.007	0.053	5597	tags=35%, list=14%, signal=40%
	HALLMARK_APICAL_JUNCTION	Details ...	200	0.50	1.54	0.000	0.008	0.065	6756	tags=39%, list=17%, signal=46%
	HALLMARK_COAGULATION	Details ...	138	0.50	1.48	0.003	0.015	0.127	6015	tags=34%, list=15%, signal=40%
0	HALLMARK_WNT_BETA_CATENIN_SIGNALING	Details ...	42	0.55	1.37	0.057	0.049	0.386	5814	tags=40%, list=14%, signal=47%

Fig. 25 Table showing the detailed enrichment results in the MYC_KO phenotype

(c) Choose the phenotype labels, it will only affect how the results are phrased (e.g., Treatment vs. Control or Control vs. Treatment).

(d) In "Collapse/Remap to gene symbols" choose "Collapse," as we provided gene IDs in the GCT file as Ensembl [37] gene IDs. GSEA [20, 21] will collapse the Ensembl [37] gene IDs into HUGO gene symbols that are required to run the analysis.

(e) Select "gene_set" as permutation type if at least one of your phenotypes has less than seven samples. Otherwise, select "phenotype."

(f) The"Human_ENSEMBL_Gene_ID_MSigDB.v7.4.chip" chip platform should be selected for this data. Select the appropriate chip platform depending on the gene IDs provided in the GCT file.

(g) Indicate that you want to save your results in */home/user/rnaseq_exp/output/gsea*.

(h) Leave all other fields as default, which is valid for most experiments. Please consult GSEA user guide for more advanced analyses (https://www.gsea-msigdb.org/gsea/doc/GSEAUserGuideFrame.html).

(i) Select "Run" from the options at the bottom of the window.

4. When the analysis finishes, a green "Success" status in "GSEA reports" will appear (Fig. 22, bottom left corner). Click on it in order to open an HTML report (Fig. 23). Alternatively, go to the results folder that has been generated inside the GSEA [20, 21] output directory and open the file *index.html*.

An example GSEA [20, 21] window with the above parameters selected is shown in Fig. 22.

"Enrichment in phenotype: MYC_KO" includes the gene sets that are enriched in MYC_KO compared to CONTROL. On the other hand, the gene sets that are enriched in the CONTROL group over MYC_KO are included in "Enrichment in phenotype: CONTROL." To determine which gene sets are enriched, GSEA sorts all the genes in the dataset based on each gene's differential expression between both phenotypes. The distribution of gene sets that are enriched will differ from the distribution constituted by all genes included in the analysis [20, 21].

Selecting "Snapshot of enrichment results" from the report home page will show the enrichment plots of the enriched gene sets (Fig. 24). In these plots, vertical black lines in the central portion represent occurrences of genes belonging to the gene set in the same positions as in the ranked list. Those occurrences increase the value of the running-ES (represented by the green line). Occurrences of other genes decrease it. The enrichment score (ES) reported for a gene set is the maximum deviation from zero that the ES takes along the ranked list. If the genes from a gene set are predominantly located in one of the two sides of the list, the ES will take significant positive or negative values. Genes correlated with each of the phenotypes are located in the red and blue areas in both ends of the ranking list [20, 21].

By selecting "Detailed enrichment results in html format," the collection of gene sets analyzed, their size, ES, NES (Normalized ES), statistical significance measures, the rank where the ES reaches its maximum value as well as parameters related to the leading edge analysis (not described in this workflow) are displayed (Fig. 25). The most relevant results to determine the significance of the gene sets are: (1) the enrichment score (ES), (2) the Normalized Enrichment Score (NES), and (3) the FDR q-val. The ES and the NES will be positive for those gene sets enriched in the first group

(MYC_KO) indicated in the phenotype labels, and negative for those enriched in the second group (CONTROL). In addition, the value of the NES serves as a measurement of this enrichment that allows comparison between gene sets. The default significance cut-off set for GSEA [20, 21] is FDR > 0.25. However, if the experiment has less than 7 replicates per condition compared and the gene-set permutation type is used, the GSEA [20, 21] user guide recommends a more conservative significance cutoff of FDR > 0.05 (https://www.gsea-msigdb.org/gsea/doc/GSEAUserGuideFrame.html).

3.11 Summary

This work describes a detailed bioinformatics workflow to allow a user to take raw RNA-seq data and develop interpretable results of cellular processes using a publicly available HUVEC dataset. Each step of the pipeline is given, along with a description and example set of commands to invoke the tools required to perform the analysis. After following the protocol, the user should be better equipped to apply the outlined template to analyze their own dataset of interest. With a base understanding of the RNA-seq analysis workflow further knowledge can be developed, exploring various tools and analyses that can build on the information extracted from the example HUVEC dataset.

4 Notes

1. If a different Linux distribution is used to run the analysis, the installation steps described in this protocol may vary, but the software usage should not differ except for possible version-related changes.
2. We tested and were able to perform the human genome indexing on a machine with 32Gb of RAM by using the option --genomeSAsparseD 2 (Subheading 3.5.1). With 64 Gb or higher memory this option should not be necessary. Read mapping (Subheading 3.5.2) was executed successfully with the default options with 32 Gb of memory.
3. Bioconductor (https://bioconductor.org/) is a free open-source and open-development project that uses the R [29] programming language for the analysis of high-throughput biological assays. The package BiocManager (available from the regular R CRAN repository) allows to install Bioconductor packages from RStudio [30].
4. The FASTQ format is used to store sequences with per-base quality scores. There are usually four lines per sequence. In the first one, an "@" character is followed by a sequence identifier. Optionally, a description can also be added to this line. The second line contains the sequence. The third line begins with

the "+" character and indicates the end of the sequence. Optionally, it can contain the same sequence identifier found in line 1. Lastly, the quality scores corresponding to each position of the sequence in line 2 are found in line 4. These values are usually Phred quality scores [44].

5. After accessing the Ensembl [37] FTP Download site (http://www.ensembl.org/info/data/ftp/index.html), scroll down to the "Single species data" section. To download the reference genome in FASTA format, click on "FASTA" under "DNA (FASTA)" in the "*Homo Sapiens*" row of the table. In the page that will open, click on "Homo_sapiens.GRCh38.dna.primary_assembly.fa.gz" to start the download. To download the genome annotation file in GTF format, click on "GTF" in the "Gene sets" column in the "*Homo sapiens*" row of the "Single species data" section. To obtain the file, click on "Homo_sapiens.GRCh38.104.gtf.gz" in the window that will open. 104 refers to the version of Ensembl [37] as of the time of writing this protocol, and it will vary in the future. The download process for both files is shown in Fig. 3.
6. The default browser for Ubuntu is Firefox. When downloading the .gtf.gz and .fa.gz files from Ensembl [37] (Subheading 3.2.10), Firefox might perform another round of compression, essentially resulting in files that have been compressed twice. In this case, STAR [14] will raise an error message when trying to generate the reference genome index (Subheading 3.5.1). Perform the below commands after having performed the first gunzip decompression command (Subheading 3.2.10) to solve the issue:

```
#Second decompression of the .fa.gz and .gtf.gz files from Ensembl:
$ cd /home/user/rnaseq_exp/resources
$ mv Homo_sapiens.GRCh38.dna.primary_assembly.fa Homo_sapiens.GRCh38.dna.primary_assembly.fa.gz
$ gunzip Homo_sapiens.GRCh38.dna.primary_assembly.fa.gz
$ mv Homo_sapiens.GRCh38.104.gtf Homo_sapiens.GRCh38.104.gtf.gz
$ gunzip Homo_sapiens.GRCh38.104.gtf.gz
```

7. Conversion of gene IDs often results in loss of information, with some IDs not having a representative ID in the desired conversion type. In the example workflow, some of the Ensembl [37] IDs do not have respective gene symbols as seen in /home/user/rnaseq_exp/output/deseq2_funcenrich/de_results.csv after ID conversion. Many of these with missing gene symbols are long non-coding RNA or

pseudogenes and may not be of interest. To minimize the loss of information having an impact on downstream analyses, the ID used in the reference genome, Ensembl [37] ID in this case, should be used throughout the workflow where possible.

For intermediate files, R code and further support see the Github page: https://github.com/vasc-bioinf/rnaseq_exp

References

1. Stark R, Grzelak M, Hadfield J (2019) RNA sequencing: the teenage years. Nat Rev Genet 20(11):631–656
2. Wang Z, Gerstein M, Snyder M (2009) RNA-Seq: a revolutionary tool for transcriptomics. Nat Rev Genet 10(1):57–63
3. Oshlack A, Robinson MD, Young MD (2010) From RNA-seq reads to differential expression results. Genome Biol 11(12):220–220
4. Conesa A et al (2016) A survey of best practices for RNA-seq data analysis. Genome Biol 17(1):13
5. Soneson C, Delorenzi M (2013) A comparison of methods for differential expression analysis of RNA-seq data. BMC Bioinformatics 14(1):91
6. Anjum A et al (2016) Identification of differentially expressed genes in RNA-seq data of Arabidopsis thaliana: a compound distribution approach. J Comput Biol 23(4):239–247
7. Hochberg Y, Benjamini Y (1990) More powerful procedures for multiple significance testing. Stat Med 9(7):811–818
8. Reimand J et al (2019) Pathway enrichment analysis and visualization of omics data using g:Profiler, GSEA, Cytoscape and EnrichmentMap. Nat Protoc 14(2):482–517
9. Andrade J et al (2021) Control of endothelial quiescence by FOXO-regulated metabolites. Nat Cell Biol 23(4):413–423
10. Kocherova I et al (2019) Human umbilical vein endothelial cells (HUVECs) co-culture with osteogenic cells: from molecular communication to engineering prevascularised bone grafts. J Clin Med 8(10):1602
11. FastQC (2015) https://qubeshub.org/resources/fastqc
12. Andrews S (2010) FastQC: a quality control tool for high throughput sequence data [Online]. http://www.bioinformatics.babraham.ac.uk/projects/fastqc/
13. Martin M (2011) Cutadapt removes adapter sequences from high-throughput sequencing reads. EMBnet J 17(1):3
14. Dobin A et al (2013) STAR: ultrafast universal RNA-seq aligner. Bioinformatics (Oxford, England) 29(1):15–21
15. Okonechnikov K, Conesa A, García-Alcalde F (2015) Qualimap 2: advanced multi-sample quality control for high-throughput sequencing data. Bioinformatics 32(2):292–294
16. Liao Y, Smyth GK, Shi W (2013) featureCounts: an efficient general purpose program for assigning sequence reads to genomic features. Bioinformatics 30(7):923–930
17. Love MI, Huber W, Anders S (2014) Moderated estimation of fold change and dispersion for RNA-seq data with DESeq2. Genome Biol 15(12):550
18. Yu G et al (2012) clusterProfiler: an R package for comparing biological themes among gene clusters. OMICS 16(5):284–287
19. Yu G, He Q-Y (2016) ReactomePA: an R/Bioconductor package for reactome pathway analysis and visualization. Mol BioSyst 12(2):477–479
20. Subramanian A et al (2005) Gene set enrichment analysis: a knowledge-based approach for interpreting genome-wide expression profiles. Proc Natl Acad Sci U S A 102(43):15545–15550
21. Mootha VK et al (2003) PGC-1α-responsive genes involved in oxidative phosphorylation are coordinately downregulated in human diabetes. Nat Genet 34(3):267–273
22. Ashburner M et al (2000) Gene ontology: tool for the unification of biology. The Gene Ontology Consortium. Nat Genet 25(1):25–29
23. Gene Ontology Consortium (2020) The Gene Ontology resource: enriching a GOld mine. Nucleic Acids Res 49(D1):D325–D334
24. Kanehisa M (2019) Toward understanding the origin and evolution of cellular organisms. Protein Sci 28(11):1947–1951
25. Kanehisa M et al (2020) KEGG: integrating viruses and cellular organisms. Nucleic Acids Res 49(D1):D545–D551

26. Kanehisa M, Goto S (2000) KEGG: Kyoto encyclopedia of genes and genomes. Nucleic Acids Res 28(1):27–30
27. Jassal B et al (2019) The reactome pathway knowledgebase. Nucleic Acids Res 48(D1): D498–D503
28. Liberzon A et al (2015) The Molecular Signatures Database (MSigDB) hallmark gene set collection. Cell Syst 1(6):417–425
29. R Core Team (2014) R: a language and environment for statistical computing. R Foundation for Statistical Computing, Vienna, Austria
30. R Studio Team (2020) RStudio: integrated development for R. RStudio, PBC, Boston, MA
31. Luo W, Brouwer C (2013) Pathview: an R/Bioconductor package for pathway-based data integration and visualization. Bioinformatics 29(14):1830–1831
32. Zhu A, Ibrahim JG, Love MI (2018) Heavy-tailed prior distributions for sequence count data: removing the noise and preserving large differences. Bioinformatics 35(12):2084–2092
33. Durinck S et al (2005) BioMart and Bioconductor: a powerful link between biological databases and microarray data analysis. Bioinformatics 21(16):3439–3440
34. Durinck S et al (2009) Mapping identifiers for the integration of genomic datasets with the R/Bioconductor package biomaRt. Nat Protoc 4(8):1184–1191
35. Wickham H (2016) ggplot2: elegant graphics for data analysis. Springer, New York. ISBN 978-3-319-24277-4. https://ggplot2.tidyverse.org
36. Leinonen R et al (2010) The European Nucleotide Archive. Nucleic Acids Res 39(Suppl_1): D28–D31
37. Howe KL et al (2020) Ensembl 2021. Nucleic Acids Res 49(D1):D884–D891
38. Hansen KD, Brenner SE, Dudoit S (2010) Biases in Illumina transcriptome sequencing caused by random hexamer priming. Nucleic Acids Res 38(12):e131
39. Evans C, Hardin J, Stoebel DM (2018) Selecting between-sample RNA-Seq normalization methods from the perspective of their assumptions. Brief Bioinform 19(5):776–792
40. The Gene Ontology Consortium (2018) The Gene Ontology Resource: 20 years and still GOing strong. Nucleic Acids Res 47(D1): D330–D338
41. Kanehisa M et al (2017) KEGG: new perspectives on genomes, pathways, diseases and drugs. Nucleic Acids Res 45(D1):D353–D361
42. Yu G (2021) enrichplot: visualization of functional enrichment result. R package version 1.12.1. https://yulab-smu.top/biomedical-knowledge-mining-book/
43. Maglott D et al (2011) Entrez Gene: gene-centered information at NCBI. Nucleic Acids Res 39(Database issue):D52–D57
44. Ewing B et al (1998) Base-calling of automated sequencer traces using Phred. I. Accuracy assessment. Genome Res 8(3):175–185

Index

Andrew V. Benest (ed.), *Angiogenesis: Methods and Protocols*, Methods in Molecular Biology, vol. 2441, https://doi.org/10.1007/978-1-0716-2059-5,

T

U

V

W

Lightning Source UK Ltd.
Milton Keynes UK
UKHW020608130223
416809UK00003B/67